Y0-BBY-550

PHYSICS IN EVERYDAY LIFE

PHYSICS IN EVERYDAY LIFE

RICHARD DITTMAN
GLENN SCHMIEG
University of Wisconsin—Milwaukee

McGRAW-HILL BOOK COMPANY

New York/St. Louis/San Francisco/Auckland/Bogotá/Düsseldorf
Johannesburg/London/Madrid/Mexico/Montreal/New Delhi/Panama
Paris/São Paulo/Singapore/Sydney/Tokyo/Toronto

PHYSICS IN EVERYDAY LIFE

1234567890 DODO 7832109

This book was set in Plantin by York Graphic Services, Inc.
The editors were C. Robert Zappa, Janice Lynn Rogers,
and Michael Gardner; the designer was Anne Canevari Green;
The production supervisor was Dennis J. Conroy.
The drawings were done by Allyn-Mason, Inc.
The cover photograph was taken by Thomas Zimmerman.
R. R. Donnelley & Sons Company was printer and binder.

Library of Congress Cataloging in Publication Data

Dittman, Richard.
Physics in everyday life.

Includes bibliographies and index.
1. Physics. I. Schmieg, Glenn, joint author.
II. Title.
QC23.D58 530 78-13381
ISBN 0-07-017056-8

CONTENTS

PREFACE

PHYSICS IN EVERYDAY LIFE is designed for use in an introductory course in physics for nonscience students. Emphasis is placed on conceptual understanding rather than on learning problem-solving techniques. Mathematics is minimized, but an important definition or law is occasionally stated in the form of an equation as well as in words.

The order and presentation of topics are traditional, proceeding from the macroscopic to the microscopic. The order roughly follows the historical progress of physics from mechanics to nuclear physics, but it is flexible. Chapter 5 (Sound) and Chapter 8 (Light) may follow Chapter 2 (Energy and Conservation Laws) with only minor adjustments. Chapter 1 (Motion and Forces) and Chapter 2 (Energy and Conservation Laws) must precede Chapter 4 (Temperature and Heat). Chapter 3 (Fluids) should precede Chapter 6 (Electricity and Magnetism).

Various learning aids appear in each chapter. Each chapter begins with a short preview of what will be covered and ends with a review of important terms, principles, and laws. In addition, the place in text where an important term, principle, or law appears is called out in the margin by its name.

Discussion questions and numerical and qualitative problems also appear in the review section. The qualitative and numerical problems are designed to reinforce the ideas presented in the chapters, whereas the discussion questions are designed to challenge the student to go beyond the information provided in the text and consider the ideas to explain other everyday experiences. Answers to all discussion questions are found at the back of the text. Additional discussion questions and corresponding answers are given in the Instructor's Manual. Also, solutions to odd-numbered problems are given at the back of the text and answers to even-numbered ones are given in the Instructor's Manual. The numerical problems are designed to give students an opportunity to use the formulas presented in the chapter. The problems are not difficult and may be skipped.

An annotated bibliography rounds out each chapter. Such bibliographies guide the interested student in learning more about selected topics and may be used as a starting point for term papers.

PHYSICS IN EVERYDAY LIFE can be used to provide background material for special topics. We find the Physics of Technology Project by the American Institute of Physics (McGraw-Hill) to be helpful in the understanding of common devices. This series of modules contains information on the analytical balance, automobile ignition system, camera, guitar, laser, pressure cooker, and so on. Use of these modules will further emphasize the role of physics in everyday experience.

The Instructor's Manual, in addition to the answers and additional discussion questions, contains information which should be helpful in organizing classroom time. Typical demonstrations are enumerated for each chapter. Titles, short descriptions, and suppliers of 16-mm motion pictures and 35-mm slides are also provided. A description of each module of the Physics of Technology Project is given and suggestions offered on the integration of the module with the material presented in the text. Finally, sample test questions with corresponding answers are given to assist in the writing of semester tests and final examinations.

We wish to thank the following reviewers for their valuable suggestions: Margaret B. Feero, El Camino College; Keith R. Honey, West Virginia Institute of Technology; Mario Iona, University of Denver; Jack E. Lynch, American River College; James Schreiber, Trenton State College; and Joseph G. Traylor, Buena Vista College. We, of course, take full responsibility for the accuracy of the text.

RICHARD DITTMAN
GLENN SCHMIEG

INTRODUCTION

PHYSICS AS A FOREIGN LANGUAGE

It is a wide-spread attitude among students and their advisors that physics is a difficult subject to study. Indeed this may be true if the subject is presented without regard for the students' capabilities. Even capable students may encounter difficulties if no guidance is provided in studying the subject matter. Below, an attempt is made to provide some reasonable guidelines for the study of physics.

Physics is the oldest branch of natural science. As such it is the most organized of the various branches of science. Because of this, you might think that physics would be easy to define or describe, but that is not the case. Dictionaries usually define physics as a study of nature or the natural world in terms of elementary laws. But that same definition could apply equally well to chemistry or biology. One wit defined physics as that which physicists do. If we accept this definition, then we can say that physics will be the material covered in this book. Of course, such a definition will not include many topics of physics. The topics in this book were chosen to reflect the title: PHYSICS IN EVERYDAY LIFE. These topics will provide a reasonably thorough and somewhat practical study of physics.

In studying, you may find it helpful to notice the similarity of physics to a foreign language. Physics not only has a special vocabulary, but there is also a "grammar" and, consequently, a "literature." The vocabulary of physics is a set of technical terms and definitions. Special words are used to define or describe the physical world in order to study physics. Such a study involves measurements and equipment, so the definitions are technical. Sometimes familiar words, such as "speed" and "work," will have to be relearned in terms that are more technical. These terms and definitions will have to be memorized, as in any foreign language, before going on to the grammar.

In physics the grammar is the group of theories, principles, and laws that describe the behavior of the natural world. The technical terms and definitions are organized to produce a general statement about some physical phenomenon.

The person who first proposes a theory, principle, or law is often honored by having his or her surname used, so we will encounter Einstein's theory of special relativity, Pascal's principle, and Newton's laws of motion. Just like the technical terms and definitions, the theories, principles, and laws will have to be learned before going on to the literature of physics.

The goal of physics is not to define as many terms or propose as many laws as possible. Rather, the goal is to select those terms and laws that describe the physical universe in the simplest way. The literature of physics is the description of what is happening and the predictions of what might happen under specific physical conditions. For example, Newton's laws of motion are relatively simple statements of a certain physical circumstance. But there are an enormous number of applications of these laws to particular situations. A discussion of these applications in physics is similar to the study of literature in a foreign language. The number of applications of the laws of physics is very large in the real world. So only applications with some amount of everyday importance will be discussed in this book.

To help you understand physics, we have included various learning aids in text. Each chapter begins with a short preview of what will be covered and ends with a review of important terms, principles, and laws. In addition, the place in text where an important term, principle, or law appears is called out in the margin by its name.

Problems and discussion questions also appear in the review section. They are designed to reinforce your understanding of the ideas presented in text and to help you apply such ideas in explaining other everyday experiences. Finally, an annotated bibliography rounds out each chapter. Such bibliographies will guide you in learning more about topics of interest to you.

THE PRACTICE OF PHYSICS

It is helpful in studying physics to recognize that definitions are built into laws and laws are then used to explain physical phenomena. But what determines whether a definition or law is "good" or "bad"? One factor is *simplicity*. If two laws can explain the same physical phenomenon, then the simpler law is the proper choice. After all, why make any concept, including definitions or laws, unnecessarily complicated?

The second determining factor for the acceptance of a law is *experimentation*. Laws are judged according to their ability to explain the experimental data. But the data itself must be the result of measurements, which in turn depend on commonly agreed upon standards. For example, distance is measured using standard lengths, while time is measured using standard clocks. In fact, by international agreement there are standard quantities for all the measurements made in physics experiments.

The international standards include the metric system of units. The United States legally adopted the metric system over a century ago. But until recently, most manufacturing and science teaching has been based on the older British system of units. Both systems will be used in the beginning of this book, because the British system is still part of everyday life in the United States. In the latter part of the book, our study will be based on only one system—the metric system.

In addition to a standard system of units, physics also uses basic mathematics. As much as possible, mathematics will be minimized in this book. But from time to time, an important definition or law will be presented in the form of a formula or equation. A formula or equation serves a useful purpose. It is a summary written in the symbols of algebra and identifies the physical quantities and their relationships to one another. The algebraic expressions should help in learning physics.

1 MOTION AND FORCES

We begin the study of physics by examining how things move. Speed, force, and inertia will be introduced as important ideas. These concepts will be put together in Newton's laws of motion and law of gravity.

1.1 SPEED

As you stand on the corner of a busy street or walk down the corridor of a crowded building, what is most noticeable? Probably you notice first the activity and noise of the various people and cars, as shown in Fig. 1-1. If you concentrate on the activity only, you recognize that many objects are in motion. The motion of objects can be discussed by comparing the speed of one object with the speed of another object. The technical definition of speed is based upon the concepts of distance and time. Both distance and time are readily measured.

Speed *Speed is defined as the distance an object moves during a certain interval of time divided by that interval of time.*

This simple statement of division can be expressed as

$$\text{Speed} = \frac{\text{distance moved}}{\text{time interval}}$$

Fig. 1-1
Rush hour in Manhattan.
(Photograph by Joel Gordon from DPI.)

or in symbols

$$v = \frac{d}{t}$$

where v = speed,
d = distance moved
t = time interval

The formula for speed is more than a shorthand notation. It can be used for calculations. Knowledge of both the distance and the time traveled on a trip allows you to calculate the speed. For example, if you drove 200 miles in 4 hours, your speed would be

$$\text{Speed} = \frac{200 \text{ miles}}{4 \text{ hours}}$$
$$= 50 \text{ miles per hour}$$

A common abbreviation for miles per hour is mph.*

Speedometers on American cars usually measure speed in units of miles per hour. But in some cases the unit of miles per hour is awkward, and other units are used. For example, in considering the dangers of high-speed driving, it is more meaningful to realize that a car is moving about 80 feet per second when it is traveling at 55 miles per hour. Although the units are different, these two speeds are equivalent. Fifty-five miles per hour equals about eighty feet per second. At this speed there is not much time to avoid hitting something which suddenly appears on the road ahead of you. If you do not apply the brakes, the car will travel 160 feet in only 2 seconds. Even if you apply the brakes, your car will travel more than 200 feet before stopping. Of course at speeds greater than 55 mph, the problem is worse. Then larger stopping distances are required. To drive safely at any speed, you must allow yourself sufficient room to come to a stop.

In either speed measurement (miles per hour or feet per second) the unit of measurement is based on standard distances established centuries ago in England. The *foot* is the distance that was based on the length of the foot of King James the First. The *mile* is 5,280 of these lengths, and the *inch* is $\frac{1}{12}$ of this length. These somewhat awkward conversion factors (5,280 and $\frac{1}{12}$), as well as other problems with other measurements, led to the development of the metric system of units. The basic unit of length in the metric system is the *meter,* which is slightly longer than the yard (39.37 inches, to be exact). The advantage of the metric system is that almost all conversions involve multiples of 10. These conversions can be performed by simply shifting the

*An average speed of 50 mph could mean, for example, that you drove for 4 hours at a speed of 50 mph or that you drove for 2 hours at a speed of 60 mph and then for the last 2 hours at a speed of 40 mph.

decimal point. For example, the kilometer (the prefix *kilo* means one thousand) is 1,000 meters, and so 1,600 meters equals 1.6 kilometers, abbreviated 1.6 km.

Several common prefixes

Prefix	Meaning	Example
kilo	One thousand (1,000)	1 kilometer = 1,000 meters
centi	One hundredth (1/100)	1 centimeter = 1/100 meter
milli	One thousandth (1/1,000)	1 millimeter = 1/1,000 meter

The only unit of measurement in the metric system not exclusively involving multiples of 10 is the second. In this case the metric system is the same as the English system. Thus 60 seconds equal 1 minute, and 60 minutes equal 1 hour. Speeds are usually stated in the metric system in meters per second.

The speeds of various objects in motion can be easily measured from the small (an ambling turtle) to the large (a satellite orbiting the earth). An even larger range of speeds is obtained when measurements are made on objects that move at speeds either too slow or too fast for direct human perception.

The importance of the concept of speed is that it gives us a means of describing and comparing objects in motion. The critical ingredients of speed are the distance and time relative *to an observer.* A case of zero speed would occur if an object is at rest relative to an observer. For example, if you measure the distance moved by a large rock to be zero, you rightly conclude that it has no speed. It is assumed that the rock does not move relative to the earth. But the earth is revolving around the sun, and so the rock does have a speed relative to the sun. In the same way, a flower vase on an Amtrak train may have *zero speed* relative to a table in the dining car on the moving train, as shown in Fig. 1-2. Notice that the vase *does have a speed* (as does the train) relative to the surface of the earth. An object's speed is always measured relative to an observer. (All motion is relative.)

Although Table 1-1 shows continental drift (the speed at which continents move with respect to the earth's surface) to be an exceedingly slow speed, this is not the slowest imaginable speed. The slowest speed would be zero, that is, absolutely no motion. It is difficult to prove that an object has zero speed by measurements. If there is no detectable distance moved in 1 second, perhaps the experimenter should wait an hour, a month, or maybe even a decade for sufficient time to elapse for the object to move a measurable distance. Nevertheless, we will accept the fact that some objects have zero speed relative to their surroundings.

The opposite situation, where something moves at very fast speeds, is not part of present-day physics. There is an upper speed limit in the universe, namely, the speed at which light travels. The speed of light is quite high,

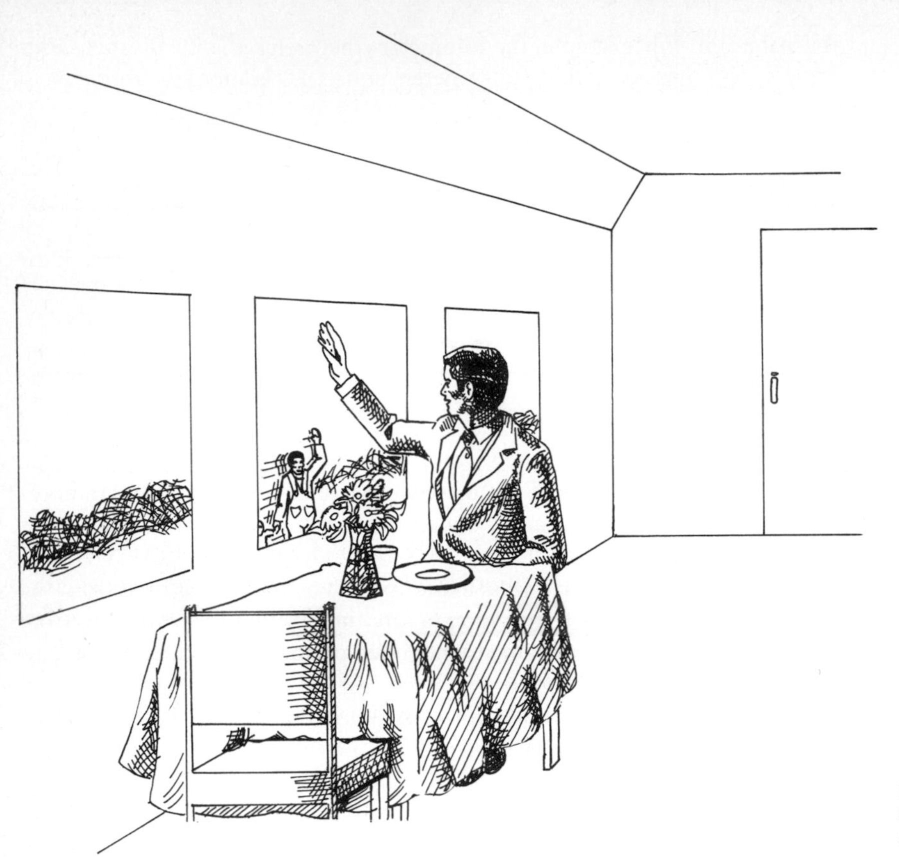

Fig. 1-2
Speed is relative to the observer.

about 186,000 miles per second, or about 300,000 kilometers per second.* This speed would allow the light beam to travel slightly more than 7 times around the earth's equator in 1 second. Fast as it is, light is not infinitely fast. Every attempt to find an object with greater speed has failed. Although nothing in the laws of physics prohibits an object from moving faster than 186,000 miles per second, such an object has not been discovered. Therefore, 186,000 miles per second now stands as nature's upper speed limit.

Speed has been shown to be one measurable means of describing motion. If the motion of objects is analyzed in terms of speed, two general conclusions are found. First, all objects in motion do not have the same speed. Some

*We will see in Chap. 8 that light travels with differing speeds in air, glass, water, and other materials. We refer here to the speed of light in a vacuum (where materials are absent), which is 186,291 miles per second, or 299,793 kilometers per second.

Table 1-1
Various average speeds

Phenomenon	Speed, miles per hour	Speed, kilometers per hour
Continental drift	0.0000000011	0.0000000018
Hour hand on watch	0.000037	0.000059
Turtle	0.015	0.024
Falling snowflake	1	1.6
Brisk walk	4	6.4
Highway traffic	55	88
Commercial jetliner	500	800
Orbiting satellite	30,000	48,000
Speed of light	670,000,000	1,100,000,000

objects move rapidly. Others move slowly. Second, a moving object can have a change in speed. While in motion, an object may increase in speed, as, for instance, when you pump harder on the pedals of a bicycle to speed up. Or an object may decrease in speed, as, for instance, when you apply the hand brakes of a bicycle to slow down.

Changes in speed can also involve objects which are temporarily at rest. A skier at rest can increase her speed by pushing on ski poles at the top of a snow-covered mountain. As she begins to move, her speed increases. Or just the opposite, a moving parachutist can decrease his speed to zero by landing in a field.

1.2 FORCE

It is obvious that changes in speed can occur for objects. What is not so obvious is the *cause* of such changes of speed. The various descriptions of the activities "pump on the pedals," "apply the brakes," and "push on the ski poles" all have a common ingredient. All these activities involve a *force*. We need to describe this new concept before combining it with other definitions to form a law.

Force *A force can be described as the everyday experience of a push or a pull.*

Force is a somewhat unusual concept in physics. It can only be described in terms of synonyms, not in terms of more fundamental quantities. Whenever you describe some muscular effort, such as a push, a pull, a tug, a lift, or a jerk, you are describing everyday examples of a force. These terms are simply other names for the concept of force. As shown in Fig. 1-3, we normally experience forces as fundamental sensory experiences and learn to recognize them in terms of muscular efforts or tactile experiences (that is, touch and feel).

This description of force is unlike our definition of speed, which in-

Fig. 1-3
When a woman applies lipstick, the force is a "push."
(Photograph by David Seymour from Magnum.)

volved the underlying fundamental quantities of distance and time. The ideas of distance and time were taken for granted. Similarly, we now take for granted the idea of force. But we can illustrate the idea of force by speaking of a push or pull, depending on the direction of the force, as shown in Fig. 1-4. Later, force will be combined with other concepts into a law that describes certain changes in motion.

Even though the concept of force cannot be defined in terms of more fundamental quantities, force can be measured. As you carry your luggage to the ticket counter of an airline, your muscular effort indicates that the luggage is capable of exerting a force on your arm. The luggage is placed on a scale and "weighed"—that is, a force is measured. The weight measured by a scale is an example of a force, in this case caused by gravity, not by any muscular effort. In the English system of units, force is measured in pounds. A scale, whether it be at an airline counter, in a grocery store, or on your bathroom floor, is simply a device that measures force (see Fig. 1-5).

Forces can be exerted in different directions. For example, you can exert a sideways force by pushing on a wall. In order to measure the force, you could hold a bathroom scale against the wall and push on the scale. Or you

Fig. 1-4
When a pickle is taken out of a jar, the force is a "pull." (*Photograph by Ted Rozumalski from Black Star.*)

could exert an upward force by *pushing* on your living room ceiling. Again, you could *pull down* on a rope tied to the ceiling, as shown in Fig. 1-6. This force could be measured by using the "spring balance" often found in older produce markets. In every case, the number of pounds is a measure of the force. In addition, every force has a direction: downward for a weight, but in general, any direction.

Several common weights

Telephone handset	$\frac{1}{2}$ pound
Gallon jug of apple cider	10 pounds
Airline baggage limit	44 pounds
Aluminum canoe	100 pounds

The pound is the fundamental unit of force in the English system of units. For convenience, a smaller unit, the ounce ($\frac{1}{16}$ of a pound), and a larger unit, the ton (2,000 pounds), are used to measure forces. In the metric system of units, forces are measured in units of *newtons,* named after Isaac Newton who proposed laws about forces in the seventeenth century. The newton is not a common unit of measurement of force in either the United States or any country using the metric system of units. In countries already using the metric system, the scales are calibrated to measure kilograms, about which more will be said later in this chapter.

Fig. 1-5
In the produce section of a market a spring scale measures the weight. Would you like some turtle soup? [(*a*) *Photograph by Robert L. Beckhard from DPI.*; (*b*) *Photograph by David Strickler from Monkmeyer.*]

Fig. 1-6
Forces can be exerted in different directions, either by pulling on the rope or by pushing on the wall. (*Photograph by Ted Rozumalski from Black Star.*)

1.3 FRICTION: A SPECIAL KIND OF FORCE

Forces have been described in terms of muscular efforts and weights. Motion has not been important in these examples. When there is relative motion between objects in contact, a new force arises—a *frictional* force.

Friction *Friction is defined as the resistance to relative motion between two bodies in contact.*

Friction is so commonplace that ordinary living would be unrecognizable if friction were to disappear. Friction keeps your shoes from sliding while you are walking. Friction is necessary for one to walk at all. Think how difficult it is to walk with very little friction, such as on a smooth, icy sidewalk. Friction also allows tires to "grip" the road when a car is starting, stopping, or turning (see Fig. 1-7). If the road is icy, sand is sprinkled on the surface of the ice to increase the friction between the tires and the road. Another example involves writing with a pencil. If there were no friction, you could not possibly leave a trail of pencil marks on paper. You could not write!

Friction is not limited to conditions of relative motion between two objects. There is friction between motionless objects in contact with one another. Nails and screws hold boards together because of friction. Also, because of friction you can stand on the side of a hill without sliding down.

Fig. 1-7
Friction is important in stopping a car. (*Photograph from The Boston Globe.*)

And yet neither of the two examples just given involves relative motion. In fact, whenever any two objects are in contact, friction is present.

In some situations friction is not desired. Friction involving moving objects wears away the surfaces. Automobile tires wear out because of friction between the tires and the road. Pants wear out at the knees and shirts wear out at the elbows from friction. Inside a car's motor, friction wears out the moving parts. The solution in this case is to lubricate the moving parts with oil or grease. The lubrication greatly decreases the amount of friction. But sooner or later the parts of even the best cared for car will wear out. These parts have to be replaced.

The cause of friction is roughness between surfaces. All surfaces are rough to some extent. As one object slides or rolls over another, the adjacent surfaces mesh with one another. The surfaces are covered with small projections. Even when highly polished, the surfaces of metals can become enmeshed or united in a type of "cold-welding." The purpose of lubricants is not only to fill in the rough surfaces between objects, but also to separate the two objects. Oil separates a pancake from a frying pan, and fine chalk dust separates fingers from a pool cue. See Fig. 1-8.

1.4 INERTIA

Although we are not ready to propose a law, experience shows that there is a connection between force and change in motion. If the motion of objects is analyzed in terms of the forces applied to them, two general conclusions are found.

Fig. 1-8
Pancakes should not stick to the pan! Frictional forces are reduced by the grease.

First, some objects are more difficult to move than others. For many solid objects, the larger the object, the more difficult it is to put the object into motion from a standstill. It is easier to move a child's wagon by pushing than it is to move a car by pushing, as shown in Fig. 1-9. And if you wish to stop an object by dragging your foot, it is easier to stop the child's wagon than to stop the car.

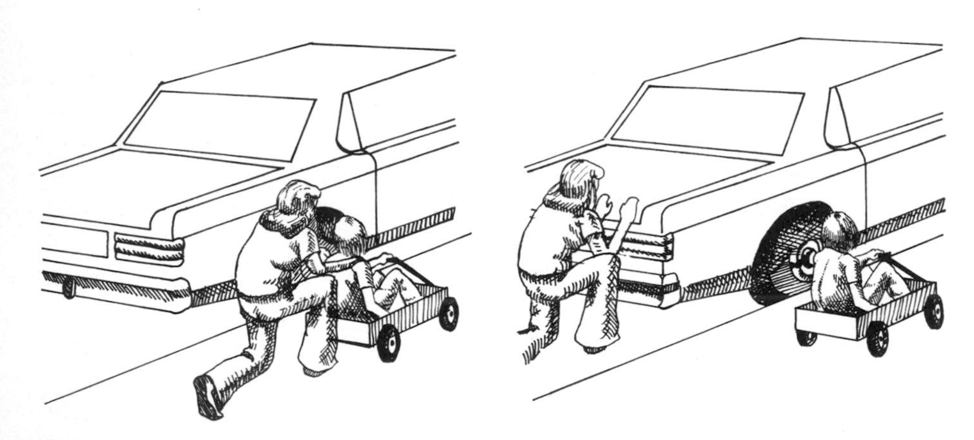

Fig. 1-9
The inertia of a car is larger than the inertia of a child's wagon.

Second, for any movable object, the greater the force exerted on an object, the more rapidly its state of motion can be changed. A race car is more easily given a large speed than a conventional car, because the race car has a much larger force applied to it. In another example, a car will stop more abruptly when it collides with a brick wall than when the brakes are applied, because the wall exerts a larger force than the brakes.

In all these cases the force is *outside* the object and is applied *to* the object. But the difficulty or ease with which different objects change their speed is due to something *inside* the object. The property of an object that indicates the relative ease or difficulty in changing its state of motion is called its *inertia.*

Inertia *Inertia is defined as the resistance offered by an object to any change in its condition of rest or motion.*

This use of the word inertia is similar to common usage. In both uses of the word there is the idea of a sluggishness or a resistance to change. If we say a committee has inertia, we usually mean it has difficulty moving ahead. For human beings to be described as possessing inertia means that their activities are sluggish, that they lack "get-up-and-go." An object with large inertia is difficult to put into motion from rest. It is equally difficult for an object with large inertia to be brought to rest from a high speed. On the other hand, an object with small inertia is easily started or stopped.

Inertia is a basic characteristic of all objects. Like force, inertia cannot be defined by a formula in terms of other physical quantities. A force is understood because of a sensory experience (that is, push or pull). Inertia is understood by comparing the difficulty encountered when forces change the motion of different objects. Forces are directly felt, speeding objects are directly seen, but the inertia of an object is indirectly experienced. It is based on experience with changing the motion of many different objects.

To correctly understand the concept of inertia, we must recognize that a force applied to an object does *not* change the object's inertia. The inertia itself is not overcome by the force. What correctly occurs is that a force changes an object's *state of motion.* A force can do this by moving a stationary object or by stopping a moving object. A force can also change the speed or direction of motion of a moving object. But in *every* case the inertia is not changed. Inertia is not reduced or overcome.

1.5 MASS

In the discussion of force it was possible to give examples and show that there are many measurements of forces (especially weights) that occur in everyday life. But when inertia was discussed, no measurements were cited. The idea of inertia, by itself, is not a quantitative concept. To remedy this situation, we introduce the word *mass.*

Mass *Mass is defined as the quantitative measure of inertia.*

Thus we may now set up standards for measurements of mass and choose a unit of mass.

A certain metal cylinder is kept at the International Bureau of Weights and Measures in order to standardize measurements. See Fig. 1-10. The mass of this cylinder is designated *one kilogram.* By recalling that the prefix kilo means 1,000, we recognize that a *gram* is a smaller unit of mass. That is, 1,000 grams equals 1 kilogram. A gram is somewhat small for many everyday measurements, but it is used in medicine and for prescriptions.

Three aspirin tablets have a mass of about 1 gram. A quart of milk in a

Fig. 1-10
Human hands never touch the nation's standard kilogram, which is protected even from air. (*Photograph courtesy of National Bureau of Standards.*)

waxed paper container has a mass of about 1 kilogram. There is also a standard unit of mass in the English system. That unit is called the *slug.* Most likely you have not heard of this unit. It is not commonly used in the United States. The slug is a larger unit than a kilogram. One slug equals 14.6 kilograms.

Many common items of food are sold in units of grams or kilograms. A box of cereal has a mass of about 300 or 400 grams. And a small container of oregano or cinnamon has a mass of 20 or 30 grams.

Sometimes packages of food are labeled with ounces or pounds. These are units of weight, not units of mass. We should be very careful when comparisons are made. When a 1-kilogram can of coffee is also labeled 2.2 pounds, we are referring to two different concepts. The can of coffee has mass. And that mass is measured as 1 kilogram. But the 2.2 pounds refers to weight. The 2.2 pounds is the *force* of attraction to the earth.

On the surface of the earth a 1-kilogram mass has a weight of 2.2 pounds. But the weight of one mass would not be the same everywhere. Even if the mass stays the same, the weight can change. If we took a 1-kilogram can of coffee to the moon, the mass would be unchanged. It would be equally difficult to shake the can back and forth on the earth or on the moon. However, the weight would change. On the moon, gravitational forces are not as strong as on the earth. On the moon a 1-kilogram can of coffee would weigh about $\frac{1}{3}$ of a pound, not nearly as much as the 2.2-pound weight on the earth.

Because of the variation in weight as objects are moved (say from the earth to the moon), we must always specify the location when making comparisons. For example, in the metric system, we say that a 1-kilogram mass has a weight *on the surface of the earth* of 9.8 newtons. Or in the English system, an object which has a weight of 1 pound *on the surface of the earth* has a mass of 0.03 slug. Each of these comparisons is only valid on the surface of the earth. If we were to move to other celestial bodies where gravitational forces are different, we would find different comparisons between weight and mass.

The important idea in inertia is that an object resists a change in its motion. In general, it might seem that the larger the object, the more the inertia. Indeed, a piano has more inertia than a gallon of milk, which in turn has more inertia than a penny. But size *alone* is not important, because a bowling ball has more inertia than a volley ball or soccer ball of the same size. Sometimes it is said that mass is a measure of the *quantity of matter.* This may help our thinking in certain applications. But it is better to associate mass with inertia than with quantity of matter.

1.6 NEWTON'S FIRST LAW OF MOTION

Up to this point the discussion of physics has been descriptive, and no reasons for the behavior of objects at rest or in motion have been given. Various terms have been used or defined: speed, force, motion, inertia, friction, and mass. In order to discuss Newton's first law of motion we must introduce one more technical definition—velocity.

The concept of speed was simply a mathematical way of describing, measuring, or calculating how fast an object moved. The definition of velocity begins with speed and adds one new piece of information: the direction of motion.

Velocity *Velocity is defined as the speed of an object in a particular direction.*

Your speedometer may indicate that your car is moving at 55 miles per hour. This is the speed. But a compass on the dashboard shows the direction of motion. When you know *both* the speed and the direction, you know the *velocity* of your car (see Fig. 1-11). Although a car can maintain a constant speed of 55 miles per hour for many hours, the velocity is not constant. The car continually changes direction as the car goes around curves, passes other cars, and goes up and down hills. The essential difference between speed and velocity is the direction.

We are now able to combine various definitions into a law about constant velocity motion (or lack of motion) in the absence of any net force.

Newton's first law of motion **In the absence of any net external forces, a mass at rest will stay at rest and a mass in motion will stay in motion at constant velocity.**

Newton's first law means that until some force disturbs a motionless object, the object will *never* move. This same law also means that until some force

Fig. 1-11
Velocity involves both speed and direction. (*Photograph from Beckwith Studios, Brooklyn, N.Y.*)

disturbs a moving object, the object will *never* slow down, speed up, or change direction. But do such statements agree with your everyday experiences? Moving objects, for example, do slow down. To reconcile experience with Newton's first law, the phrase *net external forces* must be carefully examined.

Newton's first law applies only when the net force applied to an object is zero. A net force equal to zero means either that *no forces at all* are applied to the object or that two or more forces *exactly cancel each other,* as shown in Fig. 1-12. If either of these conditions is met, then an object will not change its state of motion.

A case of no force being applied to an object could only occur in the most desolate void in outer space. In such a place the object would be away from the influence of the earth and every planet, star, or heavenly body. In such a place there would not be any force of gravity. If you could put a rock in such a place at rest, the rock would remain motionless. Or if you could put the rock in motion at such a place, the rock would move with unchanging speed in a straight line (that is, constant velocity). But perhaps it will be easier for a moment to consider a place closer to home.

Within our solar system there are forces. The largest of these forces acting over large distances is the force of gravity caused by the sun. This force causes the planets and the earth to move away from straight lines. They do not behave as the rock in our previous example. Instead, the planets follow curved paths around the sun, which means the planets change direction continuously. The motion of the planets is not one of constant velocity, because the *direction* of the motion is continually changing. Constant velocity can only occur when both the speed and direction do not change. Although planetary motion is familiar to us, it is not an example of Newton's first law.

Let us return to a case of no net force. An object can have zero net force acting on it if two or more external forces exactly cancel each other. A book at rest on a desk is such an example. The weight of the book is a force exerted downward. The table exerts an upward force that supports the book. The weight and supporting forces are equal in size but opposite in direction. Thus, they exactly cancel each other to produce no net external force. So the book remains at rest indefinitely. Any object at rest must necessarily have zero net external force acting on it, in accordance with Newton's first law. Some magicians' acts involve Newton's first law, as shown in Fig. 1-13.

Newton's first law could also be applied to a book on a desk if the book is shoved and slides across the desk top. In sliding, the book experiences a force

Fig. 1-12
The net force on the rag is zero.

Fig. 1-13
Dishes do not move unless a net force is acting on them.
(*Photograph from Joseph Perez,* The Physics Teacher, *15, 242, 1977. Copyright 1977 by The American Association of Physics Teachers.*)

of friction. If you apply a force of just the right size by pushing the book, you can keep the book moving with a constant velocity (see Fig. 1-14). In this case there would be no net force acting on the book. The force that you apply with your hand and the force of friction exactly cancel each other. Of course, the desk has to be very long if you want to give the book a constant velocity for any appreciable time.

There is always friction on the surface of the earth, but it can be very small on a smooth, icy surface. So a book tossed onto a smooth ice rink will slide quite far. If there were no friction from the air and the earth were covered by a smooth frictionless surface, a sliding book would never stop.

An object may be completely free of externally applied forces and yet still be subject to internal forces. For example, a spaceship coasting with constant velocity in deep outer space would be free of all externally applied forces. But within the spaceship astronauts could be pushing and pulling various objects. These internal forces would not alter the velocity of the spaceship. Newton's first law only specifies that net *external* forces must be absent in order to guarantee constant velocity.

1.7 NEWTON'S SECOND LAW OF MOTION

Newton's second law of motion shifts the emphasis from the inertia of an object to the net force acting on an object. Before considering this, we introduce a concept that combines two of our previous quantities: the mass and the velocity of an object. Together these two concepts make up the *momentum* of the object.

Momentum *Momentum is defined as the mass times the velocity of the object.*

Fig. 1-14
A book, whether at rest or moving, obeys Newton's first law when the net force applied to the book is zero.
(*Photographs by Mira Schachne.*)

This simple multiplication can be expressed

$$\text{Momentum} = (\text{mass})(\text{velocity})$$

or

$$\text{Momentum} = mv$$

where m stands for the mass and v stands for the velocity. Notice on the one hand that an increase in velocity results in an increase in an object's momentum. On the other hand, if two objects have the same velocity, the more massive object has the greater momentum.

From the definition of momentum we can see that any object at rest has no momentum. That is, when the velocity equals zero, the momentum equals zero. But every moving object has momentum. And that momentum is controlled by two factors. Both the mass and the velocity must be known in order to calculate the momentum. We will utilize the idea of momentum further in this section, in a statement of Newton's second law.

Newton's second law describes how an object behaves when a net force is applied to that object. Of course, the force causes a change in the state of motion. But specifically what physical quantities are changed? And by how much? Before giving answers to these questions, several facts about the interaction of forces and objects must be clearly understood.

First, it takes different applied forces to affect different objects. A small force can put a toy wagon into motion, but it takes a much larger force to begin rolling a boxcar. An even larger force would be necessary to break a building loose from its foundations and begin sliding it across the ground. In each case, for the effect described, the applied force must be greater than the force of friction (or the force of pipes and foundation holding the building). See Fig. 1-15. It is not simply the applied force that is important, but rather the *net force.* In our examples, the net force is applied force minus the

Fig. 1-15
A bulldozer exerts large forces on the building. (*Photograph by Anthony Magro.*)

opposing force (friction or a force caused by walls, pipes, connections, and so on).

Second, the same net force does not always produce the same change of motion. A certain net force applied to the less massive wagon will produce a greater change in velocity than the same net force applied to the more massive boxcar. Therefore, it is not simply the change in velocity that is important, but rather the *change of momentum* of an object. Momentum includes within it both velocity and mass.

Third, two identical objects experiencing the same net forces do not always receive the same change of momentum. If one net force is applied for a longer time than the other net force, the two objects will have different changes of momentum. Therefore, it is important to know the *time* that a net force is applied. These, then, are the three critical factors involved in Newton's second law: net force, momentum, and time.

When these factors are combined, it is possible to state a law about changes of motion.

Newton's second law of motion

Whenever a net force is applied to an object for a time, the momentum of the object will be changed.

This law, unlike Newton's first law, can be expressed in a formula, namely,

$$\text{(Force)(elapsed time)} = \text{change in momentum}$$

or

$$F\Delta t = \Delta(mv)$$

where F = net force
Δt = elapsed time* during which the force is being applied
$\Delta(mv)$ = change in momentum

The change in momentum could happen because there has been a change in mass, a change in velocity, or a change in both the mass and the velocity.

The formula for Newton's second law can be changed under certain conditions. If the mass of an object remains fixed, then the change only occurs in the velocity. Newton's second law can be written as:

$$\text{(Force)(elapsed time)} = \text{(mass)(change in velocity)}$$

or

$$F\Delta t = m\Delta v$$

In this expression the product of force and elapsed time is equal to the

*The symbol Δ which appears here is *delta,* one of the letters of the Greek alphabet. When it appears in a formula, you may read "change in." Thus, read Δt as "change in time" and read $\Delta(mv)$ as "change in momentum."

product of mass and change in velocity. Or if we divide each side of this equation by the change in time, we arrive at

$$\frac{F\Delta t}{\Delta t} = \frac{m\Delta v}{\Delta t}$$

or

$$F = m\frac{\Delta v}{\Delta t}$$

The factor $\Delta v/\Delta t$ is the *acceleration* of the object.

Acceleration *Acceleration is defined as a change in velocity divided by the elapsed time during that change.*

Acceleration is given the symbol a. That is,

$$a = \frac{\Delta v}{\Delta t}$$

It follows, then, that when the mass of an object does not change, Newton's second law can be expressed by the simple formula

$$\text{Force} = (\text{mass})(\text{acceleration})$$

or

$$F = ma$$

This formula for Newton's second law is not as general as the form involving momentum, but it is sufficient to explain many everyday events.

Whenever a net force is exerted on an unchanging mass, the mass accelerates. Consider what happens if a car or an airplane is pushed on level ground. A force greater than the opposing force of friction must be exerted in order to accelerate the car. In order to give a large acceleration to the more massive airplane, it will need a much larger force. Figure 1-16 illustrates this idea. The net force causes the mass to accelerate (speed up from a standstill). If the car or airplane is already moving and the force is in the direction of motion, then the car or airplane speeds up even more. But the force could be directed opposite to the direction of motion. Then the car or airplane would decelerate, in other words, slow down and perhaps stop. Figure 1-17 shows this idea. A mass *accelerates* when the net force is in the direction of motion (or if we begin from rest). A mass *decelerates* when the net force points opposite to the direction of motion.

1.8 NEWTON'S THIRD LAW OF MOTION

All three of Newton's laws of motion deal with forces. The first law described constant velocity in terms of inertia and the absence of force. The second law explained changes in motion in terms of momentum or acceleration in the

Fig. 1-16
The more massive airplane needs several engines to provide a large acceleration. [(*a*) *Photograph from Petersen Publishing Company;* (*b*) *Photograph by Sgt. P. R. Gannett, courtesy of U.S. Air Force.*]

Fig. 1-17
In order to have a large deceleration, a more massive object needs more parachutes. [(*a*) *Photograph from Wide World Photos;* (*b*) *Photograph by Sgt. Stephen A. McConnico, courtesy of U.S. Air Force.*]

presence of force. The third law continues the discussion of forces by focusing attention on two objects interacting with each other.

Forces are *not internal properties* of isolated objects. Instead, forces are properties of the *interaction* of objects. Whenever we speak of *force,* we are speaking of an interaction of one object with another. Without this interaction, there can be no force. For example, if you swing your hand rapidly through the air, you exert very little force on the air. If there were *no* air, the motion of your hand would exert *no* force. Forces always involve two or more objects, as shown in Fig. 1-18.

Whenever you exert a force, there is a second force that is often overlooked. For example, suppose you push with your hands against the wall of a room. You are exerting a force *on the wall.* But the fact that you can *feel* the wall indicates that the wall is also exerting a force *on your hands.* If you push quite vigorously against the wall, the wall responds with sufficient force to compress your hands somewhat. Thus, it is a property of forces to come in pairs. These two forces are called the *action* and *reaction* forces.

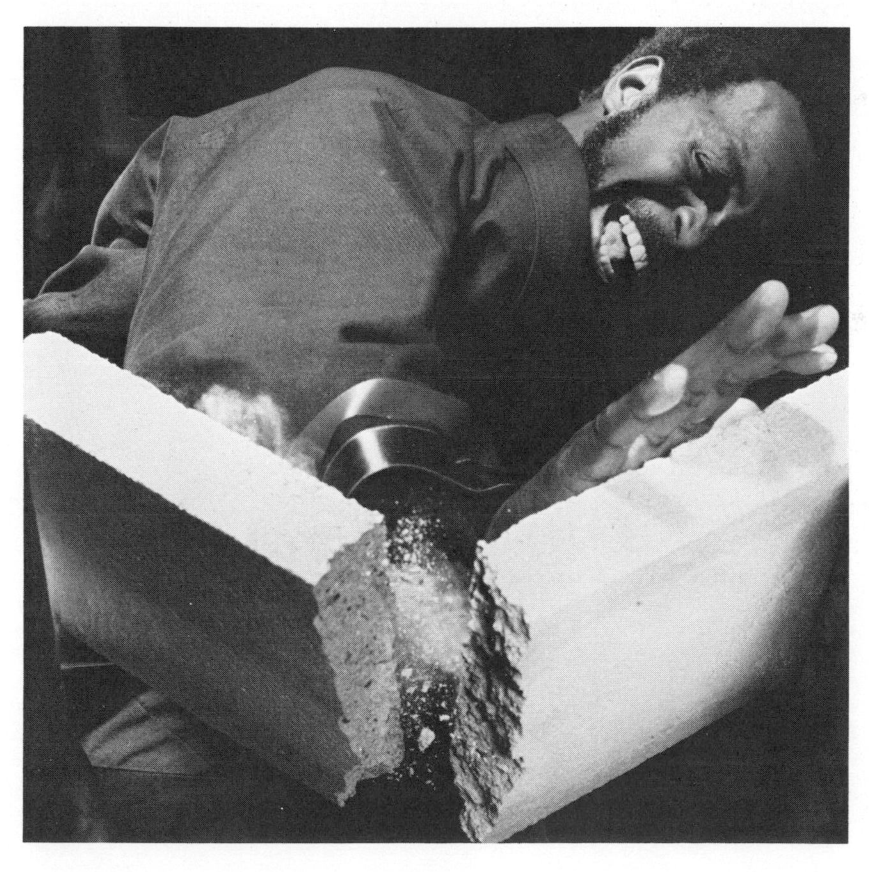

Fig. 1-18
A karate chop exerts equal and opposite forces on the hand and on the block.
(*Photograph by C. E. Miller.*)

Action and reaction forces *An action force is a force applied by one object to a second object, and the reaction force is a force exerted by the second object on the first object.*

It is now possible to state a law about forces in pairs.

Newton's third law of motion **For every action force there is an equal and opposite reaction force.**

The third law clearly recognizes two forces. They are equal in size and point in opposite directions. But what is often overlooked is that there are *two objects* as well. The two forces of action and reaction in Newton's third law always act on *different* objects. Newton's third law is illustrated in Fig. 1-19.

Every object standing or moving on the surface of the earth can be used to illustrate the third law. Each such object has weight, which is a downward force that would make the object fall if released above the ground. When the object is placed on the ground and presses against the earth, the object experiences an equal and opposite force. This upward force exerted by the earth supports the object.

Notice that the reaction force comes into existence in response to the action force. The ground does not push upward of its own accord before an

Fig. 1-19
When two lovers kiss, each feels the same force on the lips. [*Photograph of Auguste Rodin's* The Kiss (*bronze, 1897, Rodin Museum, Paris*). *From Rodin Museum, Paris.*]

object is placed on it. However, when pushed against by the presence of an object, the earth responds with a force that supports the object.

1.9 THE FORCE OF GRAVITY

Think about a golf ball sailing over a long green fairway and then falling back to the ground. After the ball is hit into the air, why does it fall back to the earth? Such a question seems almost absurd, since we are so accustomed to the fact that a golf ball always falls. Of course, golf balls are not the only things that fall. The earth attracts all material objects: people, stones, pencils, birds, or any piece of matter.

The force of gravity

The attraction of all material things for each other is called gravitation, or simply the force of gravity.

Gravitational attraction of the earth pulls all objects thrown into the air back to the earth.

Gravity is not restricted to the earth and objects falling through the air. Newton recognized that *every* material object attracts *every* other material object. Gravitational forces of attraction exist between every pair of masses in the universe, not just masses on the earth.

The gravitational force also exists when an object is at rest on the surface of the earth. The force of gravity is nothing more than the *weight* of an object. Your bathroom scale measures the earth's force of attraction on you. Your weight would be different on the moon because a different gravitational force is pulling you downward, as shown in Fig. 1-20. Nevertheless, your mass would be the same on the earth, the moon, or anywhere. Mass is the

Fig. 1-20
A scale on the moon would give a weight reading only one-sixth the weight reading on the earth. The force of gravity on the moon is only one-sixth as large as the force of gravity on the earth.

BLACK HOLES

Can you picture a really large gravitational force? So powerful that matter is squeezed and compressed into regions of extremely high density? Such a gravitational force is believed to exist. There is experimental evidence for this from astronomical observations.

When gravitation becomes the overpowering force on a stellar scale, matter is crushed into a very dense form. Calculation has shown that a mass about 3 times that of the sun could shrink to a region as small as Manhattan Island. Such a dense object has gravitational forces so strong that nothing escapes, not even light. This dark object is called a black hole.

A hot star glowing brightly in the sky can be stable. The force of gravity attracts all parts of the star toward one another. But the hot interior pushes outward and the star remains stable with a constant size. If the star cools, it will collapse. The gravitational force pulls tighter and tighter as the size lessens. If the star is not too massive, it may collapse for awhile and then become stable at some new higher density. But when a star several times the mass of the sun, or bigger, begins to collapse, it may not stop! The compression may continue indefinitely, breaking through all resistance.

Some of these features are similar to an experiment of piling books on your bed. If you cover the whole bed with one or two layers of books, the mattress will sag. Add another layer, and the mattress will be further compressed. At each stage, the mattress provides forces which resist further sagging. But if you pile the books 1 or 2 feet deep, gravitation wins out. The bed will break! The mattress cannot resist further compression.

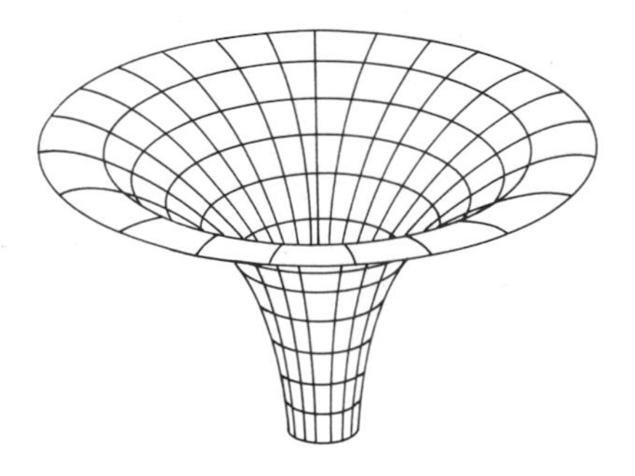

The region of space near a black hole is often mathematically pictured as a whirlpool. A black hole strongly distorts the region around it, as shown in the diagram above. When anything moves too close to a black hole, it will fall in. The intense gravitation of black holes pulls in matter for millions of miles around the black hole. The falling matter is heated up to billions of degrees. As the hot material falls toward the black hole, x-rays are emitted. By detecting such x-rays the presence of the black hole itself may be deduced.

One known source of x-rays in the sky is called Cygnus X-1. It is thought to be a black hole. The total mass is believed to be about 10 solar masses. An artist's conception of this black hole is shown in the painting below. Notice that the black hole is pulling nearby matter.

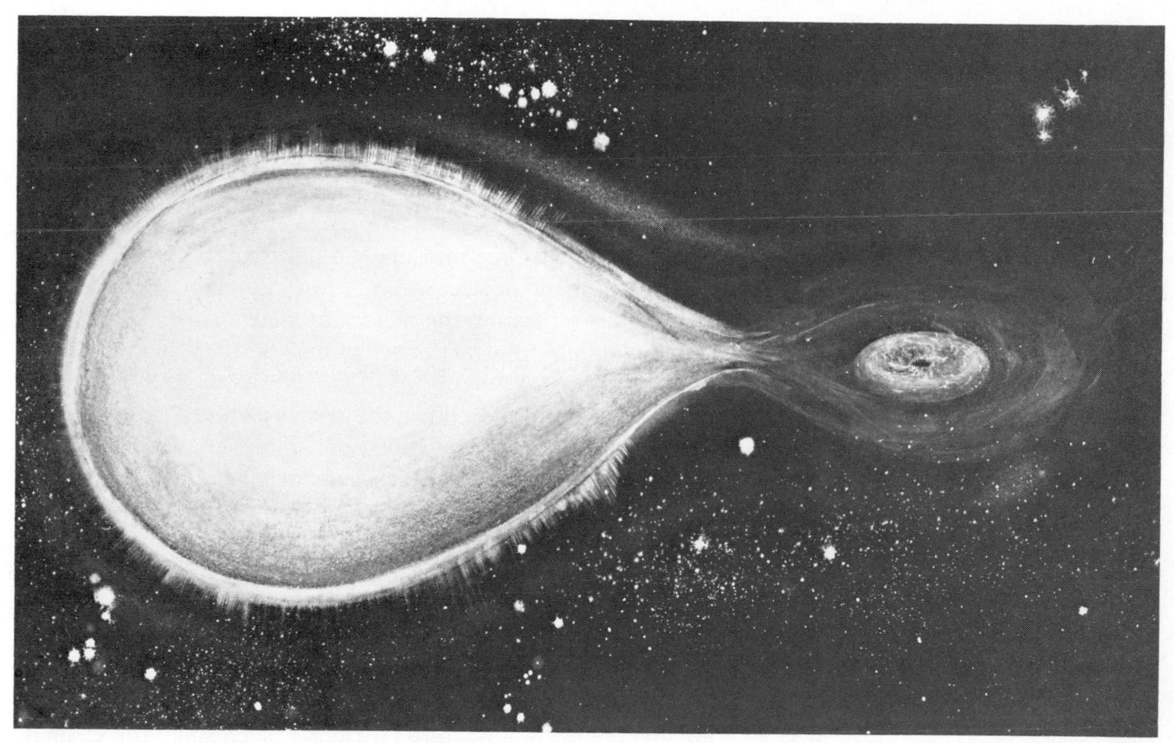

(Painting by Lois Cohen, courtesy of Griffith Observatory.)

Other black holes have been predicted on a very small scale. A mass of 10^{-5} grams may be associated with a black hole whose diameter is only 10^{-35} meters. Such tiny black holes would not occur due to the collapse of cooling stars. Instead these miniature objects may have been formed when the universe began. At such a time, matter could have been compressed into very dense primordial black holes. If so, these black holes may still exist in the universe around us. Experiments will provide new information on black holes in the years ahead.

measure of the inertia of the unchanging quantity of matter in you, but weight is a measure of a planet's force of attraction on you.

The distance between any two attracting masses affects the size of the gravitational force. The magnitude of the gravitational force decreases as the separation between two masses increases. More precisely, the gravitational force varies inversely as the square of the separation of the centers of the two masses.

The overall behavior of the gravitational force between masses is summarized in *Newton's universal law of gravitation.*

Newton's universal law of gravitation

The force of attraction acting between any two masses in the universe is directly proportional to the product of the two masses and inversely proportional to the square of the distance between them.

This law can be expressed in a formula, namely,

$$F = G\frac{m_1 m_2}{d^2}$$

where F = gravitational force measured in newtons (N)
m_1 and m_2 = masses of the objects measured in kilograms (kg)
d = distance separating the masses in meters (m)
G = gravitational constant equal to $6.67 \times 10^{-11}\ \mathrm{N \cdot m^2/kg^2}$*
See Fig. 1-21.

Since gravitational forces depend on the masses involved, as well as their distance apart, numerical values for gravitational forces are widely different. The planets Neptune and Earth are each very massive. The gravitational force of attraction between these two planets is about 20 million tons! Such large forces certainly affect the motion of Earth and Neptune. However, much bigger gravitational forces exist in our solar system. The attraction of the sun on each planet is much larger than the forces of attraction from other planets.

On a much smaller scale, two 500-gram cans of coffee placed 100 centimeters apart attract each other with a force of about 1.5×10^{-11} newtons (three-trillionths of a pound). Obviously, such small forces do not

*The number 6.67×10^{-11} is written in scientific notation, which is shorthand for 0.0000000000667.

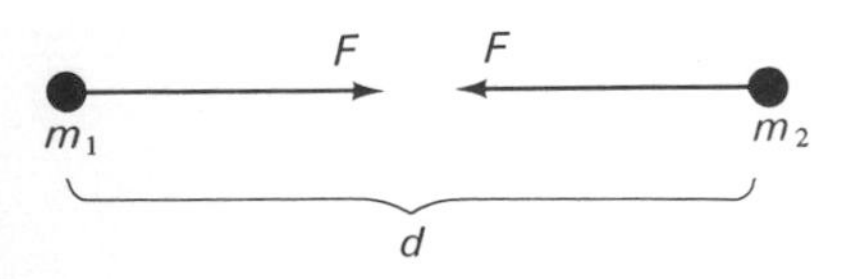

Fig. 1-21
Two masses attract each other with equal and opposite forces.

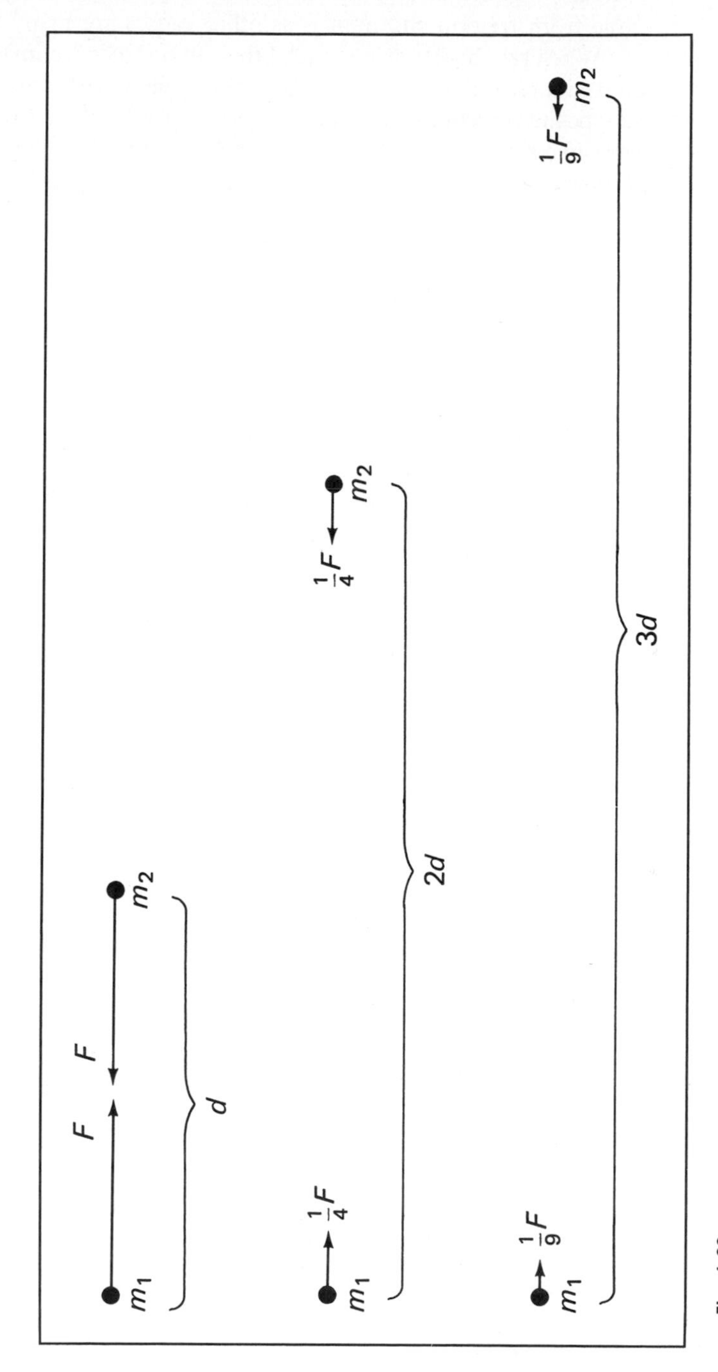

Fig. 1-22
As the distance between the same masses gets larger, the force of gravitational attraction gets smaller.

play any significant role in the behavior of coffee cans! The forces that come from friction and numerous other causes are many times larger.

When two bodies attract each other, two forces are involved. Each mass exerts a force on the other mass. Both forces are of the same size, but they act in opposite directions. All known gravitational forces act this way. Between every two masses there is a pair of attracting forces, which is another example of Newton's third law. Each pair of gravitational forces is an *action-reaction* pair.

The earth exerts a force on a communications satellite. And that same satellite exerts an equal-sized force on the earth. But these equal-sized forces produce much different accelerations. The satellite has a relatively small mass. Thus, the force from the earth causes a moderate acceleration as the satellite moves in orbit around the earth. On the other hand, the earth has a very large mass by comparison. The earth does not experience any significant acceleration caused by the attraction to the communications satellite.

One of the features of Newton's law of universal gravitation is the inverse dependence on the square of the distance. Thus, if the distance between two masses is doubled, the force of attraction decreases by a factor of 4. Or if the distance is tripled, the force decreases by a factor of 9. See Fig. 1-22. Theoretically, if we continue to separate the masses, the force gets smaller and smaller but never shrinks to exactly zero. Of course, there is a practical limit. If the gravitational force acting on a body is much smaller than some other force, the effect of gravitation will be negligible.

So far we have described gravitation in terms of forces. There is another equivalent description in terms of *fields*. A gravitational field is said to exist at any point in space where a mass would experience a gravitational force.

Thus, a gravitational field surrounds the earth. Any mass placed in this field will experience a force (directed toward the earth). And similarly a gravitational field surrounds the sun. In fact, all of space is filled with overlapping gravitational fields generated by masses throughout the universe. A mass placed at different points in a gravitational field will have larger accelerations where the field is larger.

REVIEW

You should be able to define or describe the following terms: speed, force, friction, inertia, mass, velocity, momentum, acceleration, action and reaction forces, and the force of gravity.

You should be able to state the following laws and give examples of their application: Newton's first law of motion, Newton's second law of motion, Newton's third law of motion, and Newton's universal law of gravitation.

PROBLEMS

1. A jet airliner flies 2,100 miles in 4 hours. What is the average speed?

2. A car travels 270 miles in 6 hours and 45 minutes. Calculate the average speed.

3. A snake slithers along at an average speed of 1.5 feet per second. How far does the snake travel in 30 seconds?

4. A fish swims at an average speed of 0.5 foot per second. How far does the fish travel in 1 minute?

5. How long will it take a flash of light to travel the 149,900,000 kilometers from the sun to the earth if the speed of light is about 300,000 kilometers per second?

6. How long will it take a flash of light to travel the 380,000 kilometers from the moon to the earth if the speed of light is about 300,000 kilometers per second?

7. A car accelerates at 5 feet per second per second ($5\ ft/sec^2$) when the tires exert a net force of 500 pounds on the pavement. A new powerful motor weighing the same as the old motor is installed, and the car accelerates at $10\ ft/sec^2$. What is the new net force exerted on the pavement?

8. The rubber straps in a slingshot exert a force of 5 newtons on a rock, which accelerates at 50 meters per second per second ($50\ m/sec^2$) as the rock is released. What is the force exerted on the rock when the rubber straps are pulled back far enough to give the rock an acceleration twice as large?

9. The moon is held in its orbit around the earth by the force of gravity. Suppose the distance from the earth to the moon could be doubled. What would be the new force of gravity on the moon in comparison to the present force of gravity caused by the earth?

10. Suppose the distance from the earth to the sun could be cut in half. What would be the new force of gravity on the sun in comparison to the present force of gravity caused by the earth?

DISCUSSION QUESTIONS

1. The tip of a growing cornstalk moves slowly upward. How could you calculate its speed of growth? What would you estimate to be the speed in units of feet per month?

2. Describe an experiment to measure the approximate speed of a baseball hurled by the pitcher.

3. When you are sewing with needle and thread, how fast do you pull the thread through the cloth?

4. What common objects are *pushed* when you are in the kitchen? What objects are *pulled?*

5. Describe the forces that are involved when tweezers are used to remove a splinter.

6. Think about two bricks one above another in a wall of a building. Does the mortar pull the bricks together or push the bricks apart?

7. After you eat a big Thanksgiving dinner, is your weight changed? Could the change of weight be measured?

8. When standing erect, half of your body weight is carried by each foot. How could the force on each foot be measured?

9. Is the weight of a snake when it is coiled different from its weight when it lies stretched out?

10. Is the weight of an automobile when it is moving different from its weight when it is at rest?

11. How can you increase the force measured on a scale in a gymnasium while you stand on the scale?

12. If there were no friction, would your shoelaces stay tied or would nails stay in boards?

13. If raisins are dipped in flour before mixing them in dough, the raisins will remain distributed throughout the dough. Why do the raisins not sink to the bottom of the dough?

14. Air resistance is often neglected in studying falling objects, such as rocks dropped from a bridge. Would it be reasonable to neglect air resistance in studying parachutes or falling feathers?

15. Is there a difference in the masses of 1 kilogram of lead and 1 kilogram of feathers?

16. Describe the mass changes, if any, for each of the following objects: a pencil being sharpened, an automo-

bile being driven, a book being read, and a blade of grass growing.

17. One dictionary definition for inertia is "inactivity or sluggishness." Another definition is "that property of matter by which it retains its state of rest or of uniform motion in a straight line so long as it is not acted upon by an external force." Does your dictionary definition agree with the first or second definition given above?

18. A slow steady pull unwinds a roll of toilet paper, but a sudden jerk tears a single sheet of toilet paper. What characteristic of the paper causes this difference in behavior?

19. If a race is held between a person and a horse, which one will win a 1-mile race? A 10-foot race? What factor is important in the 10-foot race?

20. A car that is stopped at a traffic light is struck from the rear. Why do occupants of the car suffer mainly from neck injuries?

21. When your head rests on a pillow, several forces are present. Identify the forces that come in pairs, according to Newton's third law.

22. What forces are involved when an ice skater starts from rest?

23. If a skyscraper were mounted on a platform placed upon a truly frictionless surface and you were standing on a fixed piece of concrete, could you move the building by pushing?

24. When you drive a car around a corner at a reasonable constant speed, why do you experience a force from the *side* of your seat belt? If you turn to the left, which side feels the force of the seat belt?

25. When jumping beans move about, do they violate any of Newton's laws of motion?

26. Can a force change the velocity of a moving object without changing the speed?

27. If your weight is a force, why does the velocity of your body not change continually?

28. Can a baseball be given a larger velocity with a sharp blow from a bat or a hefty throw from a center fielder? What are the two important factors in changing the velocity of the baseball?

29. Why does a raw egg usually break when dropped on a linoleum floor, but often does not break when dropped on a carpeted floor?

ANNOTATED BIBLIOGRAPHY

Interested in tennis, golf, or baseball? The forces involved with bouncing, rolling, and skidding are all dealt with in C. B. Daish, *Learn Science through Ball Games,* Sterling Publishing Company, New York, 1972. Also details of billiards, bowling, and squash are dealt with. Includes discussion of spin for you experts! A delightful small book.

The introduction of force which we have given is only one of several possibilities. A long book, but very clearly written, is Mary B. Hesse, *Forces and Fields,* Nelson and Sons, London, 1961. It gives the historical development of many present-day ideas of force. The important point is made that while certain of our current ideas are necessary to fit experimental facts, other ideas should be considered a matter of taste. Alternative theories and definitions could be chosen.

Chapters on the problems of movement, the role of friction in movement, and force and mass highlight R. A. R. Tricker and B. J. K. Tricker, *The Science of Movement,* American Elsevier Publishing Company, Inc., New York, 1967. Written with a pleasant British attention to detail, this is a good reference for Chap. 1.

Some critical comments which straighten out many false ideas about inertia have been given in Mario Iona, "Inertia Means All That," *The Physics*

Teacher, vol. 13, 1975, p. 242. This short article points out that inertia is a property of an object. Inertia cannot and does not need to be overcome.

One magazine which publishes regular science articles for the general reader is *Scientific American.* It covers all fields of science at a moderately high level. Many articles give some detail about recent developments. You might try two of these which deal with gravitational forces. See "Is Gravity Getting Weaker," February 1976, p. 44, and "Navigation between the Planets," June 1976, p. 58.

2 ENERGY AND CONSERVATION LAWS

In a study of motion it is possible to combine in several ways the separate ideas of distance, speed, and mass. The new combined quantities, such as energy, linear momentum, and angular momentum, are conserved. Conservation laws are fundamental and will also appear in later chapters.

2.1 CENTER OF GRAVITY

In Chap. 1 it was shown that the forces acting on an object are important when discussing motion and changes of motion. Sometimes the size and shape of an object are also important but often not. If you toss either a golf ball or a football with the same velocity, both balls will follow the same curved path through the air, both will land at the same spot, both will strike the ground at the same time, and both will have the same velocity upon striking the ground.*

The study of motion would be more general if the size and shape of various objects could be ignored and attention concentrated on the motion. This generalization is accomplished by imagining the object to be replaced by a single point known as the *center of gravity.*

Center of gravity *The center of gravity is defined as a single point associated with an object where all the forces, especially the effect of gravity, can be imagined to act on the object.*

*If you threw them very hard, air friction would make their paths different. We ignore the effect of friction here.

Sometimes even a large object can be idealized to a single point. For example, both the sun and the earth are often considered to be points in the discussion of the motion of planets in astronomy.

There is also an interesting practical reason for considering an object's center of gravity. The center of gravity is used when we wish to balance an object. For example, the weight of a ruler is a force distributed over all parts of the ruler. One way of supporting a ruler is to place it on a table. The table then exerts supportive forces on all parts of the ruler to counteract the weight. But the weight of the ruler can be imagined to act only at the center. Place the tip of one finger under the center of gravity and support the ruler at a single point. You can balance a ruler in this way with your fingertip as shown in Fig. 2-1. For this reason we will use *balance point* as another name for center of gravity.

It is easy to locate the center of gravity of many common, hand-held objects. Suppose for a moment that you did not know that the balance point of a ruler is located at the exact center. You could experimentally find the balance point by resting the ruler on top of two outstretched index fingers. By slowly drawing the fingers together, first one finger, then the other finger, will slide beneath the ruler. The two fingers will come together under the exact center of the ruler. The balance point is at the center. This same trick with sliding fingers can be repeated on nonuniform objects. Figure 2-2 shows how to find the center of gravity of a bowling pin. Try it with a pencil, a pen, or a golf club. In each case you will easily locate the balance point. But in these cases your guess ahead of time might not be so good. Make a guess for your kitchen broom. Then try the experiment.

Another technique can be used to find the center of gravity of an irregularly shaped object, such as a heart-shaped valentine card (see Fig. 2-3). If the card is loosely pinched between two fingers, it will swing back and forth, but eventually come to rest. Draw a vertical line across the card

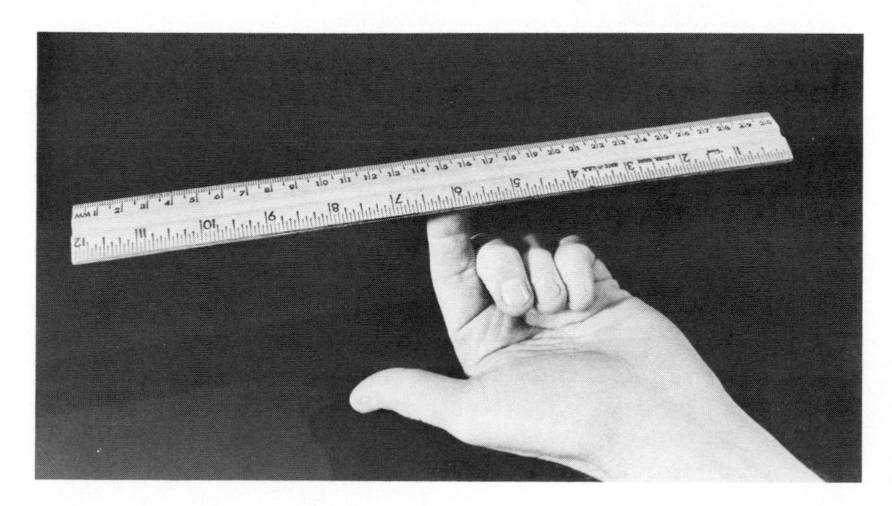

Fig. 2-1
Supporting a ruler with a fingertip under the center of gravity is equivalent to a table supporting the ruler at all points. (*Photograph by Ted Rozumalski from Black Star.*)

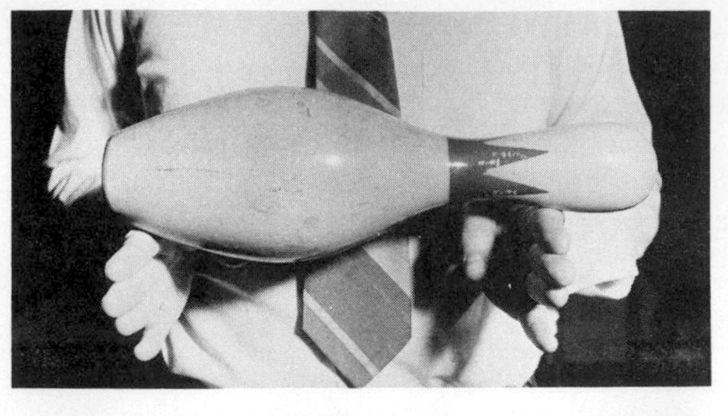

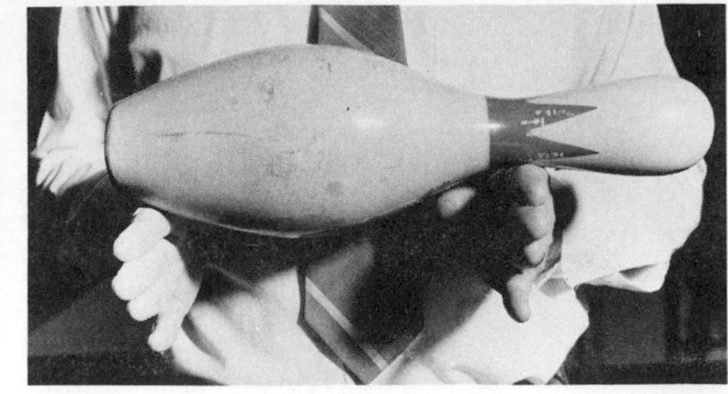

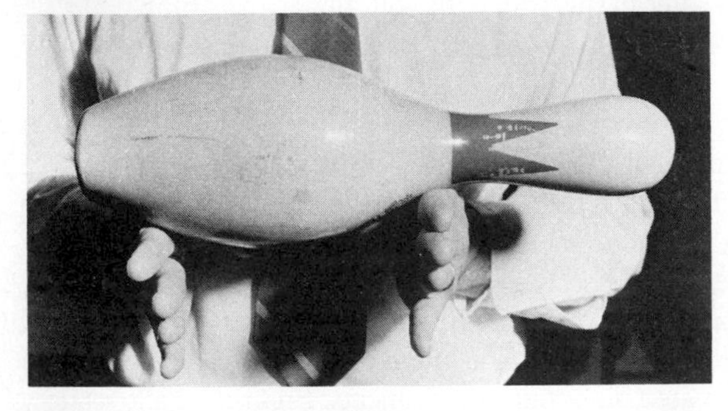

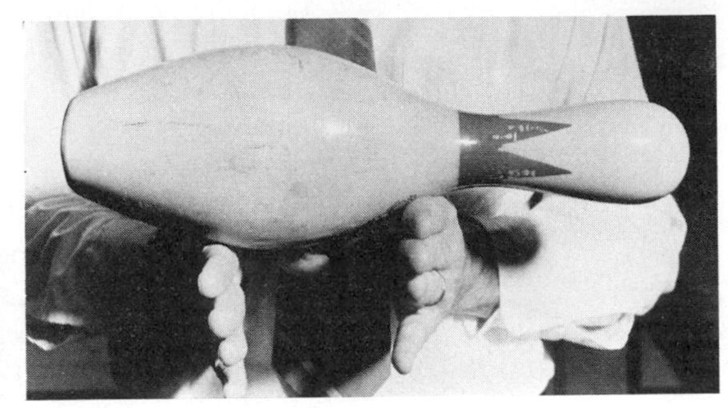

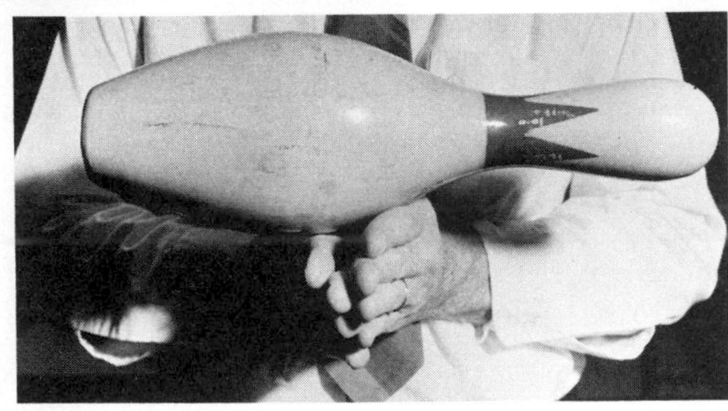

Fig. 2-2
The center of gravity of a bowling pin can be found by slowly bringing the hands together. (*Photograph by Ted Rozumalski from Black Star.*)

through the point where the card is being held. Hold the card at another point (not on the line that you have drawn) and repeat the process (including drawing a second line). Where the two lines intersect is the center of gravity, or balance point. The card can be supported by placing your finger under this point.

All objects have a center of gravity, even though the location of the center of gravity may not be obvious. For a human being the center of gravity is located approximately behind the navel, midway into the body. If a person lies on a "teeter-totter" with the navel over the support, usually the person can be balanced so that neither end of the teeter-totter touches the ground.

An even less obvious center of gravity is associated with an inner tube. In

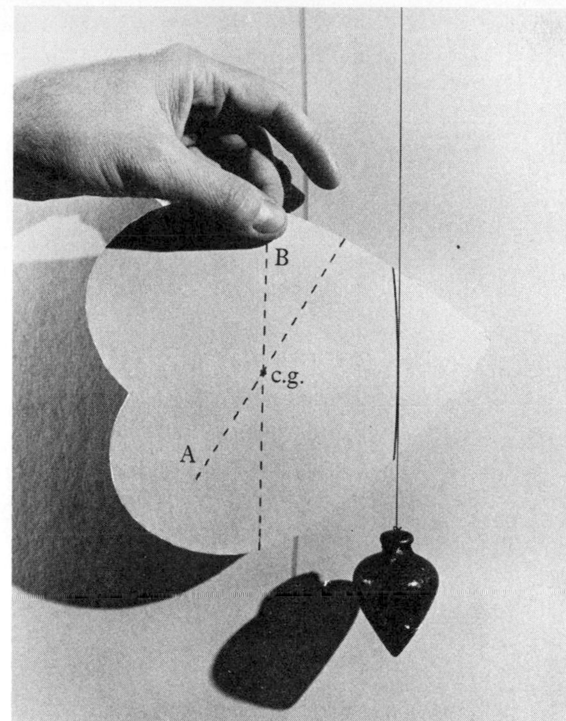

Fig. 2-3
To find the center of gravity of a valentine, draw vertical lines through two points of suspension. The point of intersection of the lines is the center of gravity. (*Photographs by Ted Rozumalski from Black Star.*)

this case the center of gravity is located in the *hole*, not in the material that makes up the object. But this is consistent with the definition of center of gravity which only stated that a center of gravity is *associated with an object*, not necessarily *in an object*. If you tried to balance an inner tube or donut, you could balance them uneasily on their edges, as shown in Fig. 2-4. But a stiff, light piece of cardboard taped to them over their holes would be necessary to support them on one finger in a horizontal position.

Any object can be altered in such a way that the center of gravity is moved. When a piece of a ruler is cut from one end, the center of gravity of the remaining piece is moved to the new center. Similarly, a ruler with a glob of putty added to one end has a new balance point. Is the balance point shifted toward or away from the putty?

2.2 EQUILIBRIUM

The definition of center of gravity determines a point for each object that we may wish to consider. The location of this balance point is independent of whether or not the object is moving. Further, the location does not depend on forces applied to the body. In this section we consider the special case of an object with no net force applied.

An example would be a broom balanced on your finger. The weight of the broom is equal and opposite to the supporting force. In such a circumstance an object is said to be in *equilibrium.*

Equilibrium *Equilibrium is defined as the condition of an object when the center of gravity experiences no net force.*

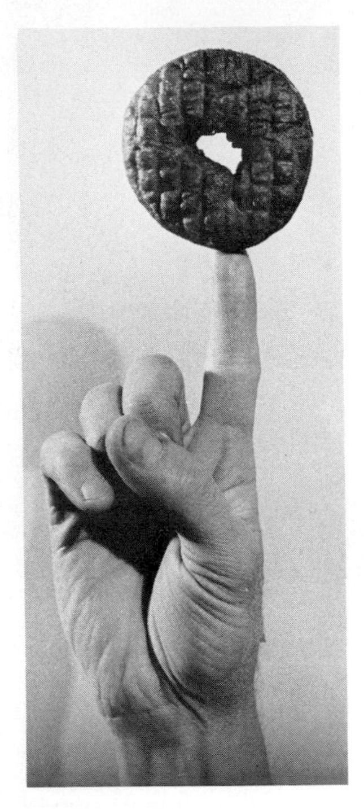

Fig. 2-4
Even though the center of gravity is in the hole, a donut can be balanced on its side. (*Photograph by Ted Rozumalski from Black Star.*)

A refrigerator standing on the kitchen floor is in equilibrium. Two forces are acting on the refrigerator. One of these forces, the weight, acts downward. The other force pushes up on the bottom of the refrigerator. This second force is provided by the floor. Both forces are the same size, but they act in opposite directions. There is no *net* force on the refrigerator. The refrigerator is in equilibrium. Any object at rest on the surface of the earth is in a condition of equilibrium.

An object being accelerated along a tabletop by an applied force is *not* in equilibrium. The weight is again opposite to the supporting force, but the applied force which produces the acceleration is directed horizontally. So the net applied force causes the object to change its momentum, in accordance with Newton's second law of motion.

The word equilibrium is also used in another way. First, as mentioned previously, equilibrium specifies a *condition* of an object—the center of gravity experiences no net force. Second, *equilibrium is used to describe a position of an object such that the object stays at rest.*

There can be one or more equilibrium positions for various objects, depending on the shape. A marble on a level table has one equilibrium position: where the marble is resting on the table, regardless of which part of the marble touches the table. A coin on a level table has two equilibrium positions, either lying on its side or balanced on its edge. (The two different sides are treated as one equivalent position, and the unlimited number of different places on the edge are treated as another equivalent position.) A cigar box on a level table has three equilibrium positions: resting on its bottom (or, equivalently, its top), its back (or, equivalently, its front), and its end (or, equivalently, its other end). The three equilibrium positions of a cigar box are shown in Fig. 2-5. Can you think of an object with four equilibrium positions?

If an object is released from any position other than an equilibrium position, the object will move toward an equilibrium position. Drop a cigar box from a few inches above a table, and it will quickly come to rest on one of the faces. It does not matter how you release it. It always reaches some equilibrium position. Or drop a coin. Again, the coin falls into one of its equilibrium positions.

Not all equilibrium positions are the same. Differences in the equilibrium positions are a matter of stability.

Law of stable equilibrium **An object is said to be in stable equilibrium when a small displacement or movement causes some oscillatory motion with the object eventually returning to its original equilibrium position.**

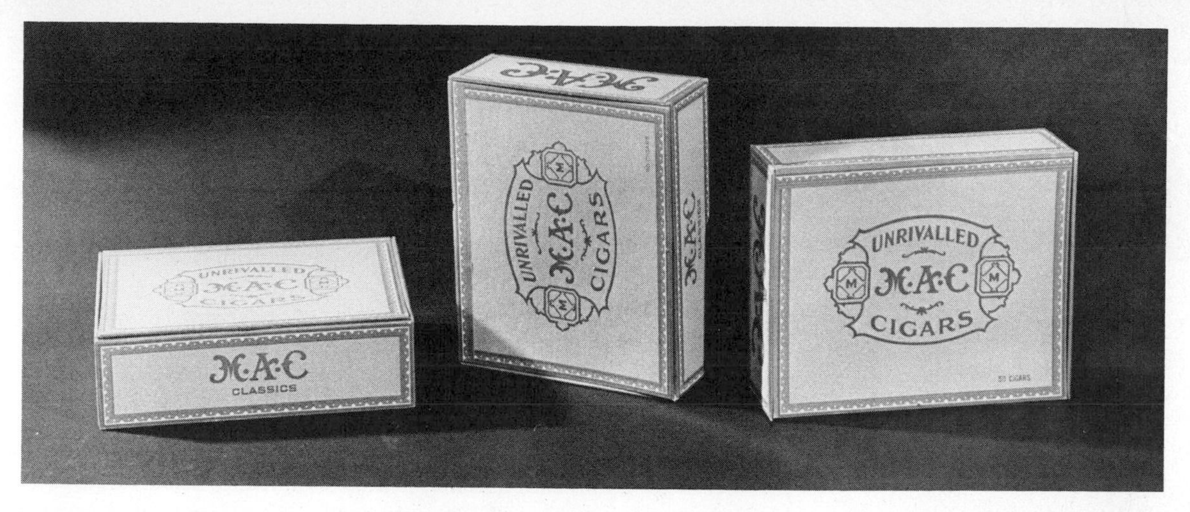

Fig. 2-5
A cigar box has three equilibrium positions. (*Photograph by Ted Rozumalski from Black Star.*)

An upright milk carton can be given a slight nudge. Even though the carton wobbles back and forth, it returns to an upright position.

The opposite of stable equilibrium is unstable equilibrium. An object is said to be in unstable equilibrium when a small displacement or movement causes no oscillatory motion. Instead, the object moves into another equilibrium position. Between the two extremes of stable and unstable equilibrium positions lies a range of positions of varying stability. For example, a coin lying on its side has more stability than a coin standing on edge (see Fig. 2-6). But even on edge, a coin is stable! A very *slight* nudge causes it to wobble

Fig. 2-6
A coin has two equilibrium positions. (*Photograph by Ted Rozumalski from Black Star.*)

back and forth, but the coin returns to its original equilibrium position. Of course, a larger force causes the coin to fall over.

If you were able to balance a pin on its point, the pin would be in an unstable position. Any force, though only barely perceptible, would cause the pin to fall over. Then it would be in an equilibrium position lying along its length.

What factors control stability? One factor is the size of the area of support. As the area increases, the equilibrium position has increasing stability. For example, a half-gallon carton of milk is more stable lying down than standing upright. Or, again, there is more stability in an ice-cream *cone* (without the ice cream) balanced upside down on the cup than right side up on the point. In fact, when the cone is balanced on the point, it is in a position of unstable equilibrium, as is any object balanced on a point or corner.

Another factor controlling stability is the position of the center of gravity relative to the *area* of support, which could be the *point* of support. An object has increasing stability as the center of gravity is lowered closer to the area of support. An example of this can be given with a common cardboard oatmeal box (in the shape of a cylinder). When the box is on end, the center of gravity is higher than when the box is lying down. (Where is the center of gravity in each case?)

Race cars at the Indianapolis 500-Mile Race have been built more "low slung" over the years in order to lower the center of gravity. As a result, the race cars are less liable to roll over because of winds from the side, collisions with other cars, and skids. In other words, low-slung race cars are less likely to roll over from an upright equilibrium position to an equilibrium position resting on the side or the top of the car.

If the center of gravity of an object is beneath the area of support, then the object is stable for all applied forces. Such a case would occur for a teacup suspended by a hook (see Fig. 2-7). Give the teacup a push—any push—and it returns eventually to the equilibrium position.

Tightrope walking on a high wire depends upon lowering the center of gravity to increase stability. Normally the center of gravity of a person is approximately at the level of the navel. A very experienced tightrope walker, by careful balancing, is able to walk on high wires without a pole, even though the body is unstable. But when carrying a pole which droops significantly at each end, a tightrope walker has a lowered center of gravity. The stability is increased.*

There is another circus act on a tightrope that looks very dangerous, but it is actually safer than a tightrope walker performing on a wire. This act involves a motorcycle with a carriage mounted under the tightrope, as shown

*The inertia of the pole is also important in tightrope walking. Inertia helps to prevent sudden motions to either side of the wire.

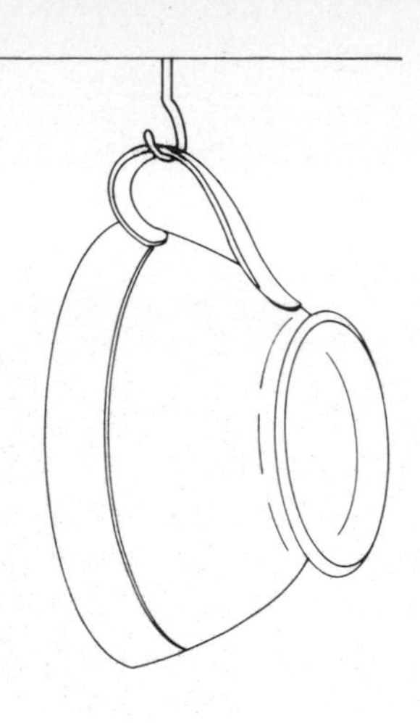

Fig. 2-7
A teacup suspended by a hook is stable because the center of gravity is beneath the support.

in Fig. 2-8. Usually one person drives the motorcycle and another person rides the carriage. This equipment is built so that the center of gravity of the motorcycle, carriage, and the two riders is *beneath* the tightrope. Such a device is perfectly stable. Both performers can lean over as much as they wish. The carriage merely swings back and forth like the teacup on the hook.

2.3 WORK

The word work is a term common to everyday life. Work is associated with activities, achievements, and fatigue. When you stop to analyze work in terms of other concepts, one factor is force. Forces are used to lift objects, to move objects, and to perform physical activities. Another factor in the everyday notion of work is time. Most employees work an 8-hour day or a 40-hour week.

In physics, work has a technical meaning with an important difference compared to the everyday usage of the word. Work is involved with the ideas of force and *distance,* totally excluding any consideration of time. Work is done whenever a force moves an object, independent of the time required.

Work *The work done on an object is defined as the force exerted on the object multiplied by the distance the object moves in the direction of the force.*

This definition can be expressed in a formula

$$\text{Work} = (\text{force})(\text{distance})$$

or

$$\mathcal{W} = Fd$$

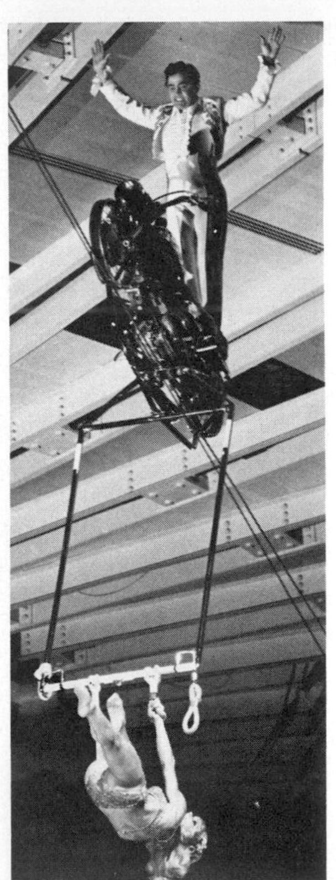

Fig. 2-8
Unstable. Stable. [(*a*) *The Metropolitan Museum of Art, Rogers Fund, 1919;* (*b*) *Photograph by K. Regan from Camera 5, Inc.*]

where $\mathcal{W}$ = work
F = force on the object
d = distance the object moves in the direction of the force.

We see that if we increase either the force or the distance moved, the work increases.

In the English system of units, force is measured in pounds and distance is measured in feet. The unit of measurement of work is the foot-pound, abbreviated ft·lb. In the metric system the unit of work is the joule, abbreviated J. When the force is expressed in newtons and the distance in meters, work is measured in joules. One newton-meter equals one joule.

As an example of the calculation of work, suppose a 5-pound bag of sugar is lifted a distance of 3 feet from the floor to the kitchen counter top.

$$\begin{aligned}\mathcal{W} &= Fd \\ &= (5 \text{ pounds})(3 \text{ feet}) \\ &= 15 \text{ ft}\cdot\text{lb}\end{aligned}$$

Various activities require that work be done, as indicated in Table 2-1. Vasily Alexeev won the 1976 Olympic weight lifting competition by lifting 532 pounds, as shown in Fig. 2-9.

Notice that the technical definition of work specifies that the distance an object moves must be in the direction of the force. In other words, the force and direction must be parallel. An example which shows this distinction is

Table 2-1
Various activities and work done

Activity	Work done, ft-lb	Work done, J
Testing a birthday cake with a toothpick to see if the cake is cooked	0.020	0.027
Lifting a 5-pound bag of sugar from the floor to the kitchen counter	15	20.4
Pushing a shovel into a pile of loose sand	30	40.8
A 90-pound boy climbing a 10-foot ladder onto a roof	900	1,224
Lifting a heavyweight barbell in the Olympics	3,230	4,390

climbing stairs. Walking up a flight of stairs is work! You use your muscles in moving your weight (a vertical force) from one floor to an upper floor. The amount of work is the product of your weight and the distance between the floors. The horizontal length of the stairs is immaterial in calculating this work. Only the vertical height between floors is important. This vertical distance is in the same direction as the (vertical) force of your weight, as shown in Fig. 2-10.

The fact that force and distance must be parallel explains why a bicycle requires so little work to keep in motion, almost irrespective of the weight of the rider. When pedaling on level ground, the rider's weight is not rising or falling, and so no work is being done in that way. But muscular effort is needed in order to move the bicycle and rider against air friction, tire friction, and bearing friction—all forces. The work expended can be decreased by decreasing all these forces: pedaling in a crouched position, using narrow tires instead of balloon tires, and lubricating the bearings. Of course, it is a different story pedaling uphill. Then your weight is being raised, which requires that work be done *in addition* to the amount discussed above.

If the force exerted on the body and the direction the body moves in are not parallel, then less work is done. In the extreme case of the exerted force being perpendicular to the direction of motion, no work is being done by that force. As an example, if a ball on the end of a chain is swung around your body in a perfect circle, the force exerted by the chain is exactly perpendicular to the ball's direction of motion at all points on the circle. See Fig. 2-11. Even though the force of the chain is continuously changing the direction of the ball's motion, the chain would be doing no work on the ball.

Fig. 2-9
The strongest man in the world. Winner of the 1976 Olympics, Vasily Alexeev lifted a mass of 241.5 kilograms (a weight of 532.2 pounds) approximately 1.85 meters (6.07 feet). It took 4390 joules (3,230 foot-pounds of work) to lift the barbell. (*Photograph from United Press International, Inc.*)

2.4 POWER

The only two variables needed to calculate work are the force and the distance moved in the direction of the force. The elapsed time plays no role in the calculation of work. Whether the work is performed rapidly or slowly does not change the work. It takes the same work to push a heavy refrigerator across a floor in 10 seconds, 20 seconds, or any elapsed time.

Although time is not a part of the technical definition of work, time is an important consideration. For example, an employer not only wants a job (work) done but wants it done in a reasonably short time. To combine the ideas of work and time, we introduce the idea of *power*.

Fig. 2-10
The work is the same in each case.

Fig. 2-11
No work is done by the force of the chain.

Power *Power is defined as the work done divided by the time interval.*

Written as an equation,

$$\text{Power} = \frac{\text{work}}{\text{time}}$$

$$P = \frac{\mathcal{W}}{t}$$

where P = power
$\mathcal{W}$ = work
t = time

In the English system of units, when the work is expressed in foot-pounds and the time in seconds, then the power is in units of foot-pounds per second (abbreviated ft·lb/sec).

Other units of power are in common use. One such unit in the English system is *horsepower.* In this system, 1 horsepower equals 550 foot-pounds per second. The power ratings of electric motors and gasoline engines are usually given in horsepower. The metric unit of power is the *watt* (abbreviated W), when work is expressed in joules and time is expressed in seconds. In the metric system, 1 horsepower is equal to 746 watts.

In the calculation of power, both the work and the time are variables. If the time is not allowed to vary, the power depends *directly* on the work done. As more work is done during the same elapsed time, more power is being delivered. Less work performed during the same elapsed time means less power is being delivered. If the work is not allowed to vary, the power depends *inversely* on the elapsed time. As the elapsed time decreases (that is, work is being done faster), the power increases. As the elapsed time increases

Fig. 2-12
In both illustrations the same work is done in pushing the refrigerator. On the left less power is used to move the refrigerator slowly. On the right more power is used to move the refrigerator quickly.

(that is, work is being done slower), the power decreases, as shown in Fig. 2-12.

When we slide a refrigerator out for cleaning, the work required might be about 120 foot-pounds. If 6 seconds is the time interval during which the work was done, we can compute the power

$$P = \frac{\mathcal{W}}{t}$$

$$= \frac{120 \text{ foot-pounds}}{6 \text{ seconds}}$$

$$= 20 \text{ ft} \cdot \text{lb/sec}$$

Suppose the 120 foot-pounds of work was done in 3 seconds. Then the calculated power would be

$$P = \frac{\mathcal{W}}{t}$$

$$= \frac{120 \text{ foot-pounds}}{3 \text{ seconds}}$$

$$= 40 \text{ ft} \cdot \text{lb/sec}$$

As the elapsed time decreased (from 6 to 3 seconds) for a fixed amount of work (120 foot-pounds), the power increased from 20 ft·lb/sec to 40 ft·lb/sec.

The concept of power can be applied to people as well as machines. How much power can people produce? The answer depends on how long we can maintain the activity. A bicycle rider can easily produce 150 watts for several hours. But if the power is raised to 750 watts, a person would be exhausted in 1 minute. And at double this power, 1,500 watts, a person would be exhausted in 5 or 6 seconds.

Another example to illustrate the relationship between work and power considers an electric sewing machine. Every sewing machine must be able to complete the job to be useful. But if the motor has insufficient power, the work will go slowly. A *powerful* sewing machine will permit the job to be completed in less time. It is not enough to have a machine which can do the *work*. A machine must have sufficient *power* to complete the work in the desired time.

2.5 POTENTIAL ENERGY AND KINETIC ENERGY

Although work is not a characteristic of an object, there can be no doubt that work does have an effect on an object. This effect is simply called *energy*. In other words, there is a cause-and-effect relationship. Work (an external activity) causes an object to gain energy (an internal characteristic). It is proper to speak of putting energy into an object by means of the activity of work. An object can be "full of energy" or "energized."

Because of the great variety of physical conditions which can exist, we speak of different kinds of energy. A rapidly moving rock has a different kind of energy than a rock that has been carried to the top of a building and is about to be dropped. A warm rock has a different kind of energy than a rock that is radioactive. Sunlight has a different kind of energy than a battery. All these different kinds of objects with different kinds of energies will be discussed in the book, but it is only natural to start with the mechanical forms of energy.

What kinds of energy can you give a Frisbee? You could pick it up from the ground and place it on a picnic table. This would be a change in the condition of the Frisbee. Your activity of work involved lifting the Frisbee some distance. Now the Frisbee is on the table. The effect is to give a *potential energy* to the Frisbee.

Potential energy *Potential energy exists whenever some object possesses energy because of its position.*

Lift a book overhead and place it on a shelf. Work has been done on the book, and it is in a new position. It is higher than it was before. You have given the book some potential energy. The book now has an energy it did not have before. Figure 2-13 shows a Frisbee thrown high in the air. At the top of its arc the Frisbee has more potential energy than when it is in the thrower's hand or on the ground.

Another example of potential energy is to stretch a rubber band out until it is taut. Now each piece of rubber has taken up a new position, and work has been done on the rubber band. Each section of the rubber band has been stretched, and each piece is farther away from its neighbors. If the tension were released, the rubber band would snap back to its original position. The stretched rubber band has potential energy.

Return now to our Frisbee. What else can be done with it? You could throw it to a friend. While it is moving, the Frisbee has a changed condition as a result of the work done in throwing the Frisbee. Now the Frisbee has some speed. The effect is to give a *kinetic energy* to the Frisbee, as shown in Fig. 2-14.

Kinetic energy *Kinetic energy exists whenever an object has energy because of its speed.*

The concepts of work and energy are abstract. Table 2-2 summarizes the cause-and-effect relationship between work and energy. On the one hand, work increases potential energy when it is observed that an object is lifted. On the other hand, work increases kinetic energy when it is observed that an object is thrown.

In principle, the energy of an object is measured by the work done on the object. But as a practical matter, there are simpler ways of determining the energy possessed by an object. In the case of lifting an object with a certain

Table 2-2
The cause-and-effect relationship between work and energy

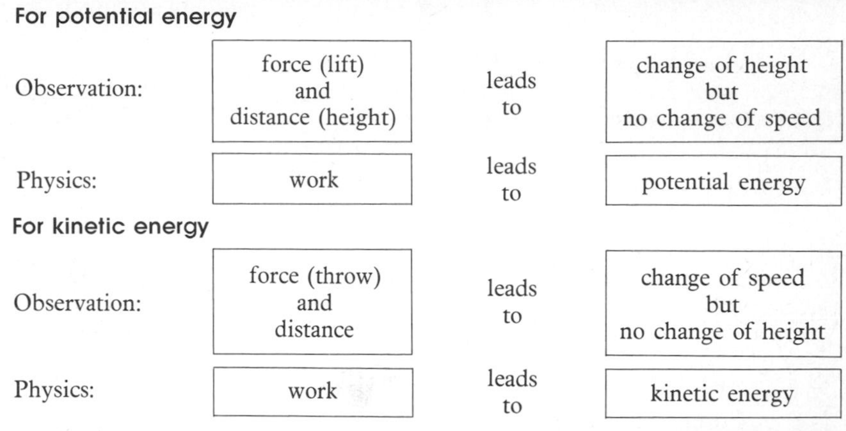

For potential energy

Observation:	force (lift) and distance (height)	leads to	change of height but no change of speed
Physics:	work	leads to	potential energy

For kinetic energy

Observation:	force (throw) and distance	leads to	change of speed but no change of height
Physics:	work	leads to	kinetic energy

Fig. 2-13
The work done in throwing a Frisbee high into the air has given the Frisbee potential energy. (*Photograph by Russ Kinne from Photo Researchers, Inc.*)

weight, the calculation of the potential energy is particularly simple, namely,

$$\text{Potential energy} = (\text{weight})(\text{height})$$

or

$$\text{PE} = wh$$

where PE = potential energy
w = weight of the object
h = height or vertical distance that the object has been raised

In the English system of units potential energy is measured in foot-pounds, when w is measured in pounds and h is measured in feet. We see that because potential energy equals w times h, an increase in either w or h produces an increase in potential energy.

Potential energy can also be computed using the metric system of units. When the weight w is in newtons and the height h is in meters, the potential energy is in joules. The metric system uses the joule for all types of work and energy. When a weight of 6 newtons is lifted 2 meters vertically, 12 joules of potential energy has been given to the object.

Potential energy increases not only because an object is lifted, but also because a person or animal jumps. Figure 2-15 shows a kangaroo at the top of its jump. The kangaroo has more potential energy in midair than on the ground. Table 2-3 lists typical values of potential energy for a boy and various animals at the top of their jump.

Fig. 2-14
The work done in throwing a Frisbee in level flight has given the Frisbee kinetic energy. (*Photograph by Russ Kinne from Photo Researchers, Inc.*)

A formula can also be given for calculating the kinetic energy associated with objects in motion. The kinetic energy involves both mass and speed. The formula is

Table 2-3
Potential energies at the top of a jump

Jumper	Potential energy, ft · lb	Potential energy, J
Grasshopper	0.005	0.0068
Bullfrog	1	1.36
Cat	30	40.8
Boy	600	816
Kangaroo	900	1,224

$$\text{Kinetic energy} = \left(\frac{1}{2}\right)(\text{mass})(\text{speed})^2$$

or

$$\text{KE} = \frac{1}{2}mv^2$$

where KE = kinetic energy
m = mass of the object,
v = speed of the object

In the metric system of units kinetic energy is measured in units of joules, when m is in kilograms and v is in meters per second. In the English system of units kinetic energy will be in foot-pounds, when m is in slugs and v is in feet per second.

Kinetic energy increases as a person, animal, or any object moves faster.

Fig. 2-15
A kangaroo at the top of its jump has potential energy because of its height above the ground. (*Photograph Courtesy of Australian Tourist Commission.*)

Fig. 2-16
A charging elephant has kinetic energy because of its speed and mass. *(Photograph from the MGM release KING SOLOMON'S MINES © 1950, Loew's Incorporated. Copyright renewed by Metro-Goldwyn-Mayer Inc., 1977.)*

Figure 2-16 shows a charging elephant, which has a large kinetic energy. The elephant's large kinetic energy is the result of both its speed and its large mass. Table 2-4 lists typical values of kinetic energy for animals moving in different ways.

In Chap. 1 it was seen that speed is a relative concept, namely, relative to the observer who measures or observes the moving object. Potential energy and kinetic energy are also both relative concepts. A person walking in a moving train has many different kinetic energies: one relative to the train, another relative to the surface of the earth, another relative to the center of the earth, another relative to the sun, another relative to our galaxy, and so on. Of course, in most applications kinetic energy is considered only with respect to the surface of the earth.

Potential energy is relative, too, but usually with respect to the level from which an object is raised. A book raised 1 foot above a table has many

Table 2-4
Kinetic energies of animals

Animal	Kinetic energy, ft · lb	Kinetic energy, J
Scurrying chipmunk	0.5	0.68
Swimming beaver	1.5	2.04
Diving eagle	2,700	3,672
Charging elephant	300,000	408,000
Fleeing blue whale	2,400,000	3,264,000

different potential energies: one relative to the table, another relative to the floor, another relative to the surface of the earth, another relative to the center of the earth, and so on. However, in most applications potential energy is considered only with respect to one level.

2.6 CONSERVATION OF ENERGY

Webster's Third New International Dictionary defines conservation as "preserving, guarding, or protecting: a keeping in a safe or entire state." In everyday speech the word conservation has an ecological meaning based on "preserving, guarding, or protecting," especially with regard to natural resources. In physics, however, the meaning of conservation is based upon the second part of the definition: "Keeping in a[n] . . . entire state."

Conservation of a physical quantity occurs when the quantity has two features. The first feature of a conserved physical quantity is that it may be transformed into other closely related quantities. It may be helpful to think of an analogy. In everyday life a certain amount of money can be changed into money of equivalent value, such as a quarter being changed into 25 pennies. The second property of a conserved physical quantity is that any transformation must yield no change in the total amount. A quarter is equal to 25 pennies, not 26 or 24. Thus there is no difference in the value of a quarter or 25 pennies.

When applied to energy, conservation means that energy can be changed from one form to another. But the total amount of energy in its various forms cannot be increased or decreased as a result of the transformation. There must be no overall change in the total amount.

A rock swinging on a string without friction illustrates a simple case of conservation of energy, as shown in Fig. 2-17. When the rock is pulled to one side, it is given a potential energy equal to the work done in lifting the rock. Before being released, the rock possesses only potential energy. Once released, the rock picks up speed as it swings downward along its arc. During the downward swing potential energy of the rock is being changed into kinetic energy. Nevertheless, the sum of the potential energy and the kinetic energy at every instant during the downward swing equals the original potential energy before the rock was released.

A unique situation occurs exactly at the lowest point of the arc of the swinging rock. All the original potential energy has been completely changed into kinetic energy for an instant. The kinetic energy at the lowest point of the arc is exactly equal to the potential energy at the highest point of the arc.

As the rock swings upward along its arc, the rock loses speed. During the upward swing the kinetic energy is being changed back to potential energy. In the absence of friction the rock will swing up to the same height from which it was released and stop. The transformation of kinetic energy back to potential energy is complete. The rock begins swinging downward back along its path, and the process repeats itself.

This simple example of the conservation of energy is part of a general *law of conservation of energy.* The statement of this extremely important law of physics is quite simple.

Law of conservation of energy

Energy can be neither created nor destroyed, only changed from one form to another.

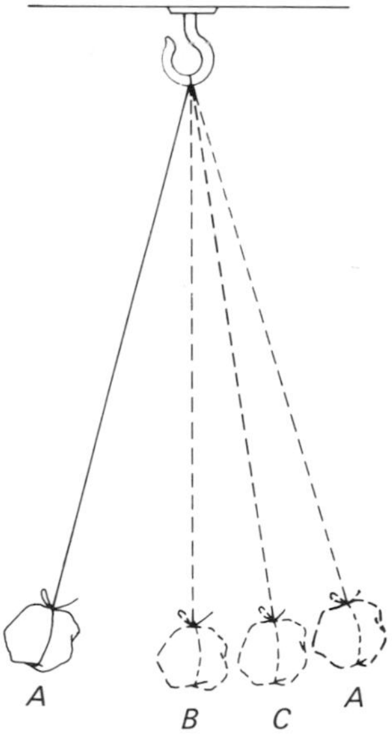

Notice that no restrictions are placed on the *form* of the energy, which may be potential energy, kinetic energy, or forms of energy not yet discussed (for example, thermal energy, electrical energy, or nuclear energy).

The law of conservation of energy has no known exceptions when *all* forms of energy are considered. However if you consider only potential energy and kinetic energy in a specific case, it may *appear* that the law of conservation of energy is violated. Take, for example, a car coming down a hill. The car possesses both potential energy and kinetic energy. If it collides with a bridge abutment at the bottom of a hill, then both the potential energy and the kinetic energy disappear. But the total energy in all forms is unchanged. One cannot ignore the fact that the car is demolished. The potential energy and kinetic energy have been converted into other forms of energy. When *all* forms of energy are identified and accounted for in any physical event, the total amount of energy is conserved.

A worldwide problem related to the conservation of energy is the "energy crisis." There is no problem with the total amount of energy available. The problem is with the economics of energy *utilization.* Coal, petroleum, and natural gas are forms of matter with large amounts of energy that can be readily processed, transported, and stored. As these sources of chemical energy are exploited, people must look for other sources of energy. A tremendous amount of energy is available in sunlight, wind, storms, and the tidal movement of the oceans. Unfortunately these phenomena are not easily and inexpensively controlled. Their energy is available, often in disastrously large amounts. But the problem is finding an *inexpensive* way to process, transport, and store energy from these sources.

A B C A

Fig. 2-17
A swinging rock possesses only potential energy at the top of its arc (position A), only kinetic energy at the bottom of its arc (position B), and both potential energy and kinetic energy at any other point (position C). The potential energy at the top of the arc is equal to the kinetic energy at the bottom of the arc.

PROJECTILE MOTION

Do you remember the last time you watered the lawn with a garden hose? As you swung the hose from side to side and up and down, the water landed at different places. Aim straight up and the water will rise, only to return upon your head. Aim slightly away from the vertical and the water will drop a few feet from your shoes. Each direction of the nozzle sends the water to a different distance.

Squirting water with a garden hose has much in common with throwing darts and even firing rockets. In each case we want one thing to land at a particular spot. With the game of darts we want each dart to end up in the bull's-eye. And every rocket should land in the planned target area. Each of these examples is a case of projectile motion. Even a garden hose sends out "projectiles." A stream of water acts like one projectile after another being sent along the same path, as shown in the photograph below.

(Photograph from Beckwith Studios, Brooklyn, NY.)

Suppose that you wish to squirt the water as far as possible. What would you do? You could turn on the faucet a little more, or open it all the way. This forces the water through the hose at a higher speed. If you hold the nozzle in one position, the more the faucet is opened, the farther the water will go. Or you could leave the faucet alone and simply aim in different directions. This will also send the water to different spots on your lawn.

Of particular concern for rockets is the energy required to achieve a certain path. Calculations are done in order to determine the path which requires the least energy.

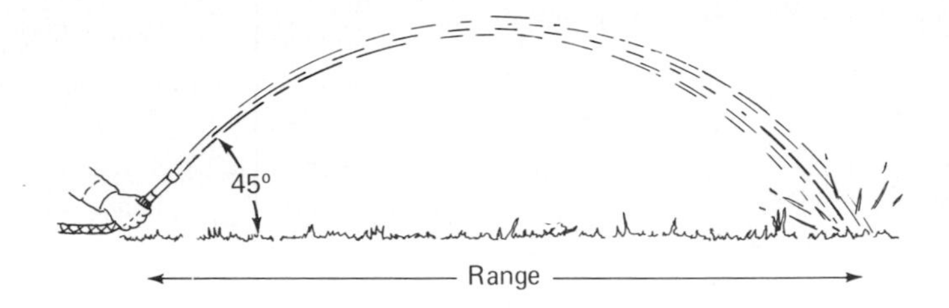

You can try a very simple case with a hose. With the end of the hose resting on the ground, try squirting the water at different angles. When the nozzle is aimed at 45° above the ground, the water will have the greatest range. It will be projected the farthest away at ground level. This is the greatest range for a fixed amount of energy given to the water (see the diagram above). If you wish to project the water stream any farther, you must give more energy to the water emerging from the hose. You must send the water out with a higher speed.

On a much larger scale, which is important to rocket travel, considerations of this kind are crucial. The path of a rocket is designed to use as little energy as possible. In this way less fuel is burned.

A typical path for a rocket sent partway around the earth is shown in the figure below. This path has been calculated to use the least amount of energy possible. It is assumed that the rocket will burn fuel for a short initial period and then coast throughout the remainder of the flight.

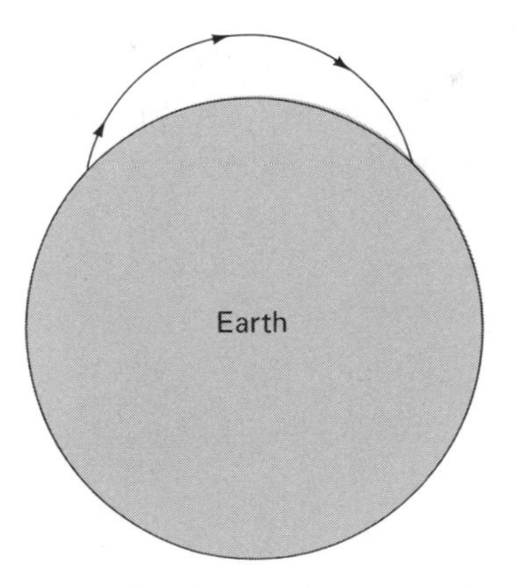

On an even larger scale such energy calculations are used to determine flight paths to the moon and planets. Newton's laws of motion enable us to minimize our use of rocket fuel. And the same laws apply to space travel as well as the streams of water from our garden hoses.

2.7 PERPETUAL MOTION MACHINES

Over the ages there has always been an energy crisis, whether it be gathering firewood for prehistoric people or finding new petroleum deposits for people now. There has been an ever-increasing demand for energy to run machines. The machines transform the energy available in various forms into useful work. For example, the kinetic energy of moving air is used to run a windmill, or the potential energy of a reservoir of water is used to generate electricity. However, machines are not 100 percent efficient in producing useful work. Some energy is always wasted as parts wear out and energy is carried out in the exhaust.

Trained engineers seek to minimize such losses of energy. But many untrained inventors ignore such everyday problems and seek to produce a *perpetual motion machine.* A perpetual motion machine is a device that is supposed to generate *more* useful work than the input of energy. In other words, a perpetual motion machine would have its output exceed its input. It would be more than 100 percent efficient. Figure 2-19 shows a hypothetical perpetual motion machine.

Many clever inventors have hoped to discover a way of conquering nature by creating energy. The fact that a perpetual motion machine would violate the law of conservation of energy is no discouragement to inventors. After all, people once said that humans would never walk on the moon. There have been laws of physics that were broken in the past, so why not the law of conservation of energy? Perhaps in time this law also will be violated or need to be corrected. But at present there are only situations in which energy is

Fig. 2-18
Disgruntled lover about to convert potential energy into kinetic energy.

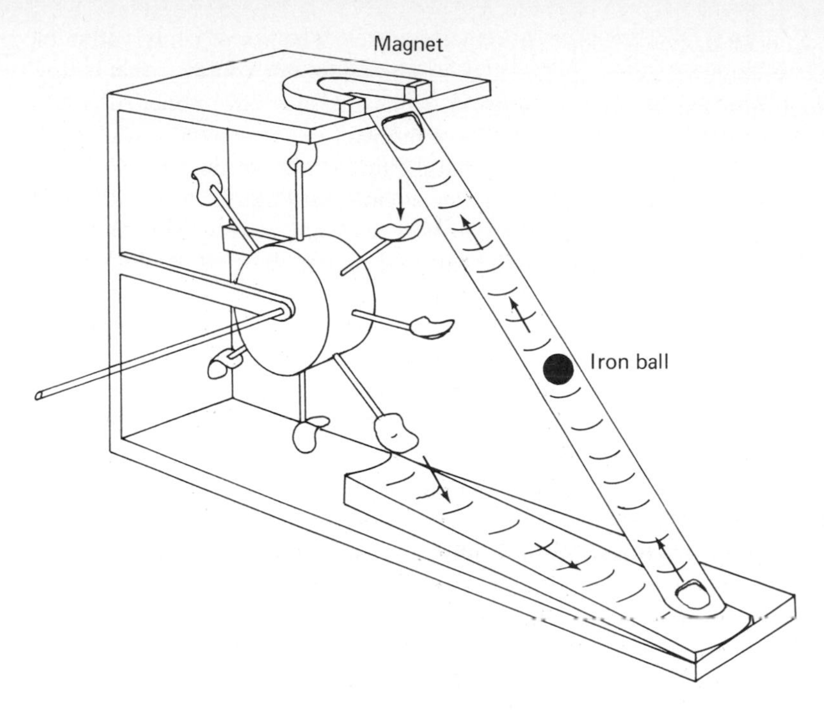

Fig. 2-19
A very strong permanent magnet is fastened to the top of the track. The magnet attracts an iron ball, which is pulled upward until it falls through the hole near the top of the track. As the ball falls, it turns the paddle wheel to produce useful work. The ball then rolls to the front of the track, and the process is repeated.

conserved. These situations are so numerous that scientists feel that it is a waste of time trying to make a perpetual motion machine. Of all the laws of physics, the law of conservation of energy is the most general statement about nature.

2.8 CONSERVATION OF LINEAR MOMENTUM

We have seen that Newton's second law of motion was based on the concept of momentum, given the symbol mv, where m stands for the mass of an object and v stands for the velocity of an object. Since momentum involves motion in a line, the symbol mv is more properly called *linear momentum.*

Linear momentum

Linear momentum is defined as the mass times the velocity of the object.

A Frisbee moving through the air was used as an example in Sec. 2.5 of an object possessing both potential energy and kinetic energy. A moving Frisbee also possesses linear momentum. The Frisbee has a mass m and a velocity v, which multiplied together produce a linear momentum mv.

Whenever two or more objects interact (such as two cars in a collision or a bowling ball striking the pins), the total linear momentum of the objects is conserved. The *law of conservation of linear momentum* is simply stated.

Law of conservation of linear momentum

The total linear momentum of a group of objects before an interaction is equal to the total linear momentum after the interaction.

Consider a bowling ball striking some pins. The law of conservation of linear momentum can be applied to this interaction. We would say that before the collision the ball has momentum, while the pins have none. After the collision, the ball and the pins all have momentum, as shown in Fig. 2-20. The sum of these momenta must equal the momentum of the ball before the collision. This is the conservation idea again—nothing lost or gained. The number of pins involved in the collision does not matter. Momentum is conserved in a collision involving one pin or ten. But we would have to consider all the pins. None could be left out. Conservation of momentum only applies to an entire interacting group of masses. (To be completely accurate for bowling ball and pins, we would need to include the effects of interactions with the floor. But we ignore that here.)

When two cars collide head on, the conservation of linear momentum again applies. Consider two identical bumper cars (same total masses including occupants) moving with the same speed v. One car has momentum mv, and the other car has momentum $-mv$ (the minus sign because it is moving in the opposite direction, although with the same speed). Before the collision there is motion, but zero total linear momentum ($mv - mv = 0$). After the collision there is no motion (no velocity), that is, zero total linear momentum, as shown in Fig. 2-21. Linear momentum has been conserved.

Another kind of collision involves two identical croquet balls. Once again each ball has the same speed, but they move in opposite directions.

Fig. 2-20
Some of the linear momentum of the bowling ball is transferred to the pins. (*Photograph courtesy of Selling Sports Goods.*)

Fig. 2-21
Two bumper cars moving at the same speed collide head on. The total linear momentum of cars and occupants is zero before and after the collision.

Before the two balls collide the total linear momentum is zero. After the collision the balls rebound with the same speed, but in opposite directions. The total linear momentum is still zero, as shown in Fig. 2-22. The linear momentum has been conserved.

A rocket uses the conservation of linear momentum to change its velocity and move through space. In order to "blast off" or accelerate, it is *not* necessary to push against the atmosphere, although rockets do fly through the atmosphere. As the exhaust gases are expelled at great velocity in the

Fig. 2-22
The system of two croquet balls moving at the same speed has zero linear momentum (mv for the ball on the left, and $-mv$ for the ball on the right before the collision). After the collision the rebounding balls still have zero linear momentum. The total linear momentum has been conserved, but each ball has exchanged its linear momentum.

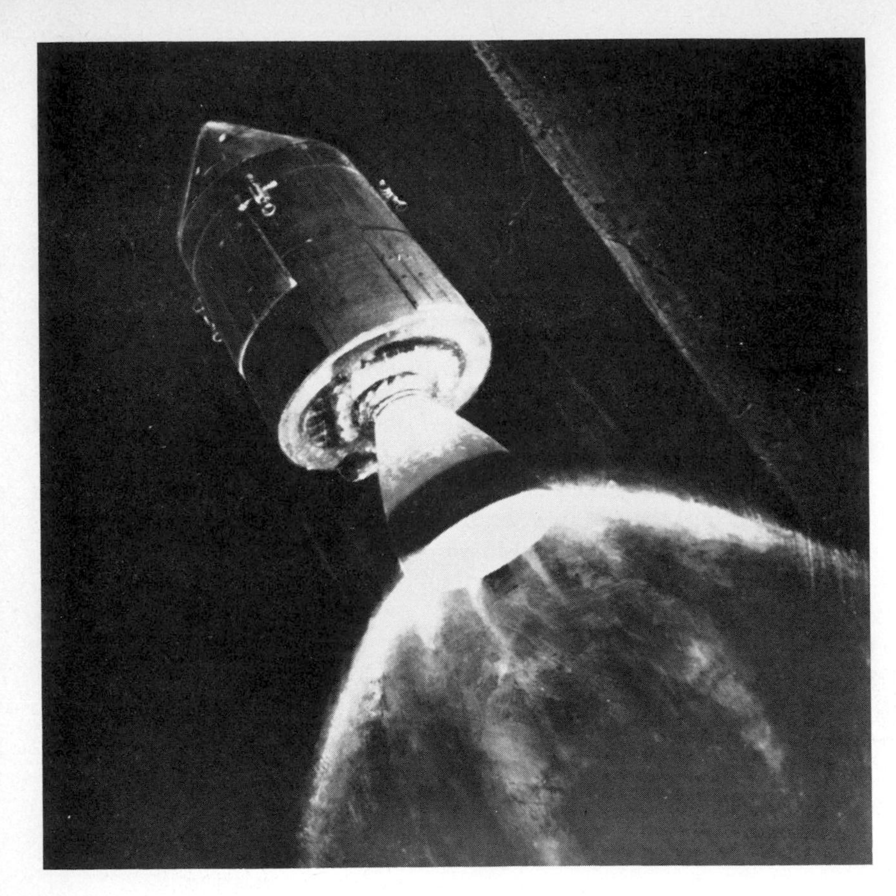

Fig. 2-23
In outer space a rocket expels its exhaust gases. The linear momentum of the gases causes the rocket to acquire an equal and opposite linear momentum. As a result the rocket is propelled forward. (*Photograph courtesy of NASA.*)

backward direction, some linear momentum exists in that direction. Because total linear momentum is conserved, the rocket and its remaining fuel must be given an equal, but opposite, linear momentum in the forward direction in order to conserve the total amount. Conservation of linear momentum for a rocket works equally well in the earth's atmosphere or in outer space where there is no atmosphere "to push against" (see Fig. 2-23).

2.9 CONSERVATION OF ANGULAR MOMENTUM

Linear momentum deals with objects moving in straight lines. But what about something spinning or revolving, such as a jar lid being unscrewed or a merry-go-round at a circus? Angular momentum is a kind of momentum used to describe objects which undergo rotation.

The definition of angular momentum is very similar to the definition of linear momentum. But angular momentum involves one extra quantity.

Angular momentum

Angular momentum is defined as the mass of the object times a speed and a radius.

This definition can be expressed as a formula

$$\text{Angular momentum} = (\text{mass})(\text{speed})(\text{radius})$$

or

$$\text{Angular momentum} = mvr$$

where m = mass of the object
v = its speed
r = radius.

But what speed and radius should be used? As a jar lid turns, some parts of the lid move faster than others. And different portions of the lid move at different radii. We will be content to say that "average" or "effective" values for these quantities must be used. Exact calculations can be made, but we will not pursue that here.

A Frisbee moving through the air is an example of an object possessing linear momentum, as noted in Sec. 2-8. The spinning Frisbee also possesses angular momentum. The Frisbee has a mass m, a speed of rotation v (not to be confused with its linear velocity), and a radius r. The product mvr is the angular momentum of the Frisbee.

Angular momentum is another physical quantity which, under the proper conditions, is conserved. Whenever a rotating object has no external forces applied to it, angular momentum is conserved.* So there is also a *law of conservation of angular momentum.*

Law of conservation of angular momentum

Whenever a rotating object experiences no external forces, the angular momentum of the object remains constant.

A spinning object, such as a child's top, would spin without change forever if there were no forces acting on it. Once given an angular momentum, it would continue to spin forever. Friction, however, eventually slows the top, and it falls. With friction present, the angular momentum of the top changes.

Another example will illustrate the effects of angular momentum conservation on a flexible object. If a ballerina is spinning, she has angular momentum that is conserved when the effect of friction is ignored. With both arms extended the ballerina has a large (average) radius and a small speed as she spins. The ballerina can change her shape by drawing her arms close to her body. In order for the angular momentum to remain unchanged, the decreased radius causes an increase in the speed. The ballerina spins faster when the arms are close to the body, as shown in Fig. 2-24. When the arms are again outstretched, the ballerina spins slower.

*If by chance an external force should be directed through the axis of rotation, the angular momentum is still conserved. Any other forces acting on the object will change the angular momentum.

Fig. 2-24
As the ballerina draws her arms close to her body, she spins faster. (*Photograph by Francis Laping from DPI.*)

This same physics can be applied to spinning on a piano stool. It has been elegantly described by Mott-Smith in *Principles of Mechanics Simply Explained* (Dover Publications, Inc., New York, 1963):

These peculiarities of angular momentum can be experimentally demonstrated by means of a revolving stool like a piano stool, only it must turn very easily on ball bearings, and not screw up or down. If a man sits on such a stool he cannot, of course, by any squirming or jerking set himself in rotation, nor when rotating stop himself except by means of an external object. If he twists his shoulders to the left, his hips will rotate to the right. . . . If he is rotating, he can diminish his speed by extending his legs and arms, and increase it again by pulling them in. If he could extend his arms a thousand miles, the rotation would become imperceptible. But however far he extended them it could never in the absence of resistances be reduced to zero. On the other hand, if he and the stool could shrivel to the size of a pinhead, . . . the rotation would increase to many millions of turns per second. Hence, although a rotation cannot be started

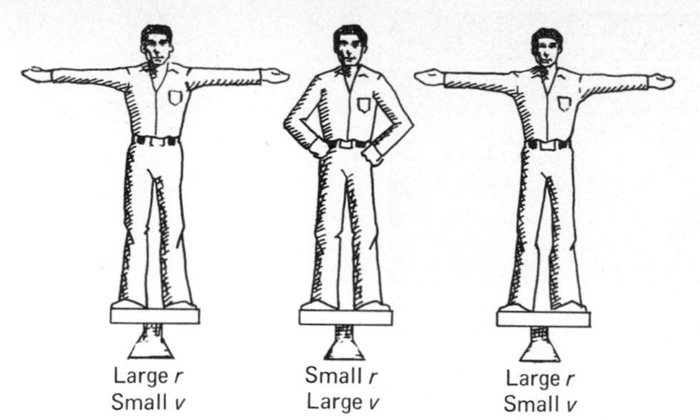

Fig. 2-25
The man spinning on a piano stool always has his angular momentum conserved. He spins slower with his arms outstretched, and he spins faster with his arms close to his body.

from nothing nor reduced to nothing, if one exists, it can be enormously increased or diminished by merely shifting masses toward or away from the axis.

Figure 2-25 illustrates the changes in motion for a man standing on a spinning piano stool.

The idea of changing the rotational speed by changing shape is used in many circumstances. Every diver uses it when doing somersaults in midair. Pull in your arms and legs and you will rotate faster. Stick them out again and the rotation will slow down. Figure 2-26 shows an acrobat using this technique.

As a final example, suppose a person stands still in the middle of a kitchen and then suddenly begins rotating by taking a number of small steps with the feet. This rotation can, of course, be continued as long as desired (barring dizziness). Since no angular momentum was present before rotation and some is present while rotating, it seems that angular momentum is not conserved. Indeed, the angular momentum *of the person alone* is not conserved. But recall that the feet provided forces that produced the rotation. If we turn our attention to the person *and the ground* on which the person is standing, then the total angular momentum of the earth and the person is conserved. As the person acquires an angular rotation in one direction, the earth underneath acquires an angular rotation in the opposite direction. The angular momentum given to the earth is insignificantly small compared to the angular momentum always present as the earth rotates on its axis each day. Therefore, the earth would not measurably change its rate of rotation. Even if all the people on the earth started spinning in the same direction, the sum of all their efforts would still not cause a measurable change in the earth's angular momentum. Nevertheless, because of the conservation of angular momentum, we must treat the earth in this way. The earth exchanges angular momentum with us as we start and stop any rotating motion by pushing against the ground.

Fig. 2-26
An acrobat jumping from the spring board has angular momentum, which is conserved. In a crouched position the acrobat spins rapidly. But when the arms and legs are extended, the speed of the spin is greatly decreased. (*Photograph by Dr. Harold Edgerton, MIT, Cambridge, Massachusetts.*)

REVIEW You should be able to define the following terms: center of gravity, equilibrium, work, power, potential energy, kinetic energy, linear momentum, and angular momentum.

You should be able to state the following laws and give examples of their application: the law of stable equilibrium, the law of conservation of energy, the law of conservation of linear momentum, and the law of conservation of angular momentum.

PROBLEMS

1. An elevator weighing 5,000 pounds is lifted from the first floor to the fourth floor. If the average distance between floors is 10 feet, how much work is done by a motor in lifting the elevator?

2. There is a force of friction equal to 60 pounds when a refrigerator is slid across the floor. What is the work done when you slide the refrigerator a distance of 7 feet?

3. An average washing machine does 1,080,000 joules of work during 1 hour of operation. What is the power used to wash clothing?

4. An electric mixer does 600 joules of work each second while whipping cream. What is the power produced by the mixer?

5. An elevator weighing 5,000 pounds is raised from

the first floor to the fourth floor. If the average height of a floor is 10 feet, calculate the increase in the potential energy of the elevator.

6. A girl pedals a bicycle up a steep hill rising 175 meters. If the rider and the bicycle together weigh 650 newtons, what is the increase in potential energy?

7. A truck of 5,000 kilograms mass moves at a constant speed of 20 meters per second. What is the kinetic energy of the truck?

8. A freight train of 8 million kilograms mass moves at a constant speed of 10 meters per second. What is the kinetic energy of the train?

9. An elevator on the fourth floor of a building has 150,000 foot-pounds of potential energy compared to the first floor. If the elevator were to fall, what would be the kinetic energy of the elevator just before it strikes the first floor?

10. A small child at the top of a slide has 500 joules of potential energy. What would be the child's kinetic energy upon landing in a sandbox if there were no loss of energy to friction?

DISCUSSION QUESTIONS

1. Try standing up straight with your feet spread apart, and then lift one foot. Why do your head and body move to one side? Could you prevent this sideward motion?

2. If a uniform yardstick is cut in two at the balance point, will each half have the same weight?

3. One way of getting up from a chair is to bring your feet back *under* the chair as you rise. Where is the center of gravity of your body as you get up? Could you get up without putting your feet under the chair or pulling on some heavy (or fixed) object?

4. Try to stand beside a wall with one foot in front of the other, so that the sides of *both* shoes touch the wall. Discuss what happens.

5. When you climb a stepladder, why are you more stable standing on the bottom step rather than on the top step?

6. Is a cat more stable standing on its feet or lying on its side? An elephant? A turtle?

7. A tall, thin flower vase is easy to tip over. List ways in which such a vase could be made more stable.

8. What is the difference between the everyday definition of work and the scientific definition of work? What do the two definitions have in common?

9. Is work done on a suitcase in lifting it from the floor to a tabletop? What about moving it back to the floor? Is work done on the suitcase when it is being carried, neither rising nor falling?

10. As a sewing needle is pushed through cloth, work is being done on the needle. Why is more work done in sewing leather than cotton flannel?

11. A coil of telephone cord stores potential energy when it is stretched. Give another example of an object that stores potential energy in this way.

12. When you step off a bathroom scale, why does the top of the scale rise?

13. When a car's brakes are released and the gears disengaged, often the car begins to roll and acquires kinetic energy. Where did the kinetic energy come from?

14. How does the kinetic energy of a motorcycle change when the speed is doubled? Tripled?

15. When the trigger of a BB gun is pulled, the potential energy of the springs is converted into the kinetic energy of the BB. Is this how a squirt gun shoots water?

16. How do the moving hands of a cuckoo clock acquire kinetic energy?

17. Discuss how the law of conservation of energy applies to using a bow and arrow.

18. What kinds of energy are involved with storing water behind a dam and then using the water to generate electricity?

19. Could a rock spontaneously rise up in the air, according to the law of conservation of energy?

20. Why does the United States Patent Office require a working model before a patent will be issued on a perpetual motion machine?

21. The jellyfish has a circular ring of nerves. When excited, these nerves carry two electrical impulses run-

ning in opposite directions until the impulses meet on the other side of the ring and extinguish one another. In a special experiment one impulse was sent traveling round and round, not losing speed or intensity for 24 hours. Could a jellyfish be made into a perpetual motion machine?

22. How are linear momentum and kinetic energy similar? How are they different?

23. How does the linear momentum of a motorcycle change when the velocity is doubled? Tripled?

24. How could you increase the linear momentum of your body?

25. Is the angular momentum of a ferris wheel constant?

26. Is it possible that the earth may have rotated on its axis at different rates during the past millions and millions of years? Could the direction of the axis of rotation change?

27. Describe what your arm feels when an electric hand-held mixer is turned on.

28. Large boats are often driven by two propellers. What differences would occur in the behavior of the boat if the propellers rotated in the same or opposite directions?

29. As a helicopter flies, why doesn't the propeller stop rotating and the body start rotating?

ANNOTATED BIBLIOGRAPHY

Perpetual motion machines have intrigued people for centuries, and there appears to be no change in sight. A thorough and easy-to-read article, which includes descriptions of several famous hoaxes, is S. W. Angrist, "Perpetual Motion Machines," *Scientific American,* January 1968, pp. 115–122. An older article which describes 12 machines and what is wrong with each is Clifford B. Hicks, "Why Won't They Work?" *American Heritage,* April 1961, pp. 78–83. For the work of current frustrated inventors (which still continues!) see *Esquire,* December 1967, pp. 142–143; *Newsweek,* Nov. 5, 1973, p. 84; *Science News,* Apr. 24, 1976, p. 262; and Arthur Ord-Hume, *Perpetual Motion: The History of an Obsession,* St. Martin's Press, New York, 1977.

Our discussion of rotating objects and angular momentum did not include a number of more subtle ideas and examples. A small paperback by John Perry, *Spinning Tops and Gyroscopic Motion,* Dover Publications, Inc., New York, 1957, has excellent examples of things which twirl and spin: hard-boiled eggs, loops of chain, toy gyroscopes, hats thrown in the air, and much more. Also, a careful discussion on the flipping of tennis racquets has been given in *The Physics Teacher,* vol. 6, 1968, pp. 118–122. (If you don't play tennis, the author also illustrates the ideas with a hammer.)

The ideas of energy and center of gravity are useful in many different subject areas. Chapter 1 of R. McNeill Alexander, *Animal Mechanics,* University of Washington Press, Seattle, 1968, has applications to animals. Some mathematics is used, but could be skipped if you wish a discussion only.

3 FLUIDS

Two of the most common materials around us are water and air. Both air and water are fluids—they flow easily from one place to another. In this chapter we will introduce the concepts of pressure, buoyant force, volume flow rate, and fluid resistance. These concepts will be used in Archimedes' principle, Pascal's principle, and Poiseuille's law.

3.1 THE ATOMIC THEORY OF MATTER

The laws of conservation of energy, conservation of linear momentum, and conservation of angular momentum were used to understand the motion of solid objects. A solid usually does not change its size or shape when small forces are applied to it. But even small forces can deform a liquid or a gas. Changes in the size or shape of a liquid or gas make it difficult to use the conservation laws already presented. The laws still hold, of course, but detailed application is difficult. Therefore, it is convenient to introduce additional definitions and laws that are especially appropriate for liquids and gases.

The simplest classification of matter is into three states: solid, liquid, and gas. Liquids and gases are both called *fluids,* which are described as substances that can flow. A substance that flows can change its size or shape.

Fluid *A fluid is defined as matter whose particles are free to move and to change their relative positions.*

But what are these particles?

An object, whether it is in the solid state, liquid state, or gaseous state, appears to your eyes to be a continuous piece of matter. The smoothness of an ice cube, the sparkle of a drop of water, or the softness of a wisp of water vapor does not suggest any granular nature. But all matter is made up of particles too small to be seen by the human eye or ordinary microscopes. These particles are *atoms* or *molecules.*

Atom *An atom is the smallest part of matter that keeps the identity of a chemical element.*

Iron, mercury, and helium are elements. Each of these materials is made up of atoms.

Molecule *A molecule is a group of atoms which are bound together as a unit to form the smallest part of a chemical compound.*

An illustration of molecules of water (two hydrogen atoms united with one oxygen atom) is shown in Fig. 3-1. All atoms and molecules are discrete entities, and they are the building blocks for the larger pieces of matter which we directly experience in the everyday world.

Some molecules in concrete, wood, or plastics contain dozens or even thousands of atoms. But even then the atoms cling together as a unit. All matter, in all forms, has this granular character. The small atomic and molecular units can be arranged in many ways, but they are always present.

The difference between a solid and a fluid is a question of the movement of the atomic or molecular particles. In a solid object the particles are not free to change their relative positions. The atoms or molecules vibrate, but do not move past one another. Thus, a solid does not change size or shape when small forces are applied to it. But a fluid, either a liquid or a gas, does change size and shape when small forces are applied to it. The atoms or molecules of a fluid are not held in fixed positions. The particles are free to change their relative positions. Thus, a fluid can flow, but a solid is rigid.

Are you thinking of some borderline cases? There are some, of course. What should we say about tar on a very cold day? Is it a solid? It seems stiff.

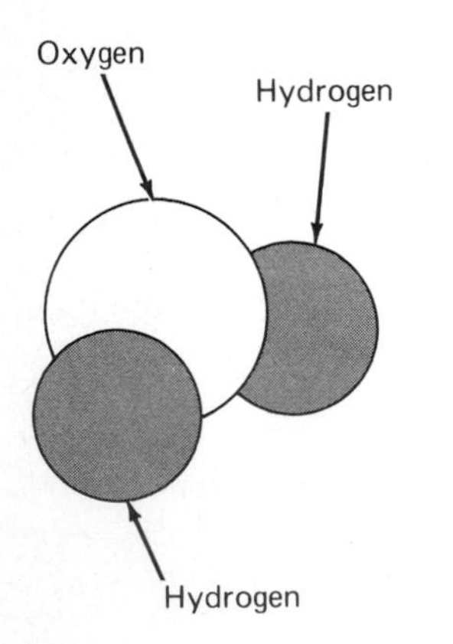

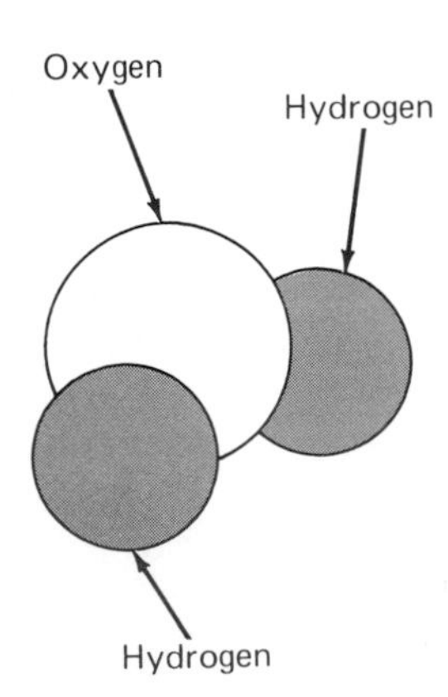

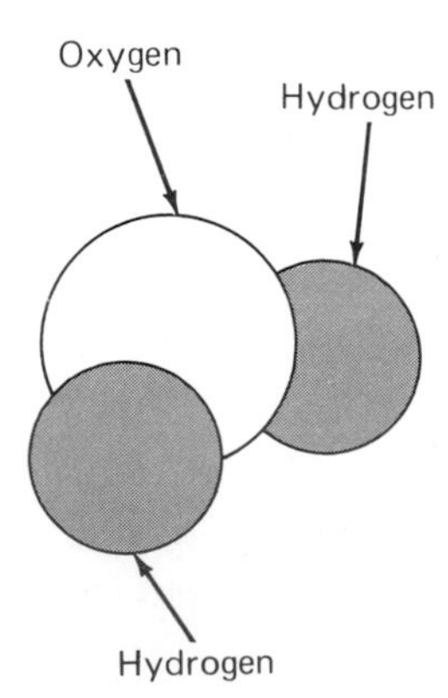

Fig. 3-1
Individual water molecules are each composed of two hydrogen atoms and one oxygen atom.

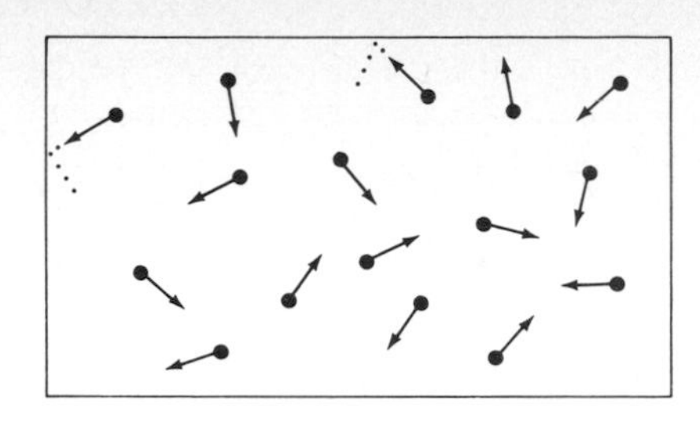

Fig. 3-2
Gas particles repel each other during brief collisions. When not colliding, the particles exert negligible forces on each other. As the temperature increases, their speed increases.

But if we watch closely for a long while, it sags. Is it a liquid? Or what about salt? The individual grains from a salt shaker are certainly solid. But a handful of salt can be poured—it flows. And we speak of sand flowing in an hourglass. In such cases it seems correct to say that "a solid flows." But we will not try to incorporate such "fluids" into our general discussion.

A gas is perhaps the simplest state of matter to describe. Figure 3-2 illustrates gas particles moving in a closed container. An oxygen molecule in the air about us moves with an average speed of 440 meters per second (1,445 feet per second) at room temperature. Only very small forces exist between the particles in a gas. And these forces hardly affect the motion unless the particles are very close together. The forces are not big enough to hold the gas together. If the lid is momentarily removed from a jar full of gas, a portion of the gas escapes. The gas does not stick together as a whole.

Air particles in the earth's atmosphere behave similarly. They do not stay together by themselves. Gravity keeps them near the surface of the earth. But if we were to remove gravity, the air molecules would drift off in every direction.

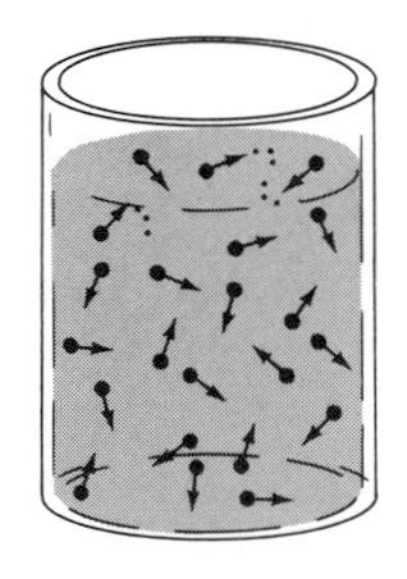

Fig. 3-3
Liquid particles are much closer together than gas particles, but are still free to move about.

In liquids, stronger forces exist which hold the particles together. There is a restricted motion of the particles within a liquid. The liquid particles are not as free as gas particles. When a liquid is in a container, the liquid takes the shape of the container, as shown in Fig. 3-3. All portions of the liquid are together. The liquid does not break up into pieces.

The flat upper surface on a glass of milk is due to the earth's gravity. If gravitational forces were much smaller, the milk would not have a flat surface. It would pull together forming a sphere of liquid. Liquids falling near the surface of the earth also form spheres or drops. Small raindrops, for example, are quite spherical in shape.

In a solid, forces are very strong between the particles. The particles are firmly bound in position and cannot move apart from each other. Figure 3-4 illustrates the atoms of a solid. The particles are free only to vibrate about their fixed positions. Thus a solid is rigid and does not flow.

3.2 PRESSURE

Hitting a piece of wood with a strong hammer blow usually produces no separation of the molecules of wood, only a shallow dent. But the same hammer blow against a nail separates the molecules of wood as the nail

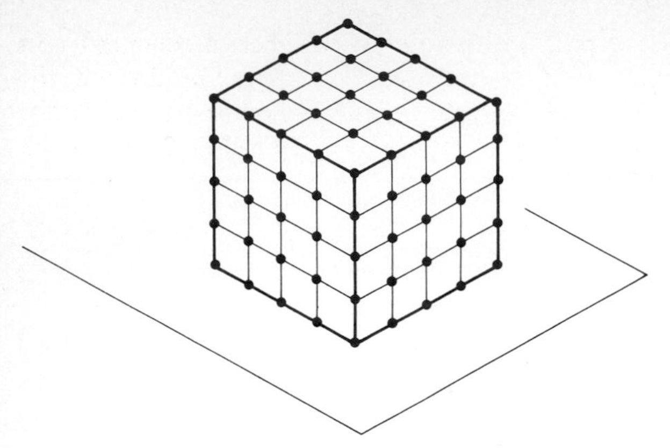

Fig. 3-4
In a solid the particles are arranged in a definite pattern. They are free to vibrate but do not change their relative positions in the pattern.

penetrates. The molecules of wood are affected differently when the same force is applied to the wood and to the nail. The applied force does not determine the effect. Instead, the different effects are caused by the unequal sizes of the hammerhead and the point of the nail.

As the area over which a force is exerted becomes smaller, the force becomes more concentrated. The technical definition which refers to this concentration of force is that of *pressure.*

Pressure *Pressure is defined as the force divided by the area.*

This definition can be expressed as a formula:

$$\text{Pressure} = \frac{\text{force}}{\text{area}}$$

or

$$p = \frac{F}{A}$$

where p = pressure,
F = applied force,
A = area upon which the force acts

We see from the definition given that if the area has some fixed value, then an increasing force will cause an increasing pressure. However, if we fix the value of force, then an increasing area will cause a decreasing pressure.

The effect of different pressures has been seen in the hammer-and-nail example. It is also evident when we consider the different sizes of heels on shoes. For example, the weight of a woman results in a *force* being applied to

Fig. 3-5
A force from a 5-pound package of sugar distributed over an area of 15 square inches produces a pressure of $\frac{1}{3}$ pound per square inch.

the floor. But wide-heeled sandals and narrow-heeled shoes produce different *pressures*. The pressure due to the heel can be calculated by dividing the woman's weight (say, 120 pounds) by the area of the heel. If we assume that in sandals she presses down on a heel area of 4 square inches (2 inches long and 2 inches wide), then

$$\text{Pressure} = \frac{\text{force}}{\text{area}}$$

$$= \frac{120 \text{ pounds}}{4 \text{ square inches}}$$

$$= 30 \text{ lb/in}^2$$

That is, she exerts a pressure of 30 pounds per square inch with each step.

If, however, she wears a narrow-heeled shoe with $\frac{1}{4}$-square-inch heels ($\frac{1}{2}$ inch long and $\frac{1}{2}$ inch wide), the pressure exerted by the heel is

$$\text{Pressure} = \frac{120 \text{ pounds}}{\frac{1}{4} \text{ square inch}}$$

$$= 480 \text{ lb/in}^2$$

Such a high pressure is sufficient to gouge some kinds of linoleum and softwood floors.

High pressures are desirable in several common circumstances. When you are sewing cloth or cutting meat, high pressures are needed to get through the materials. Can you see why *sharp* needles and knives are the best?

3.3 FLUID PRESSURE

The concept of pressure has been used to discuss the effect of one solid on another, such as a heel gouging a linoleum floor. But a fluid, whether a liquid or a gas, also generates pressure. When you swim underwater, you can feel the fluid pressure of the water pressing against your eardrums or against your face mask. This pressure is exerted on each part of your body. It squeezes you from every direction. It pushes in one direction on your eardrums and another direction on your face mask. Fluid pressures never act in one direction only.

The pressure at various depths in a fluid obeys a simple law.

Law of fluid pressures

The pressure at any depth in a fluid at rest is equal to the weight per area of a column of that fluid reaching from that depth to the top of the fluid.

Fig. 3-6
The shallow footprint is made with a large area, and so the pressure is small. The deep "pegprint" is made with a small area, and so the pressure is large.

This law allows us to find the pressure at various depths in a fluid at rest. To find the pressure at any depth, we simply use the weight per area of a column of fluid as tall as that depth.

For example, what is the pressure at a depth of 10 feet underwater? It is the same as the weight per area of a column of water 10 feet tall (see Fig. 3-7). Choose, say, a column of water with a cross-sectional area of 1 square inch. Such a column 10 feet tall weighs 4.3 pounds. The weight per area is

$$\frac{\text{Weight}}{\text{Area}} = \frac{4.3 \text{ pounds}}{1 \text{ square inch}}$$

$$= 4.3 \text{ lb/in}^2$$

The law of fluid pressures then tells us that the pressure under 10 feet of water is 4.3 pounds per square inch.

Of course, most of our lives are not spent underwater. But living on the surface of the earth means we spend our lives living under an "ocean" of air.

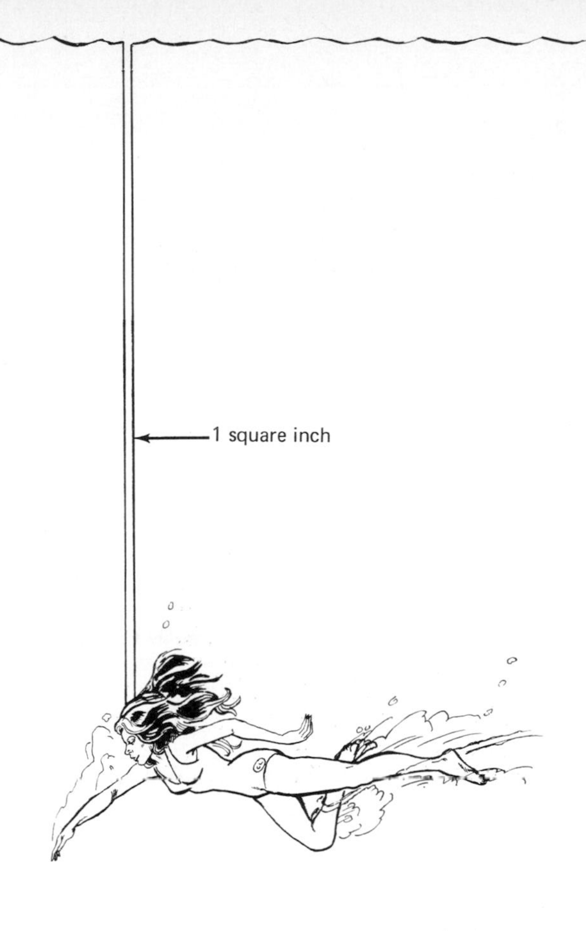

Fig. 3-7
The pressure of the water pushing in on the swimmer is equivalent to the weight of the water above her in a 1-square-inch column extending to the surface of the water.

The air is a fluid that also presses against us and everything else on the surface of the earth. The fluid pressure of the air surrounding us, the so-called standard atmospheric pressure, has been measured to be 14.7 pounds per square inch. The equivalent pressure in metric units is 1.013×10^5 newtons per square meter. This pressure arises because of the air above us. That air has weight. A column of air as deep as the atmosphere with a cross-sectional area of 1 square inch weighs 14.7 pounds. This weight is the source of the 14.7 pounds per square inch of atmospheric pressure (see Fig. 3-8).

The pressure of 14.7 pounds per square inch applies to measurements made at sea level. At higher altitudes, where there is less air above, the pressure of the atmosphere becomes less. For small changes such as walking up a few flights of stairs the pressure changes are small. But if you ride up 30 floors on a fast elevator your ears will be affected by the variation in pressure. Pressure differences caused by differences in altitude can be substantial when comparing pressure measurements made in two different cities. If one city is at sea level and the other city has an elevation of 1,000 meters (3,280 feet),

Column of air extending to the "top" of the atmosphere

Fig. 3-8
The pressure of the atmosphere pushing on your body is equivalent to the weight of the air above you in a 1-square-inch column. The column extends to the top of the atmosphere.

the pressures will vary by about 12 percent. At the high altitudes where jet airplanes fly, the pressure changes are even more noticeable. An airplane at 9,000 meters (29,500 feet) is flying in a region where the pressure is only 30 percent of that at sea level. It is flying above more than two-thirds of the atmosphere!

Atmospheric pressure would crush us if there were not a pressure inside us acting outward. By breathing, we allow air to enter our lungs and tissues. The inside pressure is normally about equal to the outside pressure, and so we feel no ill effects. If the inside pressure were suddenly removed by some unknown means, outside air pressure would crush a person. Or if the outside air pressure could be suddenly removed (such as having a pressurized space suit on the moon suddenly rip down one side), the inside pressure would cause a person to explode. It is for a good reason that the space suits of astronauts are strongly constructed.*

The standard atmospheric pressure of 14.7 pounds per square inch is something of a fiction. Atmospheric pressure is not really constant. Usually it changes slowly as storms pass overhead or as clouds disappear to reveal the sky. The change in atmospheric pressure, as measured by a *barometer* (see Fig. 3-9), is used to forecast the weather. When a weather forecaster says, "The barometer is falling," she means the atmospheric *pressure* is falling.

*The moon has no atmosphere. Therefore, a torn space suit would quickly lose its air.

Fig. 3-9
A typical household barometer. (*Photograph courtesy of Airguide Instrument Co.*)

Stormy weather usually comes with decreasing atmospheric pressure and clear weather usually comes with increasing atmospheric pressure.

Measurements of air pressure can be given in units of pounds per square inch. But another pressure unit is often used. This special unit of pressure involves a column of mercury.

Imagine that mercury is placed in a tube with 1 square inch of cross-sectional area. If the tube is 29.92 inches tall (equivalently, 760 millimeters tall), then the mercury it takes to fill the tube weighs 14.7 pounds. This weight applied to 1 square inch causes a pressure of 14.7 pounds per square inch, the standard atmosphere pressure. So measuring a distance, the height of a column of mercury, becomes a practical substitute unit of measurement for pressure. The typical range of atmospheric pressures is between 29.50 and 30.50 inches of mercury (or equivalently, between 749 and 774 millimeters of mercury). Atmospheric pressure often varies over this range in a few hours or a few days. It is seldom steady for long.

A tornado is a storm with very low atmospheric pressure inside the area of swirling winds. The characteristic funnel-shaped tornado is shown in Fig. 3-10. The pressure near the center can be from 1 to 2 pounds per square inch *less* than the surrounding atmospheric pressure. Suppose a tornado passes directly over a house. Inside the house the pressure will drop slowly, especially if all the doors and windows are closed. The pressure will be greater inside the house than outside. This difference in pressure causes a net outward force on the walls and roof. The force can be calculated by multiplying the pressure difference by the area. We see this from the formula

$$\text{Pressure} = \frac{\text{force}}{\text{area}}$$

$$(\text{Pressure})(\text{area}) = \text{force}$$

A pressure difference of only 1 pound per square inch may not seem destructive, but that amounts to a force of 144 pounds on every square foot.

Fig. 3-10
A tornado is a region of lowered air pressure. (*Photograph from United Press International, Inc.*)

For a roof 30 feet long and 50 feet wide, the force amounts to 216,000 pounds! This force can blow the roof off a house. Of course, the walls also would be blown down. The only way to decrease tornado destruction is to open all windows and doors before a tornado arrives (so you have time to safely hide in a storm cellar). This will allow the inside pressure to be closer to the outside pressure. Rain might blow through the open windows, but that is certainly better than losing the roof!

The pressure unit *millimeters of mercury* is used extensively in medicine, especially to measure blood pressure. When your heart beats, your blood pressure rises to a typical value of 120 mm for a healthy young adult. After a beat, your heart relaxes for about $\frac{3}{4}$ of a second. During this time the blood pressure falls to a typical value of 80 mm mercury before rising again during the next heart beat. These two extremes in blood pressure are usually recorded as 120/80 (read "120 over 80"). Do you know what your own blood pressure is? Next time your physician measures it, ask.

3.4 ARCHIMEDES' PRINCIPLE

It is common experience that a piece of wood floats on water. The force of gravity pulls down on the wood. But thc watcr provides a force that keeps the wood afloat. This force generated by the water is called a *buoyant force.*

Buoyant force

A buoyant force is the upward force generated by a fluid on a foreign object in the fluid.

This buoyant force acts in a direction opposite to the force of gravity. In other words, the weight of the wood is a downward force, but the buoyant force of the water is an upward force (see Fig. 3-11).

Fig. 3-11
A floating log is in equilibrium. The water provides an upward buoyant force, and the weight of the log is a downward force. (*Photograph by Joe Munroe from Photo Researchers, Inc.*)

What is the cause of the buoyant force? Recall the particles that make up the gas or liquid. When a foreign object is placed into a space previously occupied by the gas or liquid, the particles of the fluid strike the surface of the foreign object. The effect of these collisions is to impart an upward force—the buoyant force—on the foreign object.

The buoyant force does not always cause an object to float. Wood floats on water, but a rock does not. Water can generate a buoyant force large enough to support wood, but not stone. There is still a buoyant force exerted on a rock which is placed underwater. But this force is not large enough to let the rock float. Instead, this force decreases the apparent weight of the rock. The rock weighs less when lifted underwater than when lifted in air. Buoyant forces of this type exist for all fluids. We summarize by stating a principle.

Archimedes' principle is a rule by which the size of the buoyant force can be calculated.

Archimedes' principle

An object floating or submerged in a fluid at rest is pushed upward with a buoyant force equal to the weight of the fluid displaced by the object.

When an object is placed in a fluid, the fluid is pushed aside, that is, displaced, to make room for the new object. The fluid would flow back into the space occupied by the object if the object were removed. But as long as the object remains in the fluid, the fluid cannot occupy this space. The surrounding fluid exerts a buoyant force on the object.

The numerical value of the buoyant force can be found by knowing the weight of the fluid that would fill the space occupied by the foreign object, as illustrated in Fig. 3-12. For example, if the buoyant force of water on a human body had to be determined, a tub could be completely filled with water. As a person was placed in the water, all the overflow water could be collected and weighed. The weight of the overflow water would be equal to the buoyant force on the body. If a 140-pound person caused 140 pounds of water to overflow, the person would float. In this case, the buoyant force would equal the weight.

If a 50-pound suit of armor causes only 10 pounds of water to overflow, the armor would sink, as shown in Fig. 3-13. The weight would be greater than the buoyant force. The role of the buoyant force can be made particularly clear if we look at the forces after the armor has sunk and is resting on the bottom of the tub. Since the armor is at rest, there is no net applied force. The only downward force is the weight, 50 pounds for this example. One upward force is the buoyant force, 10 pounds here. What provides the other 40 pounds of upward force? The bottom of the tub! The tub pushes up with a 40-pound force.

Or you could get your hands into the act. If you reached underwater, you could hold the 50-pound suit of armor above the bottom by providing an

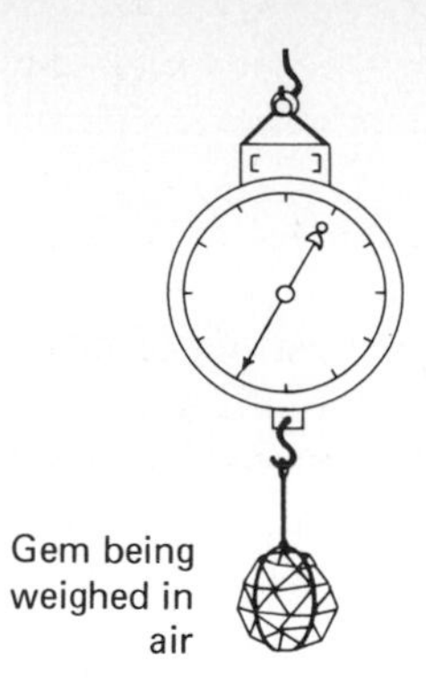

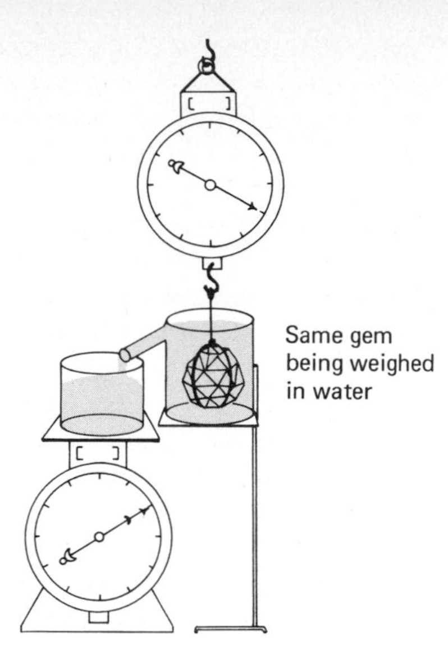

Fig. 3-12
The buoyant force is equal to the weight of the displaced water.

upward force of only 40 pounds. The water does the rest. It provides an upward buoyant force of 10 pounds. The apparent weight of the armor is only 40 pounds when it is underwater.

Often it is not practical to actually submerge an object in a liquid or gas and weigh the displaced fluid to determine the buoyant force. In such cases it is possible to predict whether an object will float or sink if the *density* of the object and fluid are known. In general, density is the quantity of a substance per unit of volume. In the case of Archimedes' principle, the commonly used quantity is the *weight density* (in the English system of units) or the *mass density* (in the metric system of units).

Weight density *Weight density is defined as the weight divided by the volume.*

Thus,
$$\text{Weight density} = \frac{\text{weight}}{\text{volume}}$$

or
$$d_w = \frac{w}{V}$$

where d_w = weight density
w = weight
V = volume

The units of weight density in the English system are pounds per cubic foot.

For example, let us calculate the weight density of concrete. To do this

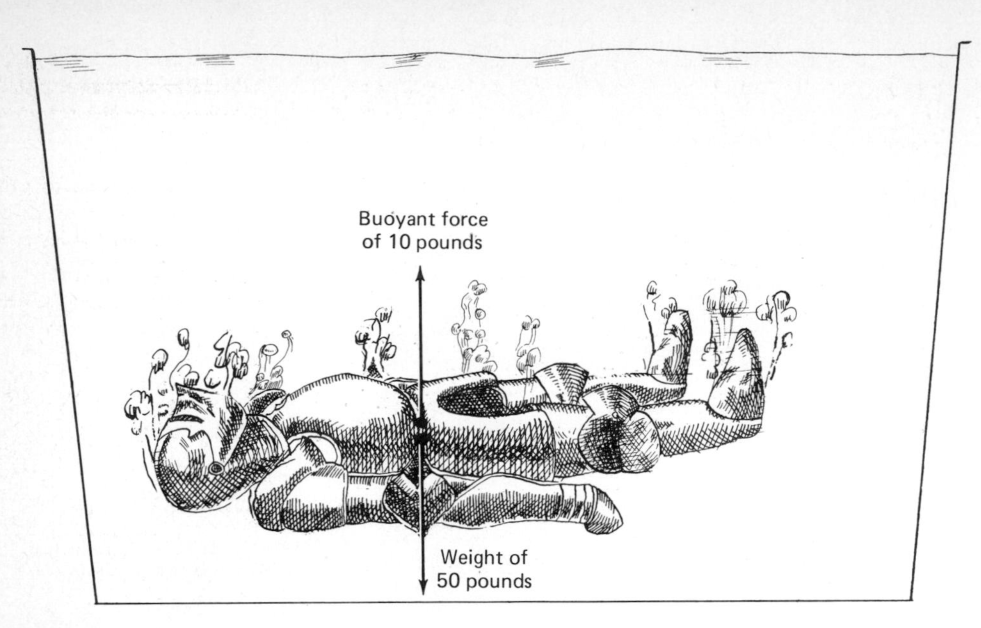

Fig. 3-13
A sunken suit of armor is in equilibrium. The water provides an upward buoyant force of 10 pounds. The weight of the armor is a downward force of 50 pounds. What provides the other 40 pounds of upward force?

we require the weight and volume of some piece of concrete. Any piece of concrete will do. It does not matter whether it is a small amount the size of your fist or a large piece the size of a bathtub. The weight density described above yields the same answer for each material, regardless of the size of the sample. Of course, the weight density does vary from one material to another.

Now a concrete example. Consider a slab of sidewalk which is 3 feet square and 4 inches thick. The total weight of this slab is 450 pounds! (Did you ever try to lift a square of sidewalk?)

The volume can be calculated as

$$\text{Volume} = (3 \text{ feet})(3 \text{ feet})(4 \text{ inches})$$

or

$$= (3 \text{ feet})(3 \text{ feet})(\tfrac{1}{3} \text{ foot})$$

$$= 3 \text{ ft}^3$$

We have found the volume to be 3 cubic feet. Now we can calculate the weight density using the formula

$$d_w = \frac{w}{V}$$

$$= \frac{450 \text{ pounds}}{3 \text{ cubic feet}}$$

$$= 150 \text{ lb/ft}^3$$

Mass density *Mass density is defined as the mass divided by the volume.*

That is,
$$\text{Mass density} = \frac{\text{mass}}{\text{volume}}$$

or
$$d_m = \frac{m}{V}$$

where d_m = mass density
m = mass
V = volume

The units of mass density in the metric system are kilograms per cubic meter.

To calculate the mass density of gasoline, for example, we require a sample of known mass and volume. The size of the sample is not important. Any size sample will yield the same result for our calculation of mass density. Therefore, let us take an entire truckload of gasoline. You have probably seen the tank truck which delivers gasoline to filling stations. Such a truck carries about 30 cubic meters of gasoline (8,000 gallons) with a mass of 20,000 kilograms. Thus,

$$\begin{aligned}\text{Mass density} &= \frac{\text{mass}}{\text{volume}}\\ &= \frac{20{,}000 \text{ kilograms}}{30 \text{ cubic meters}}\\ &= 667 \text{ kg/m}^3\end{aligned}$$

Mass densities are often used for geological and astronomical measurements. The mass density of the entire planet Earth is 5,500 kg/m^3. This is an average density. Some portions of Earth's crust have a lower density, and the core is believed to have a much higher density. The average mass density of the solar system is only 10^{-8} kg/m^3. With such a small density, it may seem that the solar system has hardly any mass. Do not be misled by the small average mass density of the solar system. The volume and the mass of the solar system are very large, even though the average mass density of the solar system is small.

Fig. 3-14
The oil and vinegar in a salad dressing separate into two layers. The oil floats on the vinegar. (*Photograph by Korrine Nusbaum.*)

By comparing numbers in Table 3-1, we can readily determine if an object floats or sinks when immersed in a fluid. Consider, for example, water with its weight density of 62.4 pounds per cubic foot. All the gases are less dense than water, and so their bubbles will rise to the surface when placed underwater. The droplets of certain liquids, such as mercury, will sink in water. Other liquids will float on water, such as gasoline or salad oil. You are probably familiar with salad oil floating on water. More familiar is salad oil floating on vinegar, as shown in Fig. 3-14. When a firm shake is given to a bottle of oil and vinegar dressing, the oil disperses in small droplets

Table 3-1
Various weight densities and mass densities*

Material	Weight density d_w, lb/ft^3	Mass density d_m, kg/m^3
Solids		
Pine wood	28	450
Ice	57.2	917
Ebony wood	76	1,220
Concrete	150	2,400
Glass	160	2,560
Aluminum	169	2,710
Iron	493	7,900
Gold	1,205	19,310
Liquids		
Gasoline	42	673
Alcohol	49.5	793
Salad oil	57.3	918
Water	62.4	1,000
Vinegar	62.7	1,005
Mercury	849	13,600
Gases		
Hydrogen	0.0056	0.09
Helium	0.011	0.18
Air	0.08	1.29
Carbon dioxide	0.12	1.98

*A few pure materials such as water or mercury have fairly precise densities. Other materials such as gasoline, wood, and glass have densities which vary from one sample to another. Such variations are not listed.

throughout the vinegar. But let the bottle stand a minute or two and the oil (which is less dense than vinegar) rises again to the top.

Finally, many solids sink in water, including ebony, but most woods will float. We could continue to compare other materials from Table 3-1. In each case, any material will float on a more dense material. For example, iron, which is very dense, will float in mercury, because mercury is even denser than iron. An iron ball floating in mercury is shown in Fig. 3-15.

A certain amount of mixing does occur when one *liquid* is put into another. In the previous discussion of floating, this was ignored. Also, we did not consider one material dissolving in another. But the problem of mixing becomes more troublesome when dealing with *gases.* Any gas mixes easily with another gas. Thus, in determining whether or not gases sink or float in air, it is advantageous to think of putting one of the gases in a very light-

Fig. 3-15
An iron ball floating in mercury. (*Photograph by Ted Rozumalski from Black Star.*)

weight balloon. Then it can be seen that a helium-filled or hydrogen-filled balloon will rise in air, but that a carbon dioxide balloon will sink.

Objects with complicated structures, such as an iron ship, can present a problem in determining the weight density. If the ship were solid iron, it would sink to the bottom of the ocean immediately. But a ship is *not* solid iron. There are passageways, rooms, and holds that are filled with air. When an unloaded ship rests in water, the sides of the ship push away, or displace, a large volume of water. This displaced water might weigh, for example, 5,000 tons. Hence, according to Archimedes' principle, there is a buoyant force of 5,000 tons supporting the ship. If the empty ship weighs only 5,000 tons, then the ship floats high on the water. However, ships are loaded and unloaded. As cargo is loaded, the ship gradually floats lower and lower in the water. More water is displaced by the ship, and a larger buoyant force is generated, just enough buoyant force to compensate for the weight of the cargo. Therefore the ship remains floating. If too much cargo is loaded, the ship can float so low in the water that waves wash over the deck and no more

Fig. 3-16
Since the red paint on the side of the upper ship is visible above the waterline, the ship is not overloaded. The lower ship is almost overloaded. (*Photographs courtesy of Sun Company, Inc.*)

buoyant force is generated. Then the ship sinks. As a precaution, engineers design freighters and passenger ships for a maximum cargo. The bottom and part of the side of a ship are painted red, as shown in Fig. 3-16. This red paint must always be visible above the waterline. If the red paint is not visible, then the ship is overloaded and may sink in rough waters.

3.5 PASCAL'S PRINCIPLE

The particles of many common gases are widely separated when the gas is at atmospheric pressure and room temperature. Applying a pressure to such a gas pushes the particles of the gas closer together. The amount of empty space between the particles is decreased. The volume of the gas is decreased (see Fig. 3-17). The volume can be decreased until the gas particles are virtually brought into continuous contact with each other. After this point is reached, further compression is very difficult. Very large pressures are needed to make the volume any smaller. But until the particles of a gas are brought very close together, a gas can be easily compressed into a smaller volume.

The particles of a liquid or a solid are not widely separated. The particles in a liquid are free to move past one another, and the particles in a solid are rigidly bound in definite pattern. Thus, a liquid or a solid cannot be easily compressed into a smaller volume under ordinary circumstances. An enormous pressure increase usually produces only an extremely small decrease in volume for a liquid or solid. Pressure changes do not affect all the states of matter equally.

Pressures are also transmitted differently for the three states of matter. A pressure at one end of a long piece of solid is transmitted to the other end without change of direction. No additional pressure is exerted to the sides. If you push a pool cue at one end, the other end pushes the ball. The sides of the cue always push outwardly against the air. But no additional pressure is directed to the side by pushing on the end.

A liquid or a gas, on the other hand, can transmit pressures around corners. The particles of a liquid or gas are free to move about within a container. A pressure applied on one side of a closed container spreads throughout the fluid and pushes against *all* the other sides of the closed container. Pushing on a tire pump handle transmits the pressure in the air throughout the pump, the connecting hose, and the tire.

It is easy to picture this distribution of pressure if you have played with water balloons. A slight squeeze on one side of a water balloon results in the transmission of pressure equally to all parts of the balloon. This kind of pressure transmission works for all fluids.

Pascal's principle is a statement based on the fact that a liquid is not easily

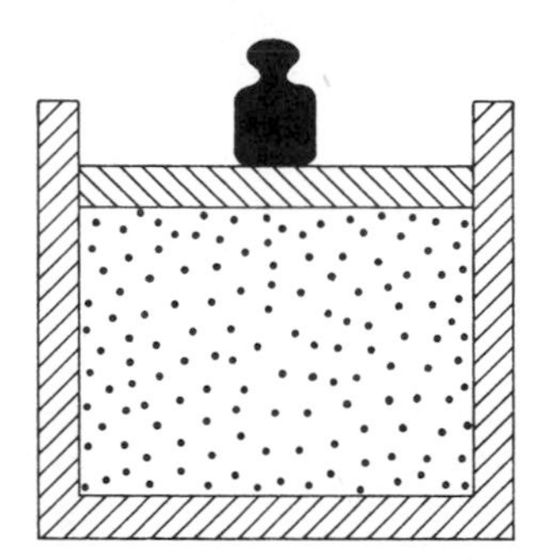

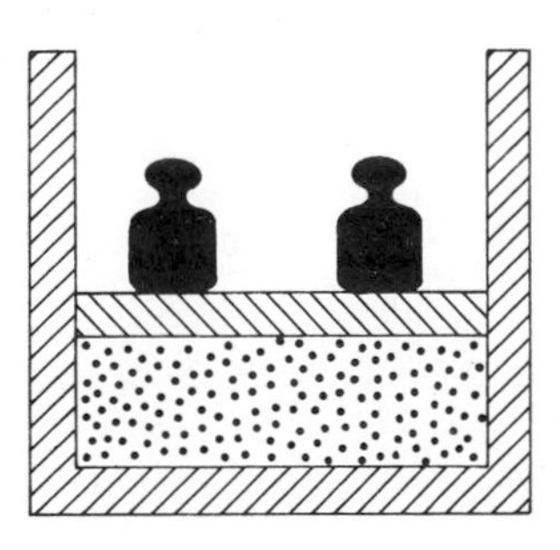

Fig. 3-17
The volume of a gas is decreased as the pressure is increased.

compressed and on the fact that an applied pressure is transmitted to all parts of an enclosed liquid.

Pascal's principle **A pressure applied to an enclosed liquid at rest is transmitted undiminished to every portion of the liquid and to the walls of the closed container.**

For example, a tube of toothpaste is a container full of liquid. The toothpaste is liquid and does flow, although very slowly. By squeezing the tube anywhere along its length, you apply a force over the area of your fingers. That is, you apply a pressure. This pressure is transmitted undiminished throughout the toothpaste. If the cap on the tube is removed, this pressure produces a force on the toothpaste. The toothpaste accelerates out of the opening. Simply stated: Squeezing an open tube of toothpaste causes it to squirt out. Children usually squeeze a tube near the opening because they believe the cause (squeeze) must be near the effect (squirt). But most adults have learned to squeeze near the opposite end of the tube and roll up the tube as the supply of toothpaste is decreased. An example of the use of Pascal's principle is shown in Fig. 3-18.

A blow to the eye also demonstrates Pascal's principle. The eye is filled with a liquid. When a pressure is applied to the eyelid with the hand, the pressure is transmitted by the liquid to the rear of the eye. There the retina experiences increased pressure. A sharp poke to the eye often causes some

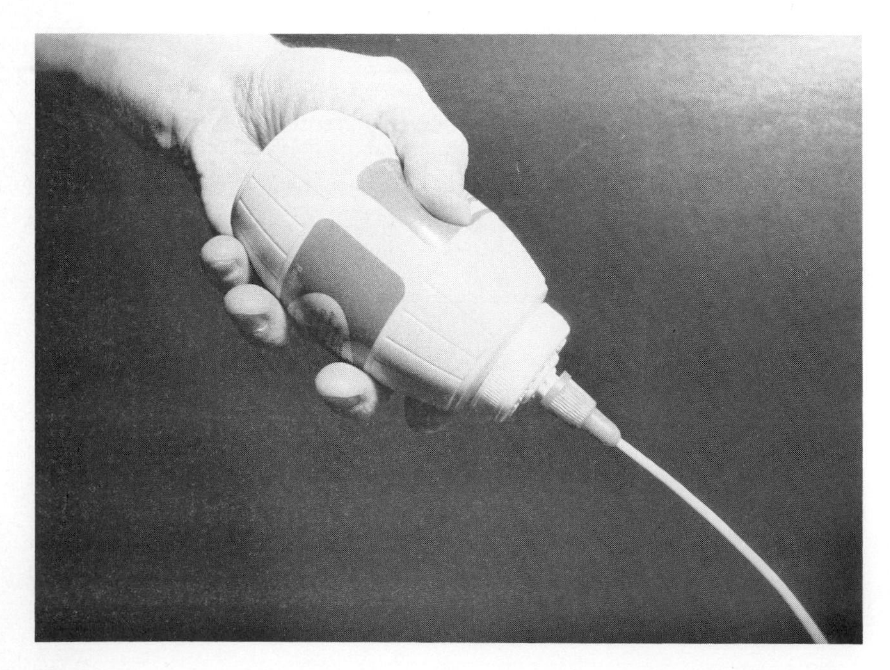

Fig. 3-18
A pressure applied in the middle of the container is transmitted throughout the mustard. (*Photograph by Ted Rozumalski from Black Star.*)

damage to the retina, blood vessels, and optic nerve, even though no damage is visible at the front of the eye.

Pascal's principle has practical application in a hydraulic lift. A hydraulic lift is nothing more than a liquid-filled vessel having two sliding stoppers (called *pistons*). The influence of gravity will cause the liquid level to be the same height in each of the tubes, even though the diameters of the tubes are not equal. Now suppose a weight is placed on one of the pistons. A pressure is exerted on the liquid, depending on the area of the piston in contact with the liquid. According to Pascal's principle, this pressure is transmitted to all parts of the liquid and consequently to all parts of the hydraulic lift. The second piston moves. In fact, it will accelerate upward (if there is little friction), in accordance with Newton's second law of motion. This will continue until the first piston reaches the bottom of its tube. If, however, we wish to hold the first piston with its weight in a fixed position, then we must counteract the increased pressure on the second piston by placing a weight on the second piston also.

But how much weight must be placed on the second piston to balance the system? Since the pressure is

$$\text{Pressure} = \frac{\text{force}}{\text{area}}$$

$$p = \frac{F}{A}$$

the *first* piston generates a pressure on the liquid equal to

$$p = \frac{F_1}{A_1} = \frac{w_1}{A_1}$$

where p is the pressure exerted by the weight w_1, acting on the area A_1 of the first piston. This same pressure is transmitted undiminished to the second piston according to Pascal's principle. Thus,

$$p = \frac{F_2}{A_2} = \frac{w_2}{A_2}$$

Notice that the two areas are not necessarily equal. The second area could be 100 times as large as the first area. If so, the second weight would also have to be 100 times as large as the first weight (see Fig. 3-19). A remarkable result occurs in a hydraulic lift. A small weight can hold up a much larger weight! By exerting a small *additional* force on the first piston, it is possible not only to support a tremendously larger weight, but also to move the larger weight upward. The small additional force is needed since all real pistons have some friction.

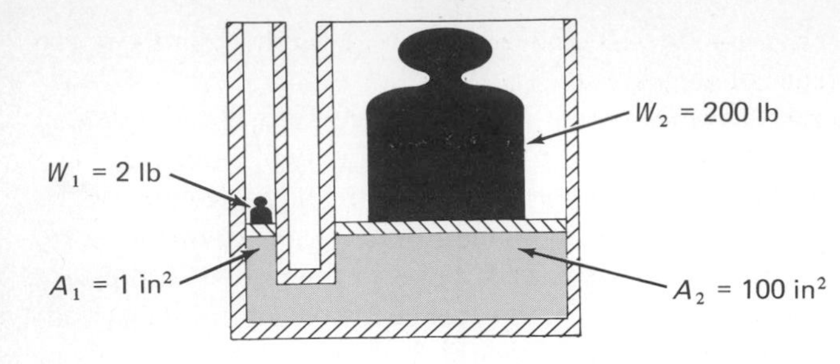

Fig. 3-19
In a hydraulic lift the pressure is the same throughout the liquid, but the forces are different on the two pistons.

There are many examples of hydraulic lifts: the grease rack in a service station (see Fig. 3-20), some automobile jacks, a dentist's chair, a beautician's or barber's chair, and some elevators. In each case the large weight to be lifted is supported by a large-area piston. The pressure that is applied to the

Fig. 3-20
The lubrication rack at a service station is raised and lowered by a hydraulic lift. *(Photograph by Korrine Nusbaum.)*

large-area piston is generated by a hand-operated or electric pump. Thus, the pump replaces the small-area piston.

A small force generating a large force in a hydraulic lift suggests "getting something for nothing." Is a conservation law being violated? Indeed, the last chapter showed that some quantities in physics, such as energy, linear momentum, and angular momentum, are conserved. But force is *not* a conserved quantity. Forces can be turned on or off, for example, simply by pushing against a wall or by not pushing. Forces can be created, destroyed, or changed in size without violating any law of physics.

The hydraulic lift, moreover, is *not* a perpetual motion machine. Although the forces on the pistons are unequal, any changes of potential energy are equal. If we consider an example of a hydraulic lift with a small piston with an area of 1 square inch and a large piston of 100 square inches, then 1 pound can support 100 pounds. If 100 pounds is lifted 1 foot, its potential energy increases by 100 foot-pounds. Correspondingly, to conserve energy, the potential energy of the single pound must decrease by 100 foot-pounds (excluding friction and kinetic energy). This means that the single pound must be lowered 100 feet. In many cases this is not practical. So a pump must do 100 foot-pounds of work on a large amount of liquid taken from a reservoir. This liquid is pumped during the operation of the hydraulic lift. Energy is not created in these devices.

What would happen if a hydraulic lift were filled with a gas instead of a liquid? If 100 pounds were placed on the larger piston and 1 pound were placed on the smaller piston, this would only serve to compress the gas and both pistons would move down. A gas can be compressed. Gas is compressed, for example, when you pump air into a tire. More air is put into the same volume. A hydraulic lift can be constructed only with a liquid.

3.6 FLUID FLOW

In a hydraulic lift the liquid is at rest (or moving very slowly) in a *closed* container. No net force is being applied to the liquid. But when a container is *not* closed, as in an open-ended horizontal pipe, fluid can flow. An externally applied pressure can exert a net force on the fluid. The fluid then accelerates according to Newton's second law of motion.

A fluid flows when a pump or reservoir exerts pressure on the fluid at one end of an open horizontal pipe. Within the fluid there will be decreasing pressure as the fluid gets farther from the applied pressure (see Fig. 3-21).

Pressure drop

The decrease in pressure in a flowing fluid is called the pressure drop and is given the symbol Δp.

The pressure drop is the higher pressure at the pump or reservoir minus the lower pressure at a point within the flowing fluid or at the open end. Recall that the symbol Δ refers to "change in." Thus Δp can be thought of as "change in pressure."

VORTEXES

We are so accustomed to living beneath an ocean of air that we may forget it exists. When you move a hand back and forth through air, you don't even feel it. But air can exert frictional forces. Sometimes these forces are small, but they can be measured. In other fluids, frictional forces are more obvious. For example, when stirring paint or moving your hand through water, you feel a resistance to the motion. These frictional forces play a role in the motion of liquids and gases.

(Photograph by Curriculum Research Group, University of Wisconsin—Milwaukee.)

One example would be stirring soup in a pan on the stove. Run a spoon round and round moving faster and faster. Friction between the soup and the spoon starts the soup rotating. Sometimes little whirlpools, or eddies, are also created in the soup by stirring, especially when you suddenly start stirring in the opposite direction. This circular pattern of fluid motion, as shown in the photograph to the left, is called a vortex.

Vortexes are also made when paddling a canoe. As the paddle moves through the water, a vortex is formed behind the blade. Since there are many variables such as the angle of the blade and the depth of the stroke, the vortexes are not all the same. Sometimes two or more will be seen. Other times a single vortex is formed which is several inches in diameter and lasts for 5 seconds or more as it drifts off behind the canoe.

Another example which you can try involves a glass soda bottle. Fill the bottle with water, cover the top, and invert. Give the water a few seconds to stop moving, and then let the water drain. It will drain out without forming a vortex. But now refill and invert the bottle. This time swirl the water and bottle vigorously and then open the end. As the water drains, a vortex forms which can be seen through the sides of the bottle.

Similar things can be done in your bathtub. When you pull the plug from a bathtub full of water, the movement of your hand and the plug sets the water in motion. The escaping water maintains this motion and sets up a whirlpool. If you slowly removed the plug from a very still tub of water, no vortex would form. However, if you deliberately stir the water, a strong vortex will form.

Much bigger vortexes are formed in the atmosphere. Hurricanes may have a diameter of 100 miles and wind speeds exceeding 150 miles per hour. Yet the basic structure of a hurricane is quite similar to the small whirlpools in your own bathtub. Of a size in between, tornadoes are the most violent storms in nature. Perhaps only a few hundred yards across, a tornado may have wind speeds reaching 300 or 400 miles per hour. Tornadoes are so powerful that wind-measuring instruments are usually destroyed if struck by a tornado. Thus accurate wind-speed measurements are hard to obtain.

Tornadoes and hurricanes are rare enough that you might live a lifetime without seeing one. Equally rare is a water spout, such as the one shown in the figure to the right. But you can see many smaller vortexes. A dust devil is a small vortex in air. The rotating winds are marked by dust or sand carried along by the air. In colder climates one may see dry leaves in the fall caught up in a similar rotating pattern. Also, there are snow devils and steam devils. The rotating air is the same in each case. But the rotation is marked by different materials.

(Photograph from Wide World Photos.)

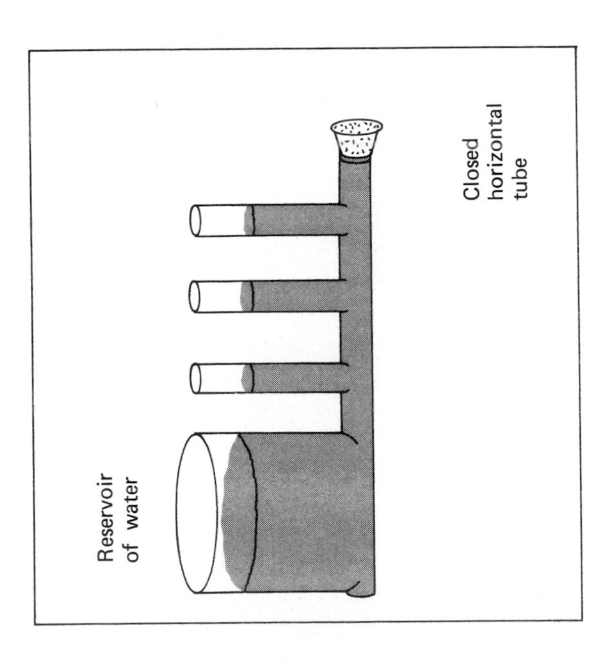

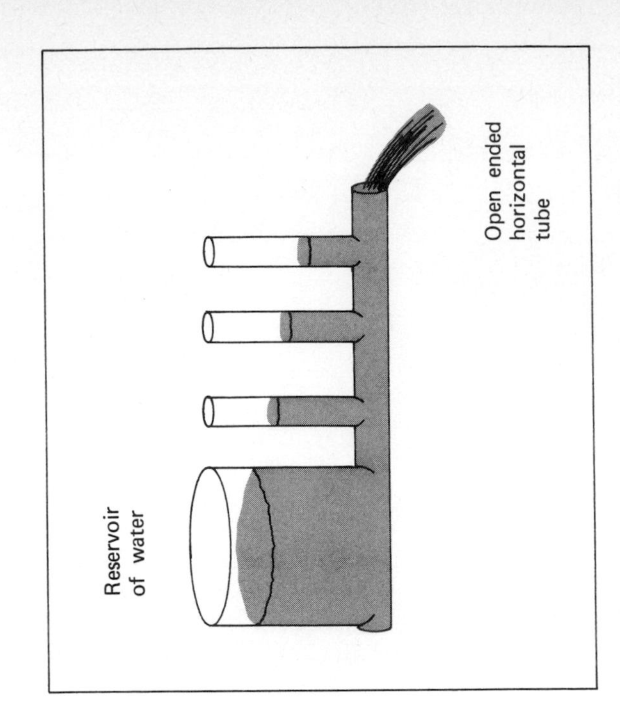

Fig. 3-21
No flow of liquid in the left illustration means the pressure is uniform in the closed horizontal tube. The liquid flowing in the right illustration means a pressure drop exists along the open horizontal tube.

As a fluid flows through a pipe, a certain volume of the fluid passes a given point during each elapsed time. This is called the *volume flow rate.*

Volume flow rate *The volume flow rate is defined as the volume of a fluid that flows divided by the time and is given the symbol* $\mathfrak{F}$.

The volume flow rate is simply a technical term for the rate at which a fluid moves through a pipe. Figure 3-22 shows a large-volume flow rate of water.

When a fluid flows, one factor that affects the volume flow rate is the pressure drop within the fluid. But there is another factor known as *resistance* that also has an effect on fluid flow. Resistance is made up of various types of friction. One type of friction depends upon the size of the pipe. As the diameter of the pipe becomes smaller, there is more friction between the fluid and the wall of the pipe. Furthermore, as the pipe gets longer, there is more friction. Another type of friction arises because of the viscosity of the fluid. Viscosity is an internal friction, or "stickiness," of a fluid. Some fluids pour more easily than others. For example, water pours more easily than syrup. The ease of fluid flow is related to viscosity. We would say that syrup is more viscous than water (see Fig. 3-23). Can you name another liquid that is more viscous than water?

Fig. 3-22
The volume flow rate is large as water flows through the discharge pipe of a dam. (*Photograph from Photo Researchers, Inc.*)

These various frictions are all combined into one term: *resistance.*

Resistance *Resistance is defined as the overall friction that decreases flow and is given the symbol R.*

If it were not for resistance in flowing fluids, the volume flow rate would simply depend upon the pressure drop in a fluid. But resistance does exist, and so we must account for it.

The three ideas of pressure drop, volume flow rate, and resistance are all combined into a law of fluid flow known as *Poiseuille's law.* In order to put these three factors together, begin by considering the volume flow rate $\mathfrak{F}$. The flow is an observable effect of moving fluids. But what controls the flow?

When fluids flow, the driving force is the pressure drop that pushes the fluids along. The larger the pressure drop, the more the flow. We say that the volume flow rate $\mathfrak{F}$ depends *directly* on the pressure drop Δp. For such a direct relationship, we could also say the smaller the pressure drop, the less the flow.

Fig. 3-23
Maple syrup flows slowly over butter and pancakes. (*Photograph by Betty Cook.*)

Another factor is also important. We must take resistance into account. When the pressure drop is held fixed, the volume flow rate can still be changed by varying the resistance. When the resistance is *increased,* the volume flow rate *decreases.* The larger the resistance, the less the flow. Thus it is said that the volume flow rate $\mathfrak{F}$ depends *inversely* on the resistance R. In an inverse relationship between two factors, one goes up while the other goes down. It is like two children on a teeter-totter, as illustrated in Fig. 3-24. They are never both up or both down.

The volume flow rate, pressure drop, and resistance can now be put together in *Poiseuille's law.*

Poiseuille's law

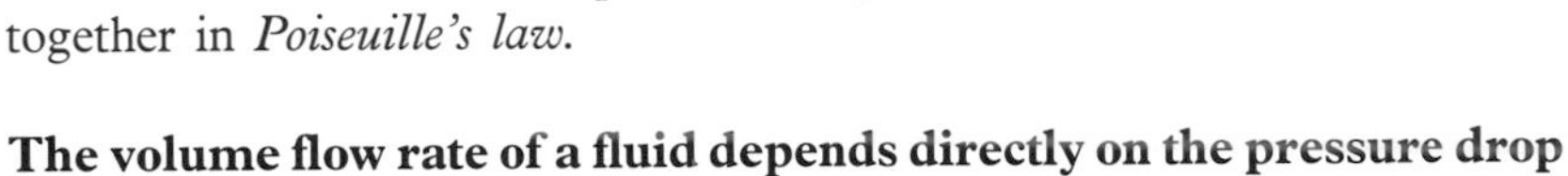

The volume flow rate of a fluid depends directly on the pressure drop and inversely on the resistance.

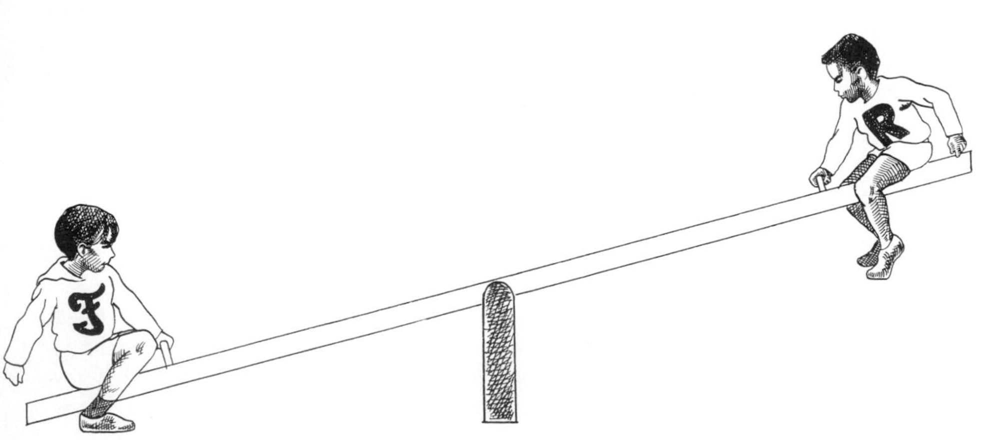

Fig. 3-24
When R is up, $\mathfrak{F}$ is down.

Poiseuille's law can be expressed in the simple formula

$$\text{Volume flow rate} = \frac{\text{pressure drop}}{\text{resistance}}$$

or

$$\mathcal{F} = \frac{\Delta p}{R}$$

Although we will not use this formula to make any calculations, it is a good shorthand summary of the important quantities and their relationships in fluid flow. As we will see in Chap. 6, this formula can also help in understanding electricity.

Poiseuille's law provides a reasonable description of the flow of fluids through pipes and tubes. Water flowing through the plumbing in a house is driven by the pressure drop maintained by city reservoirs and pumps. The volume flow rate is limited by the resistance in pipes, especially at bends or corners of the pipes.

Blood flowing through your body's circulatory system also demonstrates Poiseuille's law. The pressure drop between your heart and points along the system provides a driving force for the blood flow. As the blood flows from your heart to large arteries, then to smaller arteries, and so on, the diameter of these "pipes" decreases, thereby increasing the resistance and slowing the blood flow. This slowing of the blood flow is normal. What is not normal is when the circulatory system gets clogged with fatty deposits such as cholesterol. When fatty deposits occur, the heart must beat harder to provide higher blood pressure. Thus a physical checkup always includes a blood pressure measurement in order to obtain indirect evidence of the condition of the body's circulation system.

Poiseuille's law also applies to gases flowing through a pipe or tubing. Such an example would be the natural gas used to heat houses and to cook food. The local gas company must provide a pressurized gas line to every home in order for the gas to flow to furnaces, hot-water heaters, and stoves. The driving force of the pressure drop pushes the gas. But once again, the structure of the pipes, including bends in the pipes, imposes a resistance to the flow of gas. Engineers who design distribution systems measure or calculate the resistance before using Poiseuille's law to determine the volume flow rate. After constructing a pipeline, engineers often measure the volume flow rate to be sure that the expected gas flow is provided.

3.7 STREAMLINES

The discussion of fluid flow dealt with flow through pipes and tubes, which usually cannot be observed unless you have glass pipes. Even when a fluid is outside a pipe, such as air in the atmosphere, there may be some difficulty in observing the flow. Air, which is invisible, can be felt but not seen when the wind blows. To observe air flow we trace the motion by following the progress of dust, clouds, or smoke, as shown in Fig. 3-25. A similar situation occurs in water. The motion of clean, moving water is hard to see unless the water carries bubbles, foam, or leaves on its surface.

The pathways traced by markers in the air or water are known as *streamlines.*

Streamline

A streamline is defined as a continuous path marking the pattern of flow in a fluid.

Fig. 3-25
Smoke particles show the flow of exhaust gases. (*Photograph by John Hendry, Jr., from Photo Researchers, Inc.*)

A streamline traces smooth paths and gentle curves when the fluid meets an obstruction. Water and air flow around rocks in smooth curves. This is not like a bullet that strikes a rock. If observed from above, a bullet would be seen to follow a straight line to the rock and then bounce off in another straight line. The bullet's path has a sharp junction at the rock, unlike the path of a fluid. Fluids flow smoothly around obstructions.

The distance between streamlines indicates the speed of the fluid. If there is no change in speed, the streamlines are all equally spaced lines or pathways. Such paths can be seen traced out by leaves moving on the surface of a river if the width of the river is constant. But what happens if the river flows into a narrow channel? Then the water will flow faster in the channel or else overflow its banks. The leaves on the surface will flow faster and will also move closer together. When streamlines crowd together, the fluid is flowing faster, and when streamlines spread apart, the fluid is flowing slower. Figure 3-26 illustrates the streamlines of a fluid flowing through a pipe with a bulge in the pipe's center.

An interesting application of streamlines occurs in the design of car bodies and airplanes. A "streamlined" car or airplane is one that allows the air to flow around it with a minimum disturbance of the air. A box is not streamlined because the air follows pathways that must change direction rather abruptly in order to get around the box. But a sleek car or airplane allows the streamlines of air to gently separate and flow around the car or airplane.

The reason that large changes in direction are avoided in streamlined vehicles is that such changes cause large forces to be exerted on the vehicle by the air. When the air changes direction, there has been a *velocity* change in the air. Recall that velocity is made up of both speed and direction and a change

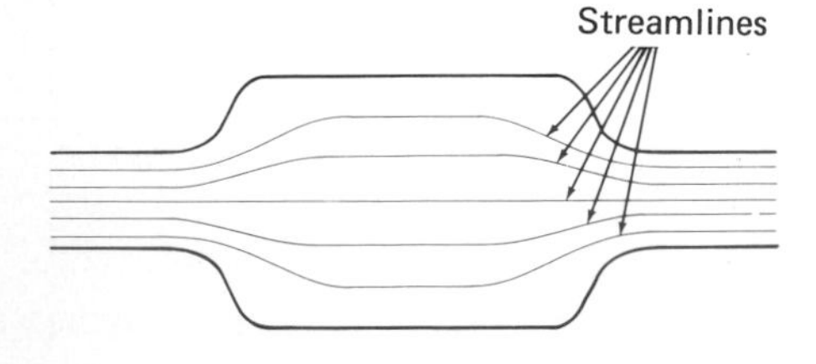

Fig. 3-26
Where the streamlines are closer together, the speed of the fluid is faster.

Fig. 3-27
The air deflected at the front end of a race car makes the car easier to steer. What is the purpose of the deflection panel behind the driver? (*Photograph by Dave Nadig from Photo Researchers, Inc.*)

in either one is equivalent to a change in velocity. A velocity change, however, must be caused by an applied force according to Newton's second law. Thus an object applies a force *on the air* to change the direction of flow of the air. But the air applies an equal but opposite force *on the object* according to Newton's third law. A streamlined car or airplane reduces the amount of deflected air. But there are other cases where deflected air is deliberately used. One example involves race cars.

Flowing air is used to exert forces on parts of a race car or airplane even though there might seem to be a loss in speed. A race car must be steered around curves and obstacles in the track. In order to increase the effectiveness of the steering, it is necessary to push the front wheels hard against the pavement without increasing the weight of the race car. The additional force comes from deflected air. The front end of a race car is sloped to deflect the air flow upward. The deflection of the air upward causes a force to be exerted downward on the front end of the race car, as shown in Fig. 3-27.

The same idea is used to increase the traction on the rear wheels of a race car. A powerful motor can easily cause the rear wheels to spin without gripping the pavement. As before, the rear wheels need to push hard against the pavement without increasing the weight of the race car. Often one can see a deflection panel mounted above and behind the driver. This panel deflects the air upward. Thus, a force on the rear of the car is exerted downward by the flowing air.

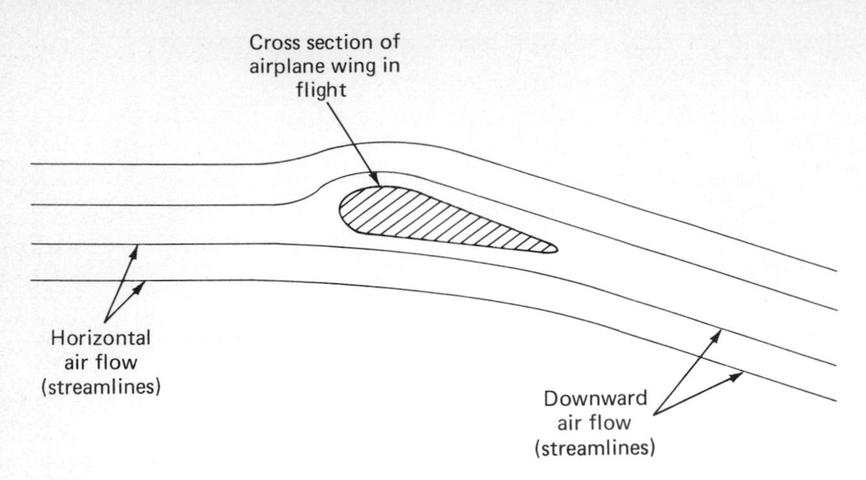

Fig. 3-28
The air being deflected downward causes the wing to be lifted upward.

The wing of an airplane provides lift because of downward deflection of air. As a result, the flowing air exerts an upward force on the wing that lifts the airplane off the ground. Once at the altitude needed for level flight, the pilot can adjust small panels known as ailerons to decrease the upward lift of the wings. The upward-lifting force is adjusted to equal the downward weight of the airplane. Later on, further adjustment of the ailerons makes the upward lift less than the weight, so that the airplane can descend to the ground.

REVIEW

You should be able to define the following terms: fluid, atom, molecule, pressure, buoyant force, weight density, mass density, pressure drop, volume flow rate, resistance, and streamline.

You should be able to state the following laws and principles and give examples of their application: the law of fluid pressures, Archimedes' principle, Pascal's principle, and Poiseuille's law.

PROBLEMS

1. An automobile weighing 3,600 pounds rests on the pavement. The area of contact of each tire is approximately 32 square inches. What is the pressure exerted by each tire on the pavement?
2. A bicycle and rider weigh 160 pounds. The area of contact of each tire is approximately 2 square inches. What is the pressure exerted by each tire?
3. The water in a small home aquarium used with tropical fish exerts a pressure of a 0.4 pound per square inch on the bottom surface. If the bottom has the shape

of a rectangle 12 inches by 20 inches, what is the total force on the bottom?

4. As a tornado passes near a house, the outside atmospheric pressure can suddenly drop by 1.5 pounds per square inch. Calculate the net outward force exerted by the air inside the house on 1 square foot.

5. A 6-inch cube of concrete weighs 18.75 pounds. What is the weight density of the concrete?

6. A 6-inch cube contains water weighing 7.8 pounds. Calculate the weight density.

7. An iron sample with a volume of 0.001 cubic meter has a mass of 7.9 kilograms. What is the mass density of the iron sample?

8. At Fort Knox, 1 cubic meter of gold has a mass of 19,300 kilograms. Calculate the mass density of the gold.

DISCUSSION QUESTIONS

1. Is syrup a fluid? Even when it is very cold? What factors must you take into account?

2. A large block of salt is quite rigid, but grains of salt flow easily in a salt shaker. Does this mean that salt is both a solid and a fluid.

3. When a container is filled with a fluid, it is sometimes said that all the space in the container is occupied. If a jar is filled with popcorn, does the popcorn occupy all the space? If a balloon is inflated with air, does the air occupy all the space?

4. Compare the roles of force and pressure for (*a*) standing flat-footed versus standing on tiptoe, (*b*) cutting with a sharp knife versus cutting with a dull knife.

5. Why are phonograph cartridges adjusted to exert very small forces on a record?

6. Why is it so easy to get a "paper cut?"

7. Compare sleeping on a regular mattress with sleeping on a water bed in terms of force and pressure.

8. Could a person really lie comfortably on a bed of nails?

9. In water, the pressure increase is *directly related* to the depth underwater. That is, at 10 feet underwater the pressure is 4.3 lb/in^2. At 20 feet underwater, the pressure is twice as much, 8.6 lb/in^2. How much is the pressure at a depth of 50 feet? At a depth of 5 feet? At a depth of 6 inches?

10. What does a barometer measure? What units are used?

11. What is the connection between tornadoes and open windows in a house?

12. Is the blood pressure in your body constant?

13. What effect would there be on your buoyancy if you went swimming with a dozen Ping-Pong balls tucked in your bathing suit? (Disregard the stares when you try this.) What would happen if you used marbles instead of Ping-Pong balls?

14. Would a teakettle float upward if it were filled with helium rather than air?

15. A cup partially filled with water resting on a scale weighs 12 ounces. If a 1-ounce marble were dropped into the cup, what would the scale read? If a 1-ounce piece of wood were dropped into the cup, what would the scale read?

16. It is sometimes asked whether a pound of feathers or a pound of lead would fall faster? What important factor needs to be considered in this question?

17. Why does a hot-air balloon rise? Would the balloon rise on a hot summer day? Would the balloon rise on a cold winter day?

18. What would it be like to swim in liquid mercury? (Disregard the toxic effects.)

19. What would it be like to swim in gasoline? (Again, disregard the toxic effects.)

20. Does Pascal's principle apply to all fluids?

21. When a tube of toothpaste is squeezed, how does it illustrate Pascal's principle?

22. Can a hydraulic press use a gas instead of a liquid?

23. List common household fluids that are highly compressible and other common household fluids that are virtually incompressible (only compressed with *great* difficulty).

24. What units could be used for volume flow rate?

25. Is the volume flow rate directly related to the resistance R? Is it directly related to the pressure drop Δp?

26. If the pressure drop is held constant and the resistance is doubled, what happens to the volume flow rate? And what would happen if the resistance is tripled?

27. What factors control the resistance?

28. If air flow under and over a wing causes an upward force on an airplane in level flight, what would happen if the airplane flew upside down.

29. Give examples of the use of different materials to observe streamlines.

30. Describe how streamlines are important to a glider pilot or a soaring hawk.

31. When a soupspoon is moved through soup, flow patterns are set up. How could these be made more visible?

ANNOTATED BIBLIOGRAPHY

A very thorough picture book exists for all of the kinds of fluid flow dealt with in this chapter. Originally taken for films, this collection is called *Illustrated Experiments in Fluid Mechanics,* Ascher H. Shapiro (ed.), MIT Press, Cambridge, Mass., 1972. Some of the topics are advanced, but many of the pictures, which range from tornadoes to whirlpools in teacups, are self-explanatory. The fourth chapter on surface tension is particularly interesting if you have an interest in soap bubbles or drops and how they form.

Another book filled with pictures of one topic is A. M. Worthington, *A Study of Splashes,* The Macmillan Company, New York, 1963. This reprint of material from 1894 is exceptionally clear and fun to read. From the preface the author says

If the present volume is so fortunate as to find many readers among the general public, as the author hopes it may, especially among the young whose eyes are still quick to observe, and whose minds are eager, it will be on account of admiration for the exquisite beauty of some of the forms assumed. . . .

One of the major areas of application of air flow is weather forecasting. James G. Edinger, *Watching for the Wind,* Doubleday & Company, Inc., Garden City, N.Y., 1967, is available in paperback (Anchor Books). Another paperback in the same series from the same publisher is Louis J. Battan, *The Nature of Violent Storms,* 1961. The book *The Weather Machine,* by Nigel Calder, Viking Press, New York, 1974, contains color pictures and descriptions. A more recent book on violent storms is John L. Stanford's *Tornado,* Iowa State University Press, Ames, 1977.

An older and somewhat obscure book, well worth tracking down, is R. S. Scorer, *Natural Aerodynamics,* Pergamon Press, New York, 1958. A master teacher, Scorer clearly deals with cloud movement, plumes from smokestacks, and the fluid behavior of sand and snow piling up near a post.

The topic of fluids in motion is a very broad one. If you have some interest in plants, you would enjoy reading "Flow and Transport in Plants" by M. J. Canny. This article is contained in *Annual Reviews of Fluid Mechanics,* vol. 9, 1977, pp. 275–296.

4 TEMPERATURE AND HEAT

There are differences in the ideas of temperature and heat. We will relate changes in temperature to the concept of internal energy. The movement of internal energy will lead to a study of the effects of transferring heat.

4.1 TEMPERATURE

On almost any day you might use the word hot or cold to describe the weather, a meal, or a beverage. Although the words hot and cold are part of our everyday vocabulary, they are not very precise. A summer day in the arctic described as hot by an Eskimo would probably be described as cold by a Tahitian. Because of differences of opinion and the need for a scientific basis, the words hot and cold evolved into a single concept—temperature, which could be measured. Temperature scales were devised so that a high temperature corresponds to the descriptive idea of hot and a low temperature corresponds to the descriptive idea of cold.

Describing physiological sensations does not, however, explain the concept of temperature. In order to understand the basis of temperature, we must consider the behavior of atoms and molecules. Such particles are continuously moving. In a gas, the individual particles move around, collide, change positions, and mix with other particles. Since the particles move, they have kinetic energy. Some particles move rapidly, and some move slowly. But even two identical particles in the same gas can have different kinetic energies. Therefore we consider an average value for the kinetic energy. Within a sample of gas the atoms or molecules possess an *average* kinetic energy which is the basis of the physical meaning of temperature.

Temperature *The temperature of a gas is defined as the quantity that is proportional to the average kinetic energy of its atoms or molecules.*

When a gas has a high temperature, the atoms or molecules move with a greater average speed than when the gas has a low temperature, as shown in Fig. 4-1.

The definition of temperature also applies to liquids and solids. The particles in a liquid move about somewhat, exchanging positions. The particles in a solid are somewhat rigidly locked in place. So the particles of a liquid or solid do not have kinetic energies in the same sense as a gas. But the atoms or molecules of a solid are *vibrating,* that is, moving back and forth about a point within the object. Associated with this vibrating motion is an average speed, which allows you to think about an average kinetic energy for the particles of the solid. The motion of particles, whether the flight between collisions in a gas or the vibrations in a solid, is the basis for the physical explanation of the concept of temperature.

In everyday life most people do not think of temperature in terms of molecular kinetic energy. Temperature is usually thought of as "what the thermometer reads." A thermometer is simply a device to measure temperature. Many thermometers contain mercury, which at ordinary temperature is a silver liquid. The mercury inside the glass tube of a thermometer expands or contracts as the temperature rises or falls. Marks along the side of the glass tube indicate the amount of expansion or contraction and so measure the temperature (see Fig. 4-2). But a practical question remains. What numbers should be associated with the marks?

The markings on the thermometer indicate the temperature in units of

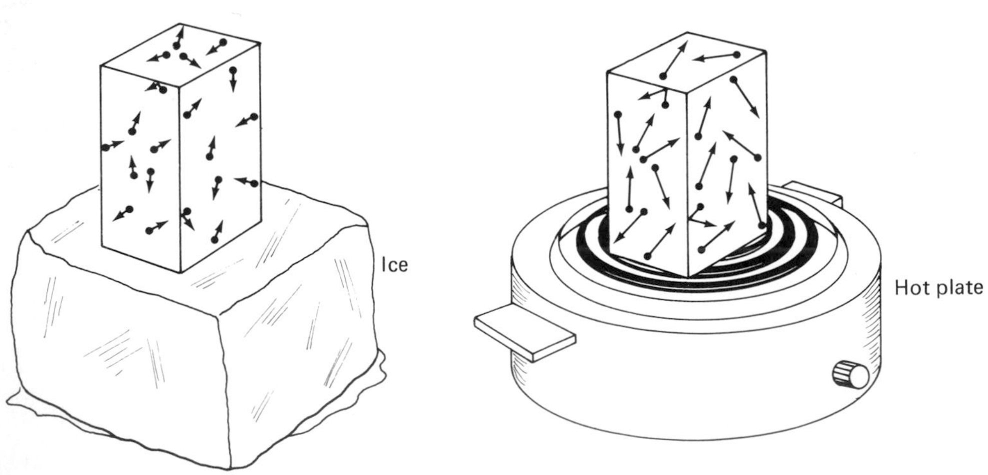

Fig. 4-1
The air molecules in a container resting on a hot plate have greater average speed than the air molecules in a container resting on a block of ice. The temperature of the air above the hot plate is higher than the temperature of the air above the block of ice.

Fig. 4-2
The thermometer measures the temperature of the meat.
(Photograph by Korrine Nusbaum.)

degrees. The numerical value of the degrees depends upon the system of units. In the English system of units the temperature of freezing water has the value of 32°F (F stands for Fahrenheit). The temperature of boiling water has the value of 212°F. Equally spaced markings are inserted between these two temperatures to fill in the temperature scale. The scale can be extended somewhat above and below these extremes to provide a practical temperature measuring device.

The common metric unit of temperature is degrees Celsius (abbreviated °C). The temperature of freezing water is given the value of 0°C, and the temperature of boiling water the value of 100°C. Again, equally spaced markings are used in the interval between. Whenever the temperature is below that of freezing water, the numbers are negative on the Celsius scale. Various experiments suggest that the lowest possible temperature is −273.15°C. This temperature is so low that it has never been reached in any laboratory. It is believed to be unachievable.* In order to eliminate the

*Many experiments are done in research laboratories at temperatures which are only slightly higher than −273.15°C. But it is thought that no process can be used to cool a substance *exactly* to this temperature of −273.15°C.

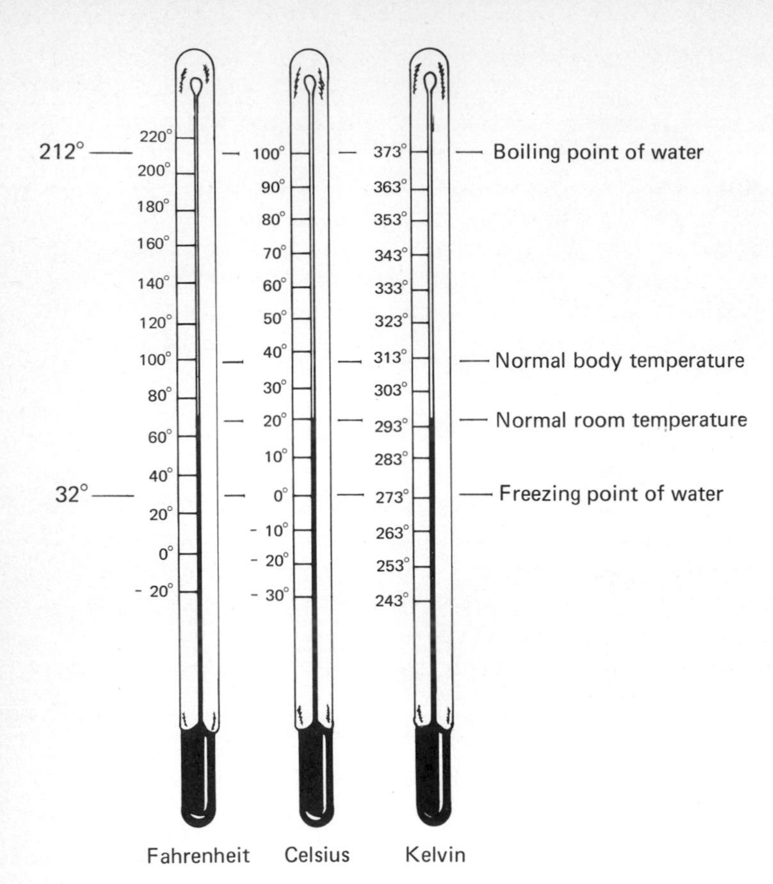

Fig. 4-3
A comparison of thermometers for the Fahrenheit, Celsius, and Kelvin temperature scales.

Table 4-1
Representative temperatures

Event	Temperature °F	Temperature °C
Electric light	3500	1930
Commercial pizza oven	550	288
Burning paper	451	233
Boiling water at sea level	212	100
Body temperature	98.6	37
Freezing water	32	0
Freezing mercury	−37	−38
Freezing alcohol	−292	−130
Absolute zero	−460	−273

negative numbers in the Celsius scale, another metric temperature scale known as the Kelvin scale has been devised. It uses degrees of the same size as the Celsius scale, but the numbering is different and the degree sign is not used. On the Kelvin scale the lowest possible temperature (−273.15°C) is given the value zero and is called *absolute zero.* The temperature of freezing water on the Kelvin scale is then 273.15K, and boiling water is 373.15K. The Kelvin temperature scale, not as commonly used as the Celsius temperature scale in everyday life, is used most often for measurements of low temperatures in scientific laboratories. Figure 4-3 illustrates the difference between the Fahrenheit, Celsius, and Kelvin temperature scales. Table 4-1 presents representative temperatures on the Fahrenheit and Celsius temperature scales.

4.2 EXPANSION AND CONTRACTION

When the temperature of an object rises, the object usually expands. Think about wires or rods, in which the expansion is most noticeable along their lengths. For example, the wire in a toaster becomes red-hot. The wire expands to a length longer than the room-temperature length. Later as the wire cools down to room temperature, the wire contracts to its original length. If the toaster were put in a deep-freeze overnight, the wire would contract to a length shorter than the room-temperature length. As a general rule, rising temperatures cause an object to expand and falling temperatures cause an object to contract.

Fig. 4-4
An expansion joint on a bridge lets the bridge expand and contract without buckling. (*Photograph by Robert A. Isaccs. From Photo Researchers, Inc.*)

Large structures, such as bridges and buildings, also expand during the summer and contract during the winter. To allow for the change in length and prevent damage to a bridge, an expansion joint is built into a bridge (see Fig. 4-4).

When the temperature changes, all objects do not expand or contract the same amount. If we wish to calculate the amount of expansion, two factors are important. One factor affecting expansion or contraction is a difference in the chemical composition of various materials. For example, an aluminum wire will change length more than a steel wire of identical size. This effect based on the type of material leads to a numerical factor known as the *coefficient of thermal expansion.* The larger the number for the coefficient of thermal expansion, the larger the expansion or contraction for equal-sized objects.

Another factor affecting expansion or contraction is the original length of the wire or rod. A long wire or rod expands or contracts more than a short wire or rod. Every part of an object contributes to this change in length, and so the longer an object, the greater will be the change in its length.

All the factors affecting expansion or contraction can be put together in one formula. We find that the change of length of a wire or rod is the product of three factors, namely, the coefficient of thermal expansion, the original length, and the change of temperature. The formula is

$$\Delta L = \alpha L \, \Delta T$$

where ΔL = change of length
α = coefficient of thermal expansion
L = original length
ΔT = change of temperature

When ΔT is positive (temperature increasing), ΔL is also positive (wire or rod expanding). Conversely, when ΔT is negative (temperature decreasing), ΔL is also negative (wire or rod contracting).

The common household thermometer measures temperature by means of an expanding or contracting column of mercury. A thermometer consists of a narrow glass tube attached to a small bulb. The small bulb acts as a reservoir for the mercury, and the glass tube is partially filled with mercury. When the temperature rises, the mercury and the glass container both expand. The mercury, however, expands more than the glass. So the mercury moves up the glass tube. A scale is etched on the glass to provide temperature readings.

As demonstrated with the mercury thermometer, all materials do not expand alike. For example, aluminum expands more than twice as much as iron. This difference in expansion is the basis of the *bimetallic strip*. Two thin strips of different metals are placed together at room temperature and welded in place along their entire length. If the temperature rises, one metal expands more than the other metal. As a result, the bimetallic strip bends. The higher the temperature goes, the more the bimetallic strip bends. If the temperature falls, the bimetallic strip becomes less and less bent. At room temperature, it would again be straight. The bimetallic strip bends in the opposite direction as the temperature goes below room temperature, as shown in Fig. 4-5.

Bimetallic strips are used in both thermometers and thermostats. A bimetallic *thermometer* is constructed with the bimetallic strip wrapped in a spiral in order to enhance the deflection. As the temperature rises, the curvature changes and a pointer deflects. Behind the pointer a scale is mounted to read the temperature (see Fig. 4-6). A bimetallic *thermostat* is a device for maintaining the temperature at a fixed value, for example, in a refrigerator, in a car's cooling system, or in a building. The bimetallic strip in

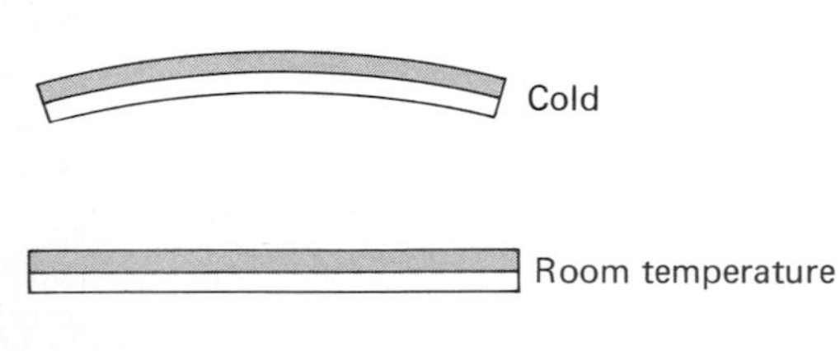

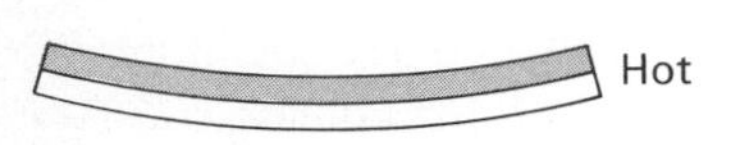

Fig. 4-5
The metals in a bimetallic strip expand unequally as the temperature is raised. The darker metal has the lower coefficient of thermal expansion.

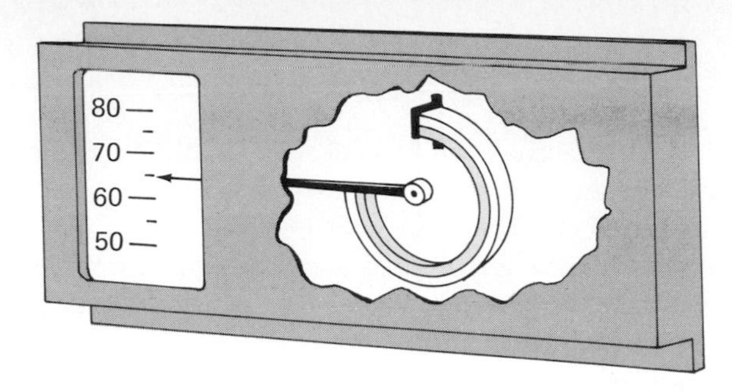

Fig. 4-6
A bimetallic strip, shown in this cutaway view, can be used to construct a wall thermometer.

the thermostat is part of an electrical circuit. As the temperature changes, furnaces, ventilators, or air-conditioning units can be switched on or off.

So far we have discussed mostly changes in length. But when an object is heated, *every* dimension changes. So a flat object will increase its area. And a three-dimensional object will increase its volume.

One way to think about expansions is to compare expansion with a photographic enlargement. When we buy an enlargement of a favorite snapshot, each dimension is bigger. The same is true when we enlarge an object by increasing the temperature. Each dimension becomes larger, including a hole, as shown in Fig. 4-7.

A cookie sheet undergoes repeated changes in size. At room temperature it has some particular length and width. One common size is $15\frac{1}{2}$ by 12 inches. When the cookie sheet is placed in an oven at 350°F, for example, the sheet expands. Both the length and the width become larger. You do not notice the changes, since they are small fractions of an inch.

Then, when the cookie sheet is removed from the oven, it cools. And it returns to its original length and its original width. You can then repeat the entire process. If you bake cookies often, you probably repeat the process several times each month!

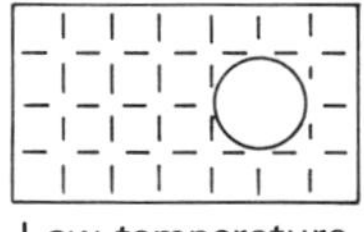

Low temperature

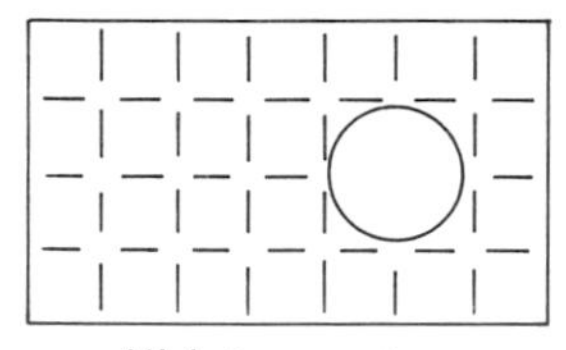

High temperature

Fig. 4-7
Thermal expansion is like a photographic enlargement. Notice that the hole also increases in size. The amount of expansion is greatly exaggerated in this illustration.

An example of volume change can be described for gasoline in the tank of an automobile. If the temperature of the gasoline rises, such as when you park in the sun, the gasoline will expand. Each portion of gasoline grows in volume. If the tank is partially filled with gasoline, the expanding gasoline will force air out of the tank. If the tank is full, the heating may cause gasoline to overflow.

We could think of mercury thermometers in the same way. As the temperature rises, the *volume* of mercury increases. This increased volume of mercury then squeezes up through the narrow glass tube. We mostly notice a change in *length* when we read a thermometer. But the *volume* of the mercury has increased.

Changes in temperature do not affect weight. The weight of a casserole dish at 450°F is the same as the weight of the dish at room temperature. But what about the weight density? If an object expands, the volume increases.

Thus by referring to Sec. 3-4, we see that the weight density decreases, even though the weight itself remains unchanged.

4.3 CHANGES IN TEMPERATURE

Temperatures change. A hot pizza cools on a plate, and a cold beverage warms in a glass. In each case there is a temperature change. Let us now consider how the temperature of an object can be raised.

There are *two* fundamentally different ways to increase the temperature of an object: either by rubbing the object or by placing the object in contact with another object having a higher temperature. As an example of the first way to raise the temperature, rub your hands vigorously together for a few seconds. The rubbing will noticeably raise the temperature.

Fig. 4-8
As work is done in rubbing the hands together, the temperature of the hands rises. (*Photograph by Korrine Nusbaum.*)

As an example of the second way to raise the temperature, put your hands in a dish of very hot water. No further activity is needed on your part, but again the temperature of your hands is noticeably raised.

These two fundamentally different activities, which you sense as "warming" and describe as raising the temperature, increase the average kinetic energy of the molecules of your hands. But how?

In the case of rubbing, small ridges on the surfaces of your hands become enmeshed, then break free. After being stretched and then released, the molecules of both hands end up vibrating more. Large-scale motion of your hands increases the small-scale motion of the molecules of your hands. When you rub, you do work on the molecules. There is an increase in the average kinetic energy of the molecules of your hands. Your hands feel warmer.

In the case of touching hot water, there are collisions between the molecules of the hot water and the molecules of your hands. Kinetic energy is transferred from the more rapidly moving molecules of the hot water to the slower moving molecules of your hands. Small-scale motion of the molecules of the hot water increases the small-scale motion of the molecules of your hands. Again, there has been an increase in the average kinetic energy of the molecules of your hands.

The case of rubbing an object to increase temperature is a specific example of the more general act of doing work on an object. Figure 4-8 illustrates hands being warmed by doing work on them. The act of doing work on an object causes the object to increase in temperature, so long as the object does not change from a solid to a liquid or from a liquid to a gas. These exceptions, which are called *changes of state*, will be discussed in Sec. 4-6.

In like manner, the case of touching a hotter object is a specific example of the more general activity of transferring *heat* to an object. Figure 4-9 illustrates hands being warmed by letting heat flow into the hands. The act of transferring heat causes the object to increase in temperature, once again so long as the object does not change from one physical state to another. The act of transferring heat usually causes an increase in temperature. Heat itself has not yet been fully described or defined, but it will be in Sec. 4-4.

4.4 INTERNAL ENERGY AND HEAT

In Sec. 4-3 it was seen that a change in temperature could be achieved either by the act of doing work on an object or by the act of transferring heat to an object. But these activities must not be imagined to "fill" the object with work or heat. The activity is external to the object. In Chap. 2 the reasoning was presented that objects do not become "full of work" as a result of work being done on them. Instead, work is the cause of an object becoming "full of energy."* In that chapter the specific kinds of energy were called potential energy and kinetic energy, which were characterized by position and speed, respectively.

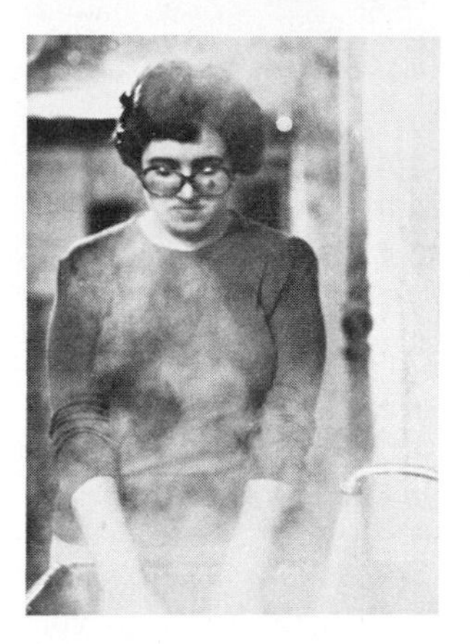

Fig. 4-9
As heat flows into the hands placed in hot water, the temperature of the hands rises. *(Photograph by Korrine Nusbaum.)*

The same reasoning applies again to the concept of heat. Both the activity of doing work on an object and the transfer of heat into an object involve something else "working on" or "heating" the object. The effect of these two processes is to increase the energy of the object. But now we find that there is a change in the *temperature,* not in the position or the speed of the object. So a new kind of energy is introduced. It is known as *internal energy.*

Some insight into the concept of internal energy may be gained by thinking about the atoms or molecules of an object. These particles possess both potential energy and kinetic energy. The energy of individual particles can be discussed, but it is a difficult task to actually calculate or measure this energy. Instead, all the individual potential and kinetic energies are imagined to be "smeared" or combined into the internal energy of the object. The internal energy represents the *total* of all the individual contributions. The potential and kinetic energies of all its atoms and molecules contribute to an object's internal energy.

Internal energy

Internal energy is defined as the energy of an object that is made up of both the potential energy and kinetic energy of the atoms or molecules of the object.

Internal energy is given the symbol U.

The question naturally arises about the difference between internal energy and temperature, since both are related to the atomic behavior of objects. On the one hand, *internal energy* is a rather abstract concept dealing with *both* the potential and kinetic energies of the atoms of an object. Internal energy is easily discussed but can only be calculated with difficulty. On the other hand, the *temperature* of an object is easily measured with a thermometer. A thermometer responds *only* to the kinetic energy of the atoms of the object.

This situation is similar to the relationship of position to potential energy and speed to kinetic energy. Position and speed are easily measured physical quantities. But potential energy and kinetic energy are a bit more abstract.

*We mean, of course, that an object has its energy increased. In every real sense, an object *never* becomes *full* of energy. If a body possesses a given amount of energy, say kinetic energy, it can always be given more kinetic energy.

Similarly, temperature is an easily measured physical quantity. But internal energy is an abstract idea. Internal energy can be calculated, although we shall not provide a formula for the calculation.

Let us return to the word heat and its usage. Common phrases suggest that heat is stored or can be stored in an object. It is typical to say, "Heat up the water," or "The hot-water bottle will heat you." These statements suggest that heat can be contained in the water, but that is false. There is only internal energy in the water. Only internal energy gives information about the state or condition of an object. However, there are *activities,* namely, doing work or transferring heat, that can change the state or condition of an object. Heat, then, is one of the two activities that changes the internal energy of objects.

Think about what happens to internal energy when you put your hand in hot water. The internal energy of the water decreases. The internal energy of your hand increases. No work is done if your hand and the water remain motionless. Yet your hand gets warmer. Thus, you speak of "heating" as a shorthand expression for internal energy moving from the water to your hand.

Heat *Heat is defined as the flow of internal energy by means of the activity of atoms or molecules.*

Whenever internal energy flows from one place to another with no accompanying work being done, the process can be described as the transfer of heat, or as simply "heating."

It is wrong to think that heat is contained in a hot object. There is no heat stored in the sun, only a lot of internal energy. Of course, there is a large amount of energy transfer (heat) between the sun and us.

The concepts of heat and internal energy are abstract. Table 4-2 summarizes the cause-and-effect relationship between work, heat, and internal energy. Work increases internal energy when it is observed that the temperature of an object is raised. Heat also increases internal energy by raising the temperature of an object.

Work and heat are activities that transfer energy. After the energy has been transferred to or from an object, work and heat no longer have meaning. But the internal energy has been changed, as expressed in the *first law of thermodynamics.*

First law of thermodynamics **A change in the internal energy of an object is the result of the activities of doing work and transferring heat.**

The first law of thermodynamics can be thought of as a statement of the conservation of energy. In this case we consider the connection between internal energy and the two activities of doing work and transferring heat. These three quantities are related. We cannot change two of them without

Table 4-2
The cause-and-effect relationship between work, heat, and internal energy

One way of increasing internal energy:

Observation:	force (rubbing) and distance	leads to	change of temperature but no change of speed or position
Physics	work	leads to	internal energy

Another way of increasing internal energy:

Observation:	interaction with hotter object	leads to	change of temperature
Physics:	heat	leads to	internal energy

affecting the third. The first law of thermodynamics describes the relationship between internal energy, work, and heat.

The first law of thermodynamics can also be stated as an equation:

$$\Delta U = \mathcal{W} + Q$$

where ΔU = change in internal energy
$\mathcal{W}$ = work done *on* the object
Q = heat transferred *to* the object

The first law of thermodynamics summarizes the relationship between the three quantities. For example, if work is done on an object *and* heat is transferred to the object, then the internal energy *must* increase.

4.5 TRANSFER OF HEAT

Whenever objects at different temperatures are allowed to interact, there will be a transfer of heat. Objects at different temperatures tend to equalize their temperatures. The internal energies can be changed by means of heat transfer in three different possible ways: conduction (*contact* of objects with different temperatures), convection (*mixing* of fluids with different temperatures), and radiation (*transmission* of electromagnetic energy between objects with different temperatures).

CONDUCTION

Some things, such as a hot frying pan or cold dry ice, are painful to touch. The pain is caused by the behavior of the molecules of the objects. The hot frying pan has greater molecular vibrations (that is, greater average kinetic energy) than your hand at normal body temperature. When you touch the hot pan, the surface molecules of your hand increase their vibrations while the surface molecules of the pan decrease their vibrations. This interaction continues (if you do not drop the pan) until both your hand and the pan have the same temperature. In order to come to the same temperature, the internal energy of the object at lower temperature increases while the internal energy of the object at higher temperature decreases.

Heat conduction

The transfer of internal energies between objects in contact is known as heat conduction.

If you touch an object such as dry ice, which is much colder than the temperature of your skin, there is again an exchange of internal energies. But this time internal energy is *lost* by your hand and *gained* by the dry ice.

All objects are capable of conducting heat, but there is great variation among different materials, as shown in Table 4-3. Materials that readily conduct heat are called *thermal conductors.* Materials that are not good conductors of heat (although there is some small heat conduction) are called *thermal insulators.*

Metals, which are very good thermal conductors, transfer heat by means of their electrons. Electrons in metals are easily separated from the atoms. These "free" electrons migrate through the interior of a metal and provide the means for distributing changes of internal energy throughout the metal. All common metals including copper, aluminum, iron, and silver are good thermal conductors (see Fig. 4-10). A frying pan made from cast iron or aluminum allows heat conduction from a stove burner to food which is being

Table 4-3
Thermal conductivities of various materials at 20°C

Material	Conductivity, W/m · K
Styrofoam	0.033
Glass wool	0.039
Body fat	0.15
Water	0.604
Glass	1.5
Concrete	1.73
Ice (0°C)	2.2
Brass	108
Aluminum	237
Silver	418

Fig. 4-10
(*a*) Heat in; (*b*) heat out. [(*a*) *Photograph by Ted Rozumalski from Black Star.* (*b*) *Photograph by Korrine Nusbaum.*]

cooked. A metal soupspoon allows heat conduction between hot soup and your fingers!

Some common thermal insulators are fabrics and plastics. We use cloth hot pads to protect our hands from hot metal pans (see Fig. 4-11). In the complex molecules of cloth, all the electrons are strongly bound to the molecules and are not free to migrate through the interior of the hot pad. It is the same with a plastic handle on an aluminum saucepan. The electrons in plastic are not free to move around. Thus plastic is a good thermal insulator. Another good thermal insulator is wood.

It is a common experience for someone in bare feet to prefer to stand on a rug rather than tile flooring in a bathroom. Both the rug and the tile have exactly the same temperature because both have been on the floor for some time. Nevertheless, the rug *feels* warmer. The mistake in your sense of the temperature is due to the unequal amounts of heat being conducted away from your feet. Tile is a better conductor than rug, and so the tile conducts more heat away from your feet than the rug does in the few seconds you might be standing on either one. Your body senses the difference in the amount of heat being conducted away as being due to a difference in temperature.

The same mistake in perception of temperature occurs when you use a cloth pot holder to carry a warm plate. Both the plate and the pot holder could have nearly the same temperature. But if you touch each one separately, the plate feels warmer because it conducts more heat to your skin. Your sense of temperature can also be misled if air moves over exposed skin. A brisk wind or a ride on a motorcycle produces the sensation of the air temperature being lowered. The temperature of the air is not actually lowered. But the heat being conducted away from exposed parts of your body is increased.

Fig. 4-11
Why do you hold a pan with a pot holder? *(Photograph by Cal Hulstein. From Leo de Wys, Inc.)*

Clothing traps warmer air near your skin, and so heat is slowly conducted away from your body. But skin unprotected by clothing loses heat faster if a steady stream of cool air makes contact with your skin. The increased conduction of heat chills your body. A chart of *wind-chill factors* summarizes the effect of the wind on how you sense the temperature. The wind-chill factor is an equivalent lower *still*-air temperature that would have the same cooling effect as the actual measured temperature and wind speed (see Table 4-4). Wind does not really lower the temperature of the air. But the wind does increase the amount of heat conduction away from a person or a warm object.

A thermometer reading might be $-5°C$ (23°F), and a weather report states that the wind-chill factor is $-15°C$ (5°F). Your car left parked on the street will not cool to $-15°C$, only to $-5°C$, the measured temperature. The effect of the wind is only to cool the car *faster* than still air would. The wind will not lower the temperature of the car below the temperature measured on a thermometer.

CONVECTION

Solids, which are rigid materials, transfer heat through their interior by means of conduction. But liquids and gases, which are fluids and do not form

Table 4-4
Windchill factors in degrees Celsius

Temperature, °C	Wind speed, km/hr 10	20	30	40	50	60	70
15	12	9	7	6	5	4	4
10	6	2	−1	−2	−3	−4	−5
5	0	−5	−8	−11	−12	−13	−14
0	−6	−13	−17	−19	−21	−23	−23
−5	−12	−20	−24	−28	−30	−32	−32
−10	−18	−27	−32	−37	−39	−40	−40
−15	−23	−33	−38	−43	−46	−48	−49
−20	−29	−40	−46	−51	−54	−57	−58
−25	−35	−47	−53	−59	−62	−64	−65
−30	−42	−54	−61	−67	−71	−72	−73

rigid objects (except at very low temperatures), transfer heat by mixing. For fluids at different temperatures there is a mixing of molecules with different internal energies.

Heat convection

The transfer of internal energies between objects by mixing is known as heat convection.

Unlike conduction, which is a transfer of internal energy between fixed atoms or molecules, convection involves an actual movement of mass from one place to another.

In the discussion of thermometers it was seen that because of expansion an object at a higher temperature usually has a larger volume than the same object at a lower temperature. So an object at a higher temperature has a lower weight density than the same object at a lower temperature. According to Archimedes' principle, the warmer (less dense) fluid will rise, while the cooler (more dense) fluid will sink.

Above a hot stove or a hot radiator there is a continual stream of rising warm air. This is an example of convection. The temperature of the air in contact with the stove increases because of heat conduction. The air expands, and becoming less dense than the surrounding cooler air, it rises. The surrounding cooler air replaces the hot air, and so a continual stream is maintained.

In a refrigerator the air in contact with the cooling coils decreases in temperature and contracts. This cooler, denser air settles downward and is replaced by warmer and lighter air. Figure 4-12 shows that the cooling coils are located near the top of a refrigerator. If the cooling coils were located near the bottom, the cool air would not circulate and not all parts of the refrigerator would be cooled.

If smoke particles are present, the flow of air can be seen. In a campfire or near a barbecue, the streamlines of the air flow can be traced by the rising

Fig. 4-12
What would happen if the cooling coils were located near the bottom of the refrigerator? (*Photograph by Korrine Nusbaum.*)

smoke. Eventually the air cools, and sometimes you can see the smoke fall back to earth.

Convection also takes place in liquids. If you keep tropical fish, convection of water takes place continually in your aquarium. When water is warmed by the aquarium heater, the water rises (see Fig. 4-13). Cooler water from other regions of the tank flows along the bottom toward the heater. Convection also occurs on a large scale in lakes and rivers. In these bodies of water convection is not the only cause of mixing water. Springs on the bottoms of lakes cause some mixing of water. Turbulent flow in rivers also mixes water at different temperatures.

RADIATION

If you bring your hand near the side of a hot object, such as an illuminated 100-watt light bulb or a hot iron, you can feel the heat warming your hands even before the object is touched. Heat is transmitted through the space between the object and your hand. This transmission of heat is not accomplished by conduction or convection of the air. You could feel the heat even if there were no air whatsoever. The sun, which is about 150 million kilometers

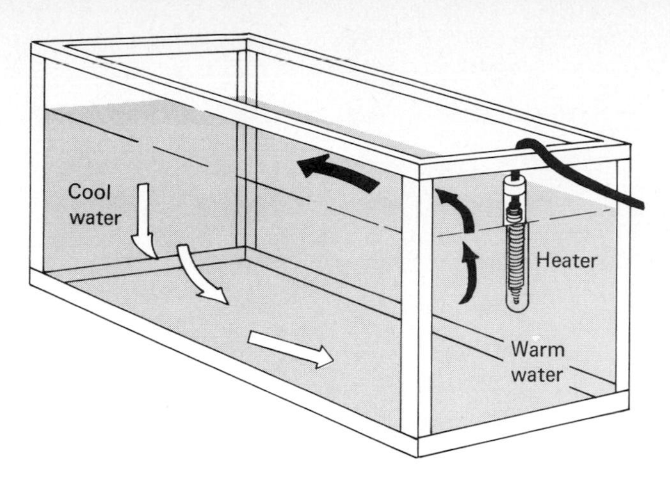

Fig. 4-13
An aquarium heater warms all the water by means of heat convection. The arrows show the circulation of the water.

away, is capable of heating you. Almost all of this distance is empty space, that is, an almost complete absence of matter. So there must be a means of heat transfer that can increase your internal energy while decreasing the internal energy of the sun.

Heat radiation *The transfer of internal energies between objects by transmission through space is known as heat radiation.*

By the action of heat radiation, heat is separated from its association with matter. It is true that heat radiation can pass through some materials such as air or glass, but not as perfectly as through empty space. In completely empty space, heat radiation is not absorbed and can travel indefinitely. But what is this heat radiation that can travel through the voids of outer space?

Heat radiation is energy traveling as electromagnetic waves. Visible light and the invisible infrared rays and ultraviolet rays are also electromagnetic waves. All these electromagnetic waves, as well as others, will be investigated in detail in Chap. 7. For the present, these rays will be identified as common forms of heat radiation. They are capable of noticeably increasing the internal energy and thus the temperature of your skin. All three of these forms of heat radiation can be felt. But only visible light can be seen with the eyes. Pure infrared rays or ultraviolet rays cannot be seen, although they can warm your eyes or cause other changes if they are intense enough.

Many fast-food restaurants use heat lamps to keep food hot (see Fig. 4-14). The heat lamps emit infrared rays, which are one form of heat radiation. The lamps also emit visible light to alert employees that the heat lamps are operating.

Heat radiation has one property in common with visible light. Just as a

Fig. 4-14
Heat radiation from the infrared lamps keeps the food warm. (*Photograph by John Harmon.*)

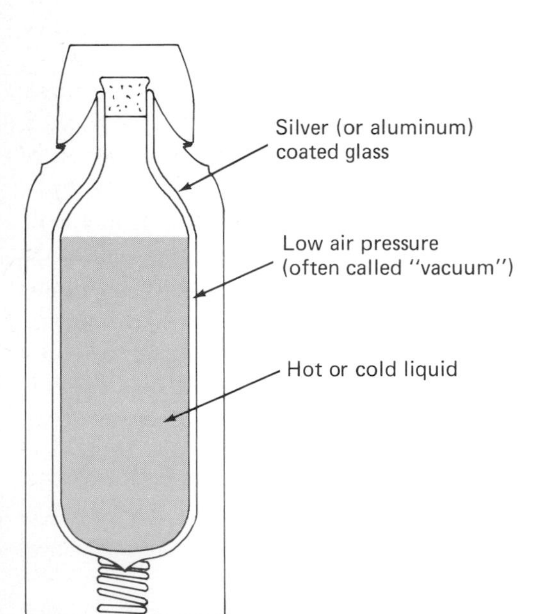

Fig. 4-15
In a thermos bottle the low air pressure between the glass walls prevents conduction and convection of heat. The silvered glass surface reflects heat radiation.

mirror (a silvered surface) can reflect visible light, a mirror can also reflect heat radiation. The reflection of heat radiation by a silvered surface is used in the common thermos bottle to keep liquids either hot or cold. A thermos bottle is a double-walled glass container with a reflecting silver layer on one of the glass surfaces, as shown in Fig. 4-15. The silver layer reflects heat radiation trying to enter or leave the glass container. Most of the air has been removed from the space between the double walls of the container, and so conduction and convection of heat are greatly decreased. Heat can only enter or leave by means of conduction through the neck of the glass walls or the stopper. So liquids remain hot or cold for several hours in a thermos bottle.

4.6 CHANGE OF STATE

According to the first law of thermodynamics, the transfer of heat to an object always increases the internal energy of the object. However, there can be two different effects on the object as a result of an increase in internal energy. One effect of the transfer of heat is to cause the object to increase in temperature. Your everyday experience of heating water on a stove is that the temperature of the water is raised. But what if the water is already boiling and you continue to increase the internal energy by continuing to transfer heat into the water? The water does not increase in temperature. Instead, another effect occurs: the water is transformed into steam. Boiling water is an example of a change of state, which is the other possible effect of the transfer of heat to an object.

The obvious difference between water and steam is a difference in state. The molecules of water are close together and attract one another quite strongly. Water is a liquid that will rest in the bottom of a closed container. But the molecules of steam have more internal energy than water molecules. Steam molecules are relatively far apart and are free to move throughout a container. Although their internal energies are different, water molecules and steam molecules can have exactly the same average kinetic energies—that is, *they have exactly the same temperature.* At 100°C, water and steam coexist together and have exactly the same temperature but different internal energies. When heat is added to water at 100°C, the increase in internal energy results in water being transformed into steam at 100°C. No increase in temperature accompanies this change in internal energy. Figure 4-16 illustrates water and steam molecules at the same temperature of 100°C.

Once all the water has been transformed into steam, any further increase in internal energy raises the temperature of the steam. Steam used to generate electricity often has a temperature of thousands of degrees. If the high-temperature steam is released to a cooler environment, internal energy is removed from the steam and enters the surroundings by means of heat transfer. Once the steam is cooled to 100°C, any further loss of internal energy causes the steam to condense into hot water at 100°C. The collected water could be used over and over. Thus, boiling water to form steam and condensing steam to form water are examples of a *reversible* change of state.

THERMAL PATTERNS

People are great pattern recognizers. We instantly recognize the path left by a person walking through snow or loose sand. And we never think of such a trail as individual marks. The collection of all the footprints together stands out in our mind as one "thing." Similarly, we recognize dozens of common patterns around us.

Any dark object which is the size and shape of a spider will cause most of us to jump. We recognize the pattern or shape first. Only later, after a moment of thought, do we see that the "spider" is made of plastic or rubber.

One interesting thing to do with patterns is to train yourself to recognize a few new ones. Perhaps some that you may have overlooked in the past. You might start with thermal patterns in the snow. These are patterns that are caused by heat transfer. Some thermal patterns result from radiation from the sun, which causes snow to melt. Other thermal patterns may be caused by conduction through the ground. In each case you can see that the snow has been unequally melted.

Sometimes the patterns of melted snow reveal intriguing details. A street or sidewalk where some snow has melted and other snow remains may indicate something underground (see photograph below). A long strip of melted snow reveals the location of an underground pipe. Perhaps the pipe carries warm air or warm water.

(Photograph by Albert Allen Bartlett. From *The Physics Teacher,* **14,** 88, 1976. Copyright 1976 by The American Association of Physics Teachers.)

Another thermal pattern is seen on roofs of houses. If you compare several houses on the same block, you will notice that snow melts quicker on some houses than others. This is the result of differences in insulation. When a house is well insulated above the living space, snow stays on the roof. Poor insulation allows heat loss through your roof. You should check your own house this way. Does it need more insulation?

Other differences can be seen on a single house. Overhanging eaves which have cold air on both sides will hold snow for a long time (see photographs below). A bare spot on a house with "blown-in" insulation indicates a part of the house that did not get any insulation. You may also be able to find a house with one of the upstairs rooms heated and another room unheated. The result will be obvious in the snow on the roof.

(Photographs by Albert Allen Bartlett. From *The Physics Teacher,* **14,** 16–17, 1976. Copyright 1976 by The American Association of Physics Teachers.)

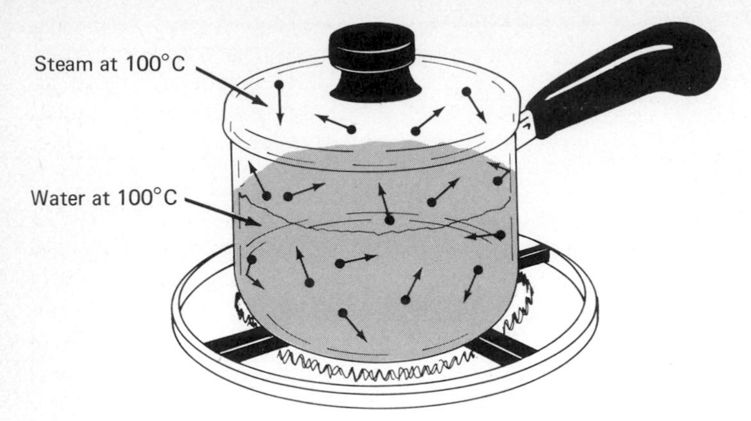

Fig. 4-16
The boiling water and the steam have exactly the same temperature of 100°C.

Another example of a change in state with no accompanying change of temperature occurs when ice melts to form water. The molecules of ice are not only close together, but locked in place as a solid. In contrast to ice molecules, water molecules have more internal energy and so are free to exchange positions and take the shape of a container. Even in their different states, ice molecules and water molecules can have *exactly the same temperature.* At 0°C, ice and water coexist and have the same temperature but different internal energies. When heat is added to ice at 0°C, the increase in internal energy results in ice being transformed into water at 0°C. No

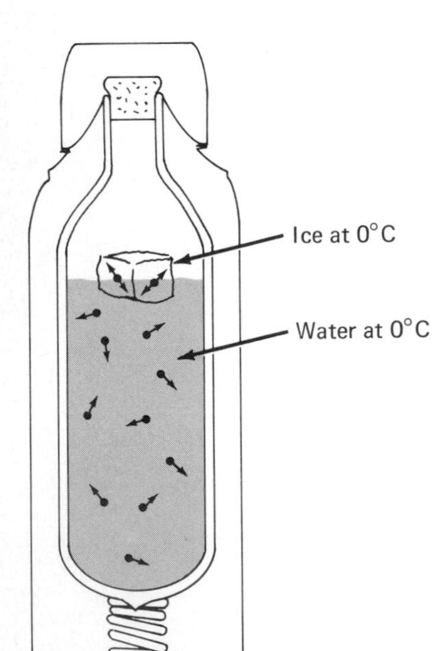

Fig. 4-17
The water and the ice have exactly the same temperature of 0°C. What is the temperature of the air in the thermos bottle?

increase in temperature accompanies this change in internal energy. Figure 4-17 illustrates water and ice molecules at a temperature of 0°C in a thermos bottle. The temperature of the air will also be 0°C if the stopper is in place.

Melting ice to produce water is also a reversible process. Water can be frozen by removing internal energy. The ice can be heated again to produce water. This process of freezing and melting occurs year after year as lakes freeze in the winter and melt in the spring.

The reversible changes of state we have just discussed involve changes from solid to liquid or liquid to gas. However, some solids change *directly* into gases without passing through the liquid state. For example, dry ice, which is solid carbon dioxide, can change directly into gaseous carbon dioxide (see Fig. 4-18). Passing directly from a solid into a gaseous state is called *sublimation.* (Carbon dioxide in the liquid state can be produced only in high-pressure vessels.) In the refrigeration industry, gaseous carbon dioxide is converted directly into dry ice, and so sublimation is reversible.

Another common example of sublimation occurs with mothballs. Several different chemicals are made into small balls or flakes, which are spread in closets to keep moths away (see Fig. 4-19). Solid mothballs do not melt but change directly from a solid to a gas by sublimation.

In some materials an increase in internal energy produces *irreversible* changes. When food is cooked, the larger molecules are usually broken down into smaller molecules that can be more easily digested. Once food is cooked, it cannot be "uncooked." The molecules will not recombine into their previous forms. Similarly, once frozen, some foods do not, upon being thawed, return to their original condition. A fresh tomato or peach can be frozen and then thawed, but there is a noticeable change in appearance. The delicate fibers and structure are ruptured, so that thawing a frozen tomato or peach produces a mushy pulp. These molecular rearrangements are irreversible. The pulp cannot be restored to the fresh, firm tomato or peach.

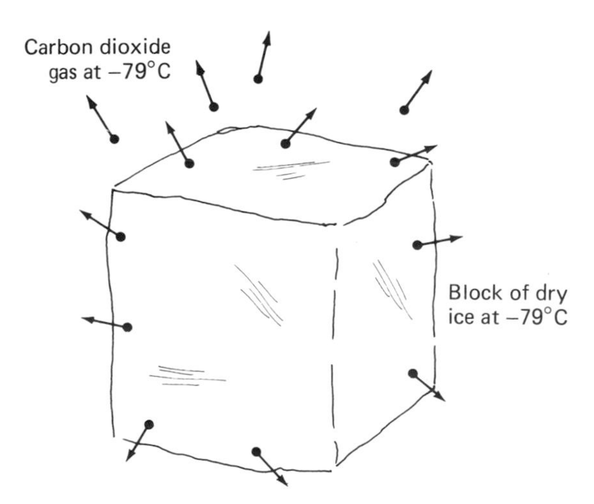

Fig. 4-18
Dry ice exposed to the atmosphere changes directly from a solid to a gas without producing a liquid.

Fig. 4-19
Mothballs do not melt. (*Photograph by Korrine Nusbaum.*)

4.7 HUMIDITY

In the act of boiling water the internal energy is increased. Molecules of water are driven from the liquid by the transfer of heat to the water at its boiling temperature of 100°C. At temperatures less than 100°C, however, molecules of water can and do spontaneously leave the liquid and enter the atmosphere as a gas by means of *evaporation.*

Evaporation can be explained by considering the kinetic energies of

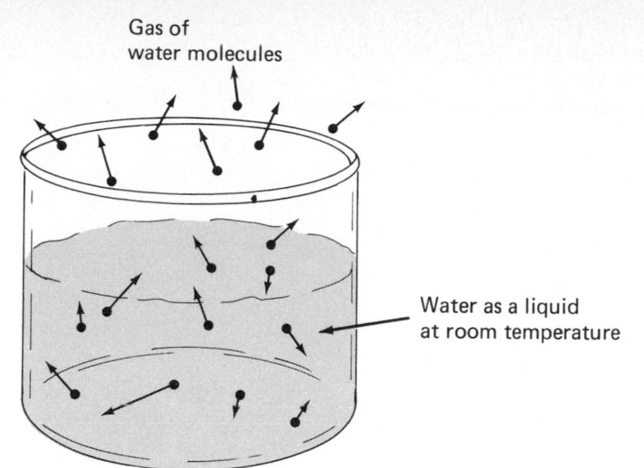

Fig. 4-20
Water at room temperature evaporates to form a gas of water molecules.

individual water molecules. The molecules do not all have the same kinetic energy. Some of the molecules on or near the surface of the liquid may have more kinetic energy than other molecules. These more energetic molecules will break away from the surface of the liquid and escape into the atmosphere by evaporation (see Fig. 4-20).

Evaporation *Evaporation is defined as the action of molecules spontaneously escaping from a liquid environment to a gaseous environment at temperatures below which the liquid boils.*

The water that escapes from the liquid into the atmosphere is a gas consisting of individual molecules. Another name for this gas of water molecules is *humidity.*

Humidity *Humidity is defined as the gaseous water density of the atmosphere.*

Humidity, like the various gases mentioned in Chap. 3, has a density. Unlike the densities of these gases, the density of humidity varies widely, depending upon local weather conditions and geography.

The greatest density that the humidity can have at a given temperature is known as the *saturation density,* as shown in Table 4-5. If the temperature of saturated humidity is lowered, the excess water molecules combine into liquid droplets of fog, clouds, or dew (see Fig. 4-21). But if the temperature of a fog should be increased by warm sunshine, the allowed saturation density may be greater than the actual humidity present. Then the fog evaporates.

Table 4-5
Saturation water-vapor densities at various temperatures

Temperature, °C	Water-vapor density, kg/m^3
−10	0.0022
0	0.0049
5	0.0068
10	0.0094
15	0.0128
20	0.0173
25	0.0228
30	0.0304
35	0.0396
40	0.0511

In everyday life one does not speak of humidity as a *density* of water in the atmosphere in order to describe the weather or the air. Instead, the concept of *relative humidity* is introduced.

Relative humidity

Relative humidity is defined as the percentage of the saturation density present at a given temperature.

That is, the relative humidity is given by the expression,

$$\text{Relative humidity} = \left(\frac{\text{actual humidity density}}{\text{saturation density}}\right)(100\%)$$

As an example, if the actual humidity density were 0.01 kg/m^3 and the saturation density were 0.02 kg/m^3, the relative humidity would be 50 percent.

Changing the temperature of the humidity will change the relative humidity. Cold humidity brought into a warm house may have a large relative humidity outdoors but actually a low density. When the humidity is warmed indoors, the possible saturation density increases, but the actual humidity density is unchanged. So the *relative* humidity decreases in warm buildings during the winter. Humidifiers are put in rooms or added to the heating systems in order to moisten dry winter air. During the summer there is just the opposite problem. Outside air with a high relative humidity is almost saturated with moisture. If this air is cooled, the humidity becomes saturated and moisture condenses on cool objects, such as ice-filled water glasses (see Fig. 4-22) or cold pipes in a basement. Dehumidifiers are put in basements or attached to air-conditioning systems in order to dry out moist summer air.

Fig. 4-21
Fog appears if the humidity in the atmosphere condenses to form small droplets of water. (*Photograph by Fred Lyon. From Rapho Photo Researchers.*)

Fig. 4-22
The humidity in the air condenses on the cold side of a glass. (*Photograph courtesy of The Seven-Up Co.*)

One of the effects of evaporation of water is to transport individual molecules of water into the atmosphere, thereby increasing the relative humidity. A second effect of evaporation of water is to remove those molecules on the surface of the water with the greatest kinetic energy. The molecules remaining in the water have less kinetic energy on the average, or in other words, a lower temperature. Thus, evaporation cools a liquid (usually water) or wet objects. Our bodies are cooled by the evaporation of perspiration, which is mostly water. The cooling effect can be increased by fanning ourselves. The moving air sweeps away energetic molecules just above our skin. Although these energetic molecules have escaped from the surface of the liquid, collisions might cause them to be directed back to the liquid. Fanning prevents the return of the molecules to the liquid by sweeping the molecules away.

REVIEW You should be able to define the following terms: temperature, internal energy, heat, heat conduction, heat convection, heat radiation, evaporation, humidity, and relative humidity.

You should be able to state the first law of thermodynamics and give examples of its application.

PROBLEMS

1. A 100-meter-long bridge is anchored in place on a hot summer day when the temperature is 35°C. During the winter the temperature drops to -15°C. Assuming the bridge has a coefficient of thermal expansion of 11×10^{-6} per degree, calculate the change in length.

2. The tungsten filament in a light bulb is about 1 centimeter long. The temperature of the hot filament increases about 1900°C when the electric light is lit. Assuming that the filament has a coefficient of thermal expansion of 4.5×10^{-6} per degree, calculate the change in length when the filament expands.

3. A gas is compressed by doing 1350 joules of work on the gas. Simultaneously 650 joules of heat are added to the gas. How much does the internal energy change as a result of these two activities?

4. A gas is compressed by doing 1350 joules of work on the gas. Simultaneously 650 joules of heat *leave* the gas. Calculate the change of internal energy.

5. On a windless day a motorcyclist travels at a speed of 50 kilometers per hour. If the temperature of the air is 15°C, what is the wind-chill factor?

6. A skier stands outside a warming hut. The wind is blowing at 30 kilometers per hour and the thermometer reads -15°C. What is the wind-chill factor?

7. At a temperature of -10°C the actual humidity density is measured to be 0.0020 kilogram per cubic meter. What is the relative humidity? If this humidity is brought into a living room at 20°C, what will be the relative humidity?

8. At a temperature of 5°C the actual humidity density is measured to be 0.0064 kilogram per cubic meter. What is the relative humidity? If this humidity is brought into a house at 25°C, what will be the relative humidity?

DISCUSSION QUESTIONS

1. Glass thermometers are filled with either mercury or alcohol. Which thermometer could measure lower temperatures?

2. What special property of Pyrex makes it well-suited for use in ovenware?

3. Does a giraffe's neck contract in winter because of the lower temperature?

4. A piece of quartz can be raised to a white-hot temperature and then plunged into cold water without breaking. What thermal property of quartz permits this harmless change of temperature?

5. Why does smoke rise from a cigarette? What would happen if the cigarette smoke were cool?

6. Some self-winding table clocks use temperature changes of the air to wind the clocks. Is this an example of a perpetual motion machine?

7. Why are the walls of test tubes used in a chemistry laboratory made thin?

8. Suppose a jeweler were to make a solid gold statue with a loosely fitting solid gold bracelet. Would the bracelet fit snugly if the statue were placed in a freezer overnight?

9. How would the temperature readings of a glass thermometer change if a different liquid were put in the thermometer?

10. If you open a refrigerator door wearing only your bathing suit, why do you first feel the cool air on your feet?

11. When water flows over a waterfall, what temperature change would you expect to occur in the water?

12. Could a refrigerator with its door left open be used as a room air conditioner?

13. Why does defrosting a refrigerator save money for the owner?

14. How does a hot-water bottle raise the temperature of nearby cooler objects?

15. Two rods of identical size, one of glass and the other of iron, can each be red-hot on one end and not visibly hot on the other. Why can the glass rod be held with the bare hand, but the iron rod cannot?

16. Why does a rug at the same temperature as a tile floor feel warmer to your bare feet than does the floor?

17. Why are double-paned windows or storm windows important in cooling your home in the summer or heating your home in the winter?

18. How does a deep snowfall protect some plants from damage in cold weather?

19. How does a cup full of hot coffee cool? Compare a porcelain cup with a Styrofoam cup.

20. Why do the more expensive aluminum saucepans have a plastic or ceramic handle?

21. A piece of red-hot metal has a high temperature. As it cools, it appears black, but how can you determine whether or not it is quite hot without touching it?

22. Cupcakes fresh from the oven can be picked up. Would you grab the pan they were cooked in?

23. What advantage does a double boiler have over a single-walled boiler?

24. What advantage does a pressure cooker have over a single-walled boiler?

25. When ice melts, is heat transferred to the ice or away from the ice? Give an example where this heat transfer is important.

26. When water freezes, is heat transferred to the water or away from the water? Give an example where this heat transfer is important.

27. Under equal conditions, does hot water or cold water freeze faster?

28. Some people have their bodies stored in liquid nitrogen after their deaths in order to be thawed out after certain diseases are cured. Comment on this idea.

29. A friend claims to have sat naked in a sauna bath at a temperature of 160°F for 1 hour. Would she suffer burns?

30. How does a fan cool you? Would a fan be effective in a very humid room?

31. Why do you feel cold when coming out of a shower in the winter, even though the room temperature may be 75°F?

32. If the temperature in the winter is 10°F and the wind causes a wind-chill factor of −40°F, should you cover the radiator of your car with a blanket to keep it warm?

33. How can the internal energy of air be increased?

ANNOTATED BIBLIOGRAPHY

Not many popular books have been written about heat and temperature, but one very good book is M. Mott-Smith's *Heat and Its Workings,* Dover Publications, Inc., New York, 1962. This book is clearly written, although it may be difficult for the beginner.

Many practical examples of the applications of thermal physics to industry and technology are given in *The Way Things Work,* Simon & Schuster, Inc., New York, 1967. The book also contains material on many other areas of physics.

M. Zemansky has written a clear discussion of the common mistakes made in confusing the ideas of heat and internal energy. See his article "The Use and Misuse of the Word 'Heat' in Physics Teaching," which appeared in *The Physics Teacher,* vol. 8, 1970, pp. 295–300.

An interesting application of heat convection to geology is given in "Convection Currents in the Earth's Mantle," *Scientific American,* November 1976, p. 72.

Much current research in physics deals with very low temperatures near absolute zero. Two interesting review articles are "Large-Scale Applications of Superconductivity," *Physics Today,* July 1977, p. 34, and "Superfluid Helium 3," *Scientific American,* December 1976, p. 56

A philosophical book on the phenomena of heat and temperature is P. W. Bridgman's *The Nature of Thermodynamics,* Harper Torchbooks, New York, 1961. In this book Bridgman presents a detailed analysis of the concepts of thermodynamics.

5 SOUND

Many phenomena in physics involve wave motion. In this chapter we will first discuss wavelength, amplitude, frequency, and speed as they apply to visible water waves. Then we will use these ideas to understand sound, which is an invisible but audible wave phenomenon.

5.1 WAVES

Chapter 3 ended with a discussion of fluid flow and streamlines. A fluid, whether a gas or a liquid, flows from one place to another through the movement of its molecules. In such movement the molecules can transport kinetic energy, or they can transport internal energy (as in the process of heat convection, discussed in Chap. 4).

Fluids can also transport energy by means of waves. Waves are caused by vibration in a fluid. For example, drops of falling water send waves in the form of ripples across the surface of a lake (see Fig. 5-1). The falling drops collide with the surface of the lake and cause the molecules of water to vibrate. The smooth surface is disturbed by the vibration. The disturbance carries away the energy of the falling drops as ripples on the water. After the ripples have moved over the water, the surface is again smooth.

All waves, whether as ripples on water or as sound in air (or as electromagnetic waves discussed in Chap. 7), carry energy from place to place. But unlike fluid flow or convection, waves do not permanently move the molecules of a fluid. The molecules vibrate as the wave passes, but after the disturbance the fluid returns to its original condition.

Fig. 5-1
Waves carry energy from one place to another. (*Photograph by Allan D. Cruickshank. From National Audubon Society.*)

Wave *A wave is a disturbance that carries energy from place to place without permanently changing the medium.*

The definition of a wave does not restrict waves to fluids, for as we shall see, sound can also travel through a solid.

In order to discuss waves in various materials, it is necessary to define certain features found in all waves. Since water shows the features of waves most clearly, we shall examine water waves before discussing sound waves.

WAVELENGTH

Although waves on the surface of water are always in motion, imagine a "snapshot" taken at one instant. Such a picture can record the cross section of a series of waves or ripples. Figure 5-2 shows that a wave is a succession of crests and troughs. The distance or length between two adjacent crests is the *wavelength* λ (Greek letter lambda).

Wavelength *Wavelength is the distance between two successive like points of a wave.*

In Fig. 5-2 the wavelength is measured between two adjacent crests. It could have been measured between two adjacent troughs. In either case we measure the same wavelength for this particular wave.

All waves do not have the same wavelength. In fact, wavelengths vary greatly from wave to wave. The wavelengths of small water ripples are only a few centimeters, whereas that of ocean waves found near hurricanes are many meters long.

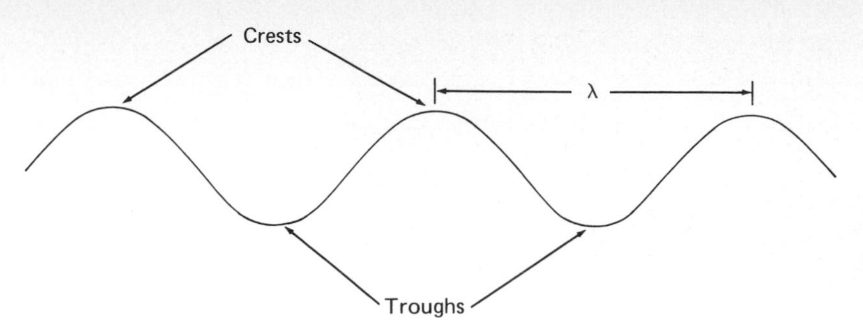

Fig. 5-2
The wavelength λ is measured between two adjacent crests.

AMPLITUDE

The distance between the top of the crest and the bottom of the trough is independent of the distance from crest to crest. Some waves are "tall" and others are "short," so to speak. The height of a wave is related to a feature of waves known as the *amplitude.*

Amplitude *Amplitude is the maximum value of the displacement of a wave.*

The height of a wave is the distance from the bottom of a trough to the top of a crest. The displacement, however, is half the height of a wave, as shown in Fig. 5-3.

The reason the displacement (not the height) defines the amplitude of a wave is based on the behavior of the medium. The disturbance that transfers energy in a wave is caused by the vibration of the medium. In a material medium, molecules move back and forth as the wave passes. The distance that the molecules move from their original position of rest to their maximum displacement is the amplitude. Figure 5-3 shows a series of ripples and the level of the still water. The height of the wave is 6 centimeters (about 2.5 inches). But the amplitude of the wave is half the height, or 3 centimeters. Wave amplitudes are not all the same. Figure 5-4 shows a series of ripples with different amplitudes made by a water-skier.

FREQUENCY

Waves involve motion. The disturbance moves over the surface of water. A motion picture, not a snapshot, is needed to describe additional features of waves. Imagine sitting on a pier and watching the waves pass in front of you. It takes several seconds for a few waves to move past the pier. The rate at which wave crests pass a point is the *frequency.* Instead of dealing with wave crests, we can call a complete wave a cycle.

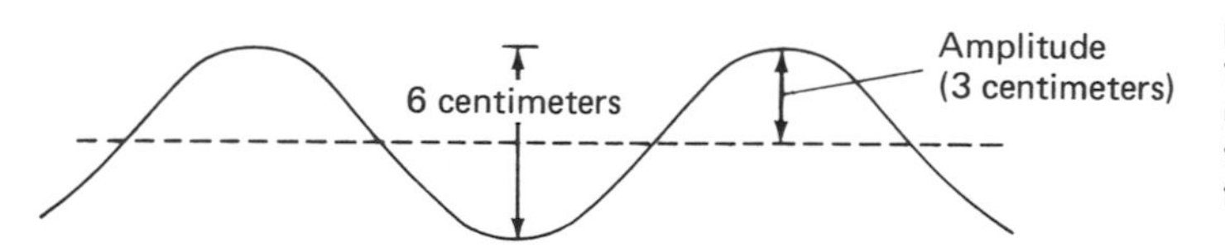

Fig. 5-3
The amplitude is always measured from the undisturbed position of the medium carrying the wave. Amplitude is not measured from trough to crest.

Fig. 5-4
A water-skier sends waves of many different amplitudes away from the skis. (*Photograph from The New York Library Collection.*)

Frequency *Frequency is the number of cycles occurring during a certain interval of time divided by that interval of time.*

Using the symbol f for frequency, we write a formula for frequency

$$\text{Frequency} = \frac{\text{number of cycles}}{\text{time interval}}$$

or

$$f = \frac{\text{number}}{t}$$

where f = frequency
t = time

Frequencies can be measured in units of cycles per second. This unit has been given the name *hertz* (abbreviated Hz) in honor of Heinrich Hertz, a pioneer in the study of radio waves. One hertz equals one cycle per second, no matter what kind of a wave is being considered. Ripples of water typically have a frequency of a few hertz. A hummingbird has a higher frequency for its beating wings (see Fig. 5-5). Higher yet is the buzzing sound of a mosquito's wings, which have frequencies from 300 to 600 Hz. The higher frequencies of electromagnetic waves exceed billions of hertz.

SPEED

The concept of frequency concerns the *number* of wave crests passing a given point each second. As the waves pass, however, the waves move along with a certain speed. In other words, the concept of speed concerns the *distance* a wave crest moves each second.

Speed of a Wave

The speed of a wave is the distance traveled by a wave during a certain interval of time divided by that interval of time.

The definition of the speed of a wave is very similar to the definition of the speed of an object as given in Chap. 1. In both definitions speed is

Fig. 5-5
The rapidly beating wings of a hummingbird have a frequency of about 50 hertz. (*Photograph by Alexander Lowry. From Photo Researchers, Inc.*)

measured in units of distance per time. But in the case of wave speeds, we must be careful to use only the *forward* distance moved by the crest of a wave. We ignore any oscillatory motion, such as the vibration of a water wave up and down. Only the forward motion of any wave is used to determine the speed of that wave. A surfer riding a breaker can be easily carried forward with the wave at a speed of 13 meters per second (about 30 miles per hour).

5.2 THE WAVE FORMULA

The wave formula relates the speed of a wave to the frequency and wavelength of that same wave. Consider what happens to a breaker that carries a surfboarder over the surface of the ocean at a speed of 13 meters per second. A reasonable value for the frequency of such a wave might be one crest passing a stationary observer every 10 seconds. Since the frequency is usually expressed as the number of waves passing *each* second, the frequency is 0.1 cycle per second, or equivalently 0.1 Hz.

Now look at the role of wavelength in a moving wave. The distance between crests in large breakers might be 130 meters. In other words, the wavelength is 130 meters for the wave being considered.

If we combine the concepts of frequency and wavelength, we see that every 10 seconds one complete wave passes in front of us and moves 130 meters. The wave has a speed of 130 meters per 10 seconds, which is the same as 13 meters per (each) second. The same number for the speed of the wave could be obtained by multiplying together the frequency and the wavelength. In this way we obtain the *wave formula*,

Wave formula

$$\text{Speed} = (\text{frequency})(\text{wavelength})$$

or

$$v = f\lambda$$

where v = speed
f = frequency
λ = wavelength

If the wavelength is measured in meters and the frequency in hertz, the speed calculated from the wave formula will be in meters per second. For the wave carrying the surfboarder

$$\begin{aligned} v &= f\lambda \\ &= (0.1 \text{ hertz})(130 \text{ meters}) \\ &= 13 \text{ meters per second} \end{aligned}$$

The wave formula is applicable to many different types of waves. The wave formula can be applied not only to the analysis of water waves, but also to the analysis of sound waves. We will also use the wave formula in Chap. 7 for the discussion of electromagnetic waves.

5.3 TRANSVERSE AND COMPRESSIONAL WAVES

All waves possess the features of wavelength, frequency, speed, and amplitude. However, amplitudes are measured differently for different types of waves. There are two main wave types, *transverse* waves and *compressional* waves.

A transverse wave has its amplitude measured transverse, or crosswise, to the direction of motion of the wave. Water waves, whether small ripples or large breakers, are examples of transverse waves. The transverse water wave is a disturbance that moves across the surface. The water molecules vibrate mainly up and down, which is transverse to the direction of motion of the wave. Figure 5-3 shows the amplitude of a ripple of water being measured from the plane of the level surface of water.

There are other examples of transverse waves. All strings, ropes, and cords carry transverse waves. A guitar string, for example, moves back and forth crosswise to the length of the vibrating wire, while the wave itself moves along the length of the wire. Every stringed musical instrument relies on the transverse waves of its strings to produce sound. Sound, however, is not an example of transverse waves, as we shall see later in this chapter.

A transverse wave can be made with a Slinky—the toy spring that somersaults down stairs. A Slinky can be stretched out on a smooth floor and one end shaken back and forth perpendicular to the length. A series of transverse waves will move away from the hand. But there is another way of shaking the stretched Slinky. Instead of moving the hand back and forth, the hand is moved to and fro along the length of the spring. As the comparison shows in Fig. 5-6, the second way of vibrating the spring produces a fundamentally different type of wave, namely, a compressional wave.

A compressional wave is a series of compressed regions (called *compressions*) and expanded regions (called *expansions* or sometimes *rarefactions*) that move through a medium. The compressions and expansions are the disturbances that carry the energy of a compressional wave. The Slinky shows that in a compressional wave vibrating particles move along the direction of the wave itself. Another example of a compressional wave is found in an earthquake. The first vibrations which arrive from a distant earthquake alternately compress and expand the ground beneath our feet.

Compressional waves have the features of wavelength, amplitude, frequency, and speed like transverse waves. The wavelength of a compressional wave is still the distance between two successive like points, such as two compressions shown in Fig. 5-7. A compressional wave in a Slinky typically has a wavelength of about 1 meter.

The compressional waves of an earthquake on the surface of the earth have much longer wavelengths. Within a few kilometers of the fault that generates the earthquake the wavelength ranges from half a kilometer to a

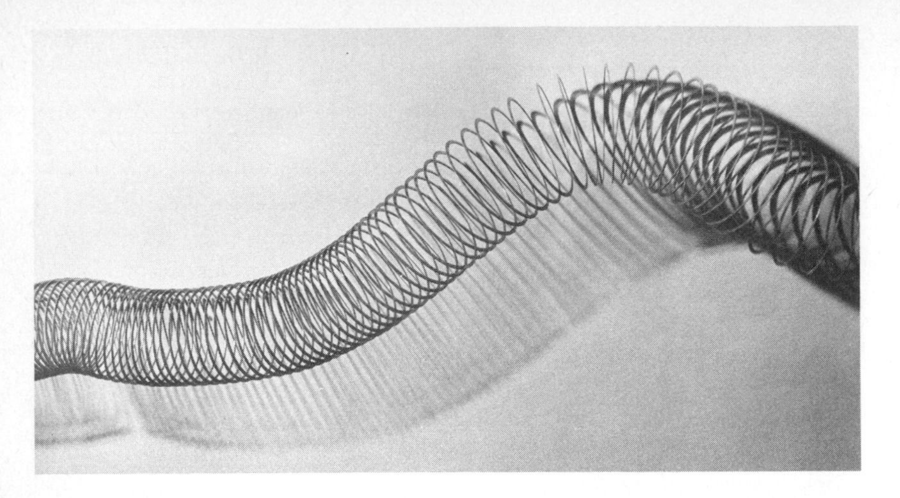

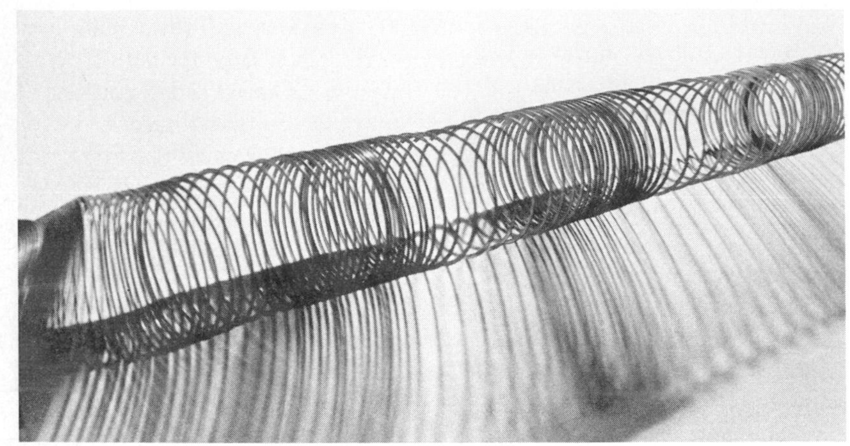

Fig. 5-6
A Slinky swung back and forth generates a transverse wave, but a Slinky moved to and fro generates a compressional wave. (*Photographs by Ted Rozumalski. From Black Star.*)

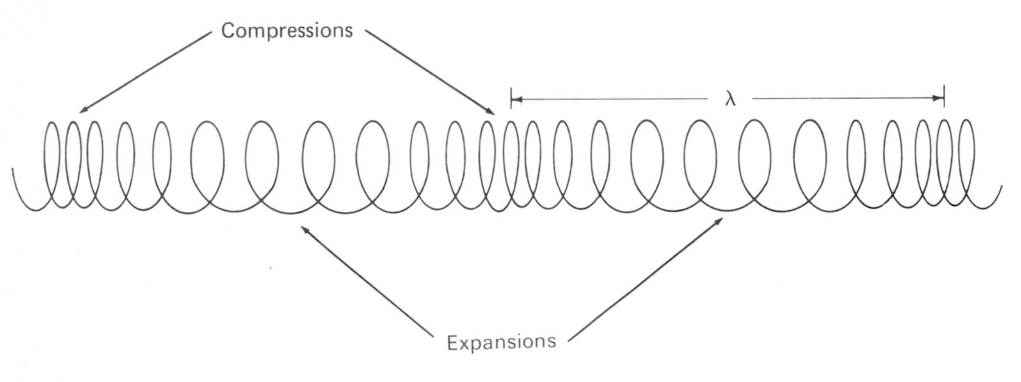

Fig. 5-7
The wavelength λ is measured between two adjacent compressions.

few kilometers. At a distance of several hundred kilometers from the epicenter of an earthquake the wavelength is about 10 kilometers.

The amplitude of a compressional wave is more difficult to identify than the amplitude of a transverse wave. Both types of waves define amplitude in the same way, namely, the maximum value of the displacement. In the compressional wave, however, the displacement is measured in the same direction as the wavelength. (Recall that in the transverse wave the amplitude and wavelength are crosswise to each other.) The compressional wave amplitude is the maximum distance that a portion of the material moves from the equilibrium position, while the wavelength is the distance from one compression to the next. Figure 5-8 compares a stretched spring with and without compressional waves. The amplitude of the compresssional wave is $\frac{3}{4}$ centimeter, the maximum distance the material has been displaced from its equilibrium position.

The amplitude of a compressional wave is quite different from that of a transverse wave. Otherwise, the features of the two types of waves are very similar. Not only is the definition of wavelength the same for both types of waves, but so too is the definition of frequency and speed. The frequency of a compressional wave is the *number* of cycles occurring in an interval of time divided by that interval. Similarly, the speed of a compressional wave is the *distance* traveled during an interval of time divided by that interval. Lastly,

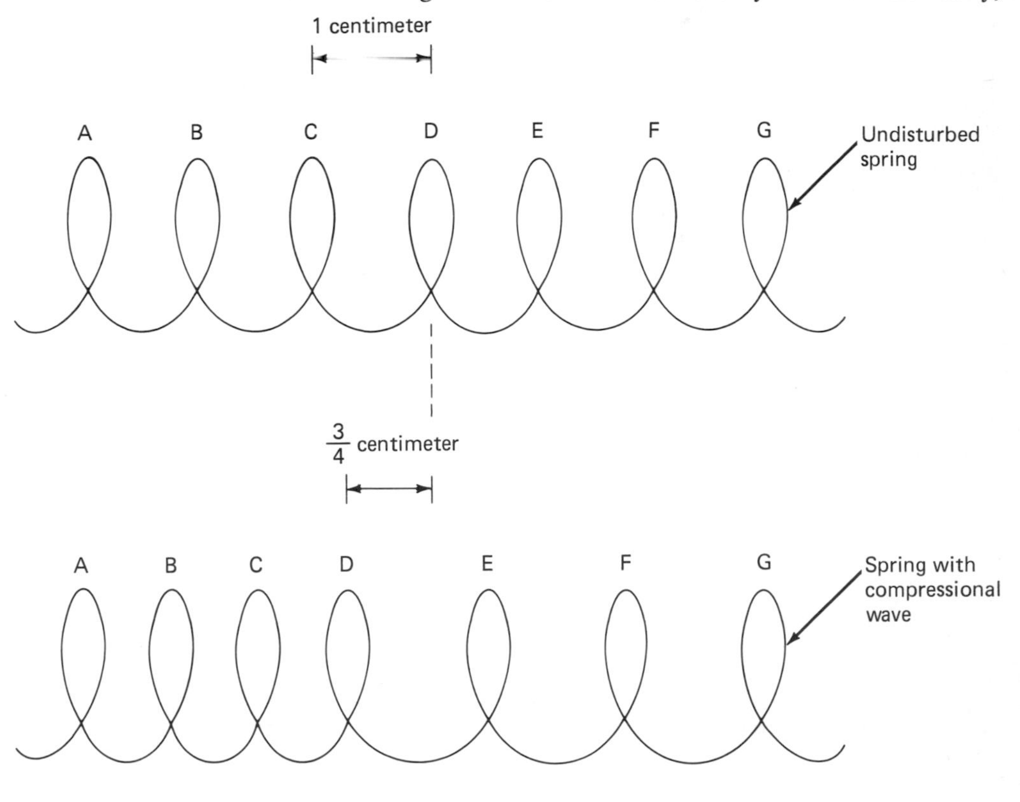

Fig. 5-8
The maximum displacement of coil *D* is $\frac{3}{4}$ centimeter. This is the amplitude of the wave.

compressional waves are governed by the same wave formula $v = f\lambda$ as are transverse waves.

5.4 SOUND WAVES

All sounds around us, whether pleasant music or irritating noise, are generated by something vibrating in the air. The vibrating speaker cone in a radio and the garbage can lid striking the pavement both send sound waves through the air. These sounds, although common, are quite complicated. So in order to understand how sound is produced and what it is, let us first consider a simple sound generator.

A tuning fork is a metal device consisting of a handle and two prongs free to vibrate. If the prongs are struck with a pencil, they generate a sound heard as a pure tone. As one vibrating prong moves forward, it compresses the air to form a compression that moves away. As the prong moves backward, the air is left less compressed to form an expansion that follows after the compression. In other words the vibrating tuning fork sends out compressional waves.

Sound waves *Sound waves are compressional waves usually transmitted through air.*

The series of alternate compressions and expansions cannot really be seen. Nevertheless Fig. 5-9 indicates with dots where the air is being compressed and expanded at one instant of time. Corresponding to each of these compressions and expansions are pressure changes above and below normal atmospheric pressure. These pressure variations are also shown in Fig. 5-9.

Besides the simple tuning fork you can consider simple musical instruments. A drum consists of a sheet of rubber or animal skin which is stretched over a housing. The vibrating drumhead alternately compresses and expands the air to transmit a compressional wave. Of course, as Fig. 5-10 shows, the drum is struck repeatedly to keep the drumhead vibrating and set the musical beat.

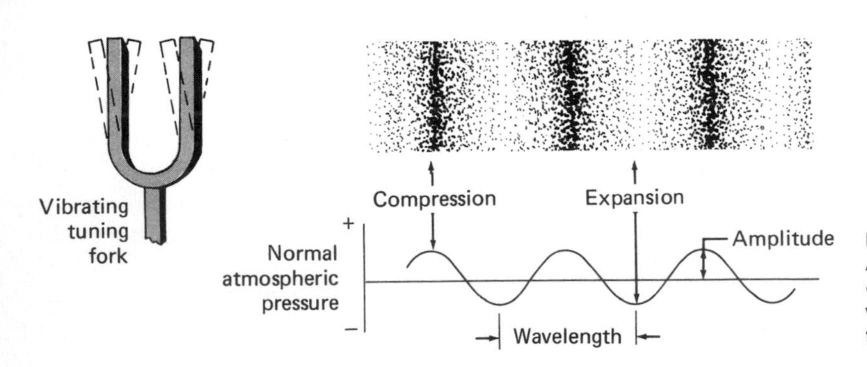

Fig. 5-9
A tuning fork sends out a wave of compressions and expansions, which are very small variations in the air pressure.

Fig. 5-10
The vibrating drumhead emits a sound wave, which is shown as a series of visible compressions and expansions of the air.

A more complicated musical instrument is a string instrument, such as a guitar, violin, or piano. The strings are set into vibration by plucking, rubbing, or striking the individual strings. The vibrating string then sends transverse waves back and forth along the string. Very little sound is produced by the striking itself. Rather, the string makes the wooden housing (so-called soundboard) of the instrument vibrate. The soundboard is much larger than a string and is more effective in producing compressional waves in air. The musician controls the strings in order to produce the sound of a particular note. The string, in turn, forces the soundboard to vibrate so that the note can be heard.

Probably the most complicated "string instrument" is the human voice box (larynx). Your voice box is located inside the windpipe of your throat. Two flexible vocal cords are stretched across the voice box. Air blowing through your throat causes the vocal cords to vibrate. High or low notes are produced by muscles stretching or loosening the vocal cords. The sound is changed into more complicated speech patterns by varying the shape of the lips, closing the teeth, and moving the tongue. The complexity of human speech is possible in part because of the complexity of sounds produced by the vocal cords.

5.5 SOUNDS IN AIR

Like every other wave, sound travels at a certain speed from one place to another. In air the speed of sound is about 340 meters per second (760 miles per hour). At this speed sound travels about $\frac{1}{5}$ of a mile each second. Although the speed of sound is fast, the speed of light is even faster. The speed of light is about 300,000 kilometers per second (186,000 miles per second), which is about a million times faster than the speed of sound.

As a result of the tremendous speed of light, you see a lightning flash almost instantly. However, the sound of the thunder is heard a split second or longer after the flash is seen. Knowing that sound waves take about 5 seconds to go a mile, you can calculate the distance from you to the lightning bolt. Count the number of seconds between a flash of lightning and the first sound of thunder. Divide the number of seconds by 5 to find how many miles away the lightning struck. If the number of seconds is less than 5, the lightning struck within 1 mile of you. Less than 1 second means the lightning is too close for comfort.

Apart from the question of loudness, there is a difference in the sound of distant thunder and nearby thunder. Distant thunder rumbles for a few seconds, but nearby thunder quickly dies away. The more persistent sound of distant thunder is due to echoes.

Sound waves in air do not easily penetrate a solid. In fact, most of the

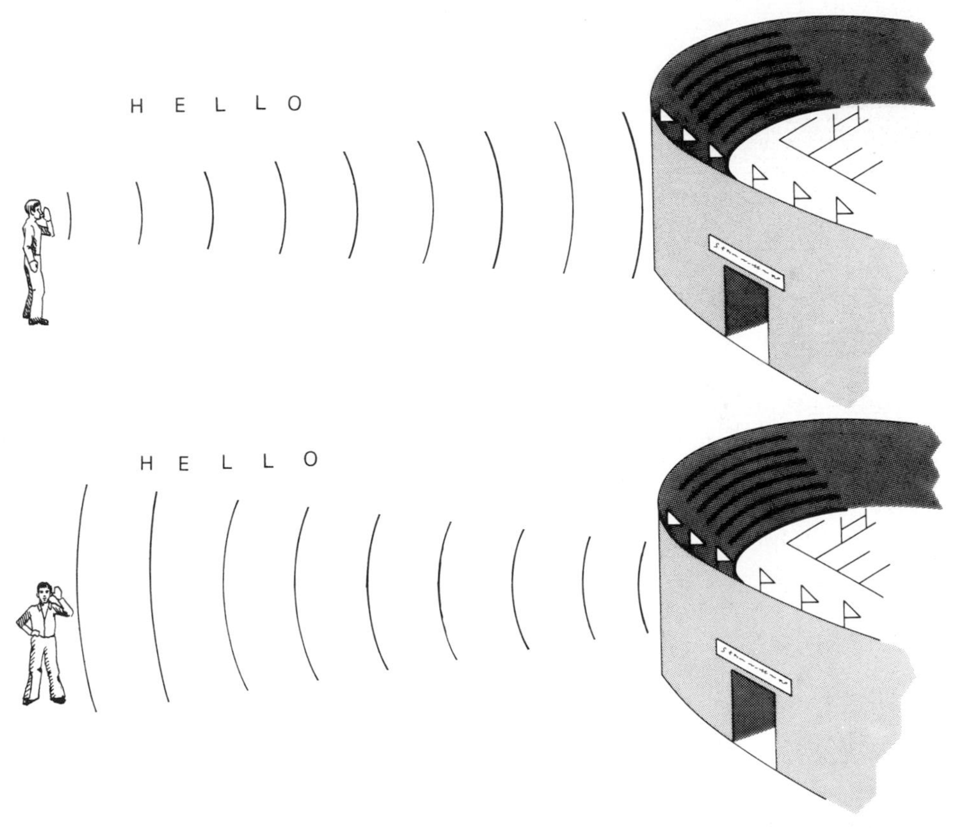

Fig. 5-11
If you shout from the parking lot toward a stadium wall about 170 meters (560 feet) away, you will hear an echo 1 second later.

sound waves that strike a smooth solid surface bounce off the surface. An echo is a sound wave that has been *reflected.*

Reflection *Reflection is defined as the return of a wave from the smooth surface of an obstacle.*

The reflecting surface must be smooth if a clear echo is to be heard. Surfaces with many irregularities, such as an acoustical tile on the ceiling or a curtain across a stage, scatter the sound in all directions. In addition, some of the energy of the sound wave is absorbed. The net result is to prevent echoes from being formed. But the smooth surface of the wall of a building or a tunnel easily reflects the sound. The rumble of distant thunder is partially the result of hearing the thunder coming directly from the lightning bolt as well as echoing off buildings, walls, and hills. Nearby lightning usually produces thunder that comes directly to your ears without any echoing.

If you shout toward the smooth wall of an isolated building, you can hear an echo. Suppose the building is 170 meters away, which is the length of nearly two football fields. At a speed of about 340 meters per second it takes the sound 1 second to make the round trip from you to the wall and back, as shown in Fig. 5-11. As a rule of thumb, every second it takes an echo to be heard means the reflecting surface is another 170 meters away.

A bat flying in the darkest cave navigates using echoes. The bat cannot see the obstacles or insects in front of it. So the bat sends out very high frequency squeaks that humans cannot hear. The echoes of these squeaks inform the bat of the distance to whatever reflects the sound. As the time of arrival for the echo becomes shorter, the bat knows that it is nearing an obstruction or its prey.

Fig. 5-12
Sound cannot be heard outside the jar if the air is removed from the jar. (*Photograph by Ted Rozumalski. From Black Star.*)

The sounds emitted by a bat, lightning bolt, or human voice box all depend upon air to transmit the compressional wave. The compressions and expansions of a sound wave need a medium in which to travel. The importance of air for sound transmission can be demonstrated by placing a noisemaker in a glass container and pumping out the air. Figure 5-12 shows a metronome placed inside a glass bell jar. As air is pumped from the glass container, the clicking sound of the metronome becomes softer until the sound disappears. When most of the air has been removed, there is no material medium to transmit the sound waves. As air is admitted back into the bell jar, the sound of the metronome can again be heard.

5.6 SOUND IN VARIOUS MATERIALS

Sound waves expand in all directions in the air. Consequently the energy supplied to the wave is distributed over a surface increasing in size. As a result, the amplitude of the sound wave decreases. The decrease in amplitude of an expanding sound wave can be visualized as being similar to the decreasing amplitude of an expanding ripple on the surface of water.

SHOCK WAVES

Think of a loud noise. One that starts suddenly without warning. One that might startle you. What comes to mind? Perhaps the loud crack of thunder as you have heard it on a warm summer night. Maybe the noise caused by a friend unexpectedly popping a balloon. Although these two loud noises have very different sources, they have something in common. They are both examples of shock waves.

A shock wave is a very narrow disturbance in the air. On each side of the shock wave there are different values for the pressure and density. The pressure differences and density differences across the shock wave are shown in the diagram below. In air, the thickness of a shock wave is about a millionth of an inch.

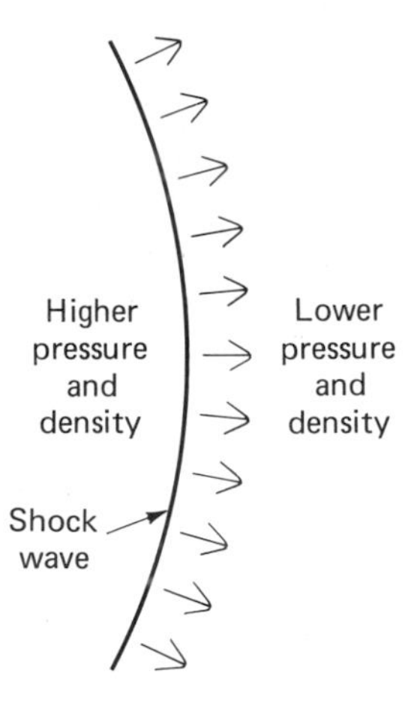

An abrupt release of energy in a small space is the cause of every shock wave. The crash of a meteorite causes a shock wave. This wave travels through the ground. The explosion of a firecracker causes a shock wave. In this case the chemical energy of gunpowder is converted to the energy in the shock wave. The shock waves from firecrackers and meteorites are both short-lived. A single shock wave is sent out and then nothing more. There is no additional energy release.

Other circumstances produce shock waves continuously. Speedboats do this. The evidence is left behind in V-shaped trails on the water. And airplanes traveling faster than the speed of sound also form continuous shock waves. The evidence is a sonic boom. A sonic boom is the noise heard when a shock wave passes by. The shock wave behind an airplane is shown in the diagram. These waves are only generated by supersonic aircraft.

Several commercial aircraft fly at supersonic speeds. The Russian TU-144 and the British-French Concorde both do. The Concorde is designed to cruise at 64,000 feet at a speed just over twice the speed of sound. It produces loud sonic booms heard on the ground underneath the flight path. Because of this, the Concorde has been involved in several legal disputes. Many people have strong objections to the noise.

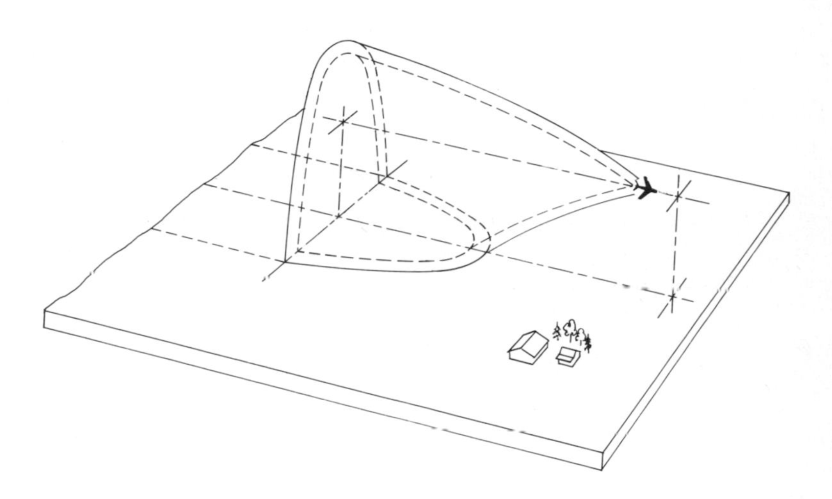

The diagram above shows that a supersonic aircraft produces a cone-shaped shock wave. This cone intersects the ground underneath in the shape of a broad letter U. As the airplane moves ahead, the sonic boom sweeps over the ground. If you are standing on the ground, you hear the sonic boom as the shock wave sweeps over you. Both ahead and behind the sonic boom are regions of undisturbed air. If you were at one of these spots, you would not hear any sound. The sonic boom is only heard briefly at each spot. A military fighter plane produces a sonic boom lasting about $\frac{1}{10}$ of a second. The sonic boom of a supersonic airliner lasts about $\frac{3}{10}$ of a second. These are the lengths of time that the noise is heard by any one person. But remember that the shock wave is spread out. It will be heard by other people who are in other positions. Each individual hears the sonic boom as the shock wave sweeps over <u>him</u> or <u>her</u>.

At present most supersonic aircraft generate sonic booms over the oceans. Considerable controversy exists over whether or not these planes should be allowed to generate sonic booms over land. There are many open questions. For example, the effects of sonic boom on wildlife behavior are unknown. Much research remains to be done.

It has been quipped "Why must every advance in science be accompanied by an increase in noise?"

When sound waves are prevented from expanding in all directions, the energy is not dissipated as much. The amplitude of a sound wave restricted to the inside of a pipe or a speaking tube does not decrease very much. Before there were telephones, voice communication in department stores and ships was possible using speaking tubes between floors or decks. The sound waves spoken into one end of the tube are confined to moving inside the tube. Some of the energy in the waves is absorbed by the walls of the tube, but most of the sound reaches the other end of the speaking tube.

If the length of the speaking tube is shrunk and the tube is capped on the lower end, it looks like a tin can. The energy in the sound waves striking the bottom of the can sets it into vibration. If a string is stretched tightly from the bottom of one tin can to the bottom of another tin can, then a child's tin-can-and-string telephone is made (see Fig. 5-13). The children talk into and listen to the tin cans, which act as senders and receivers of the sound. The sound travels back and forth mainly through the string. In other words, sound is able to travel through material media other than air.

When you think of all the places you hear sounds, you may soon realize that sound waves travel through many different materials. Sound waves, for example, move through the walls of houses and apartments and through the glass windows of buildings and cars. Although the sounds are clear when they come to you directly through the air, the sounds are different after passing through other materials. For example, most of the sounds coming through walls are muffled and distorted.

Fig. 5-13
Sound is transmitted through a tight string connected to tin cans. *(Photograph by Ted Rozumalski. From Black Star.)*

Some distortion of sound occurs when we hear the same sound under different conditions. As an example, listen to someone eating celery. The sound you hear comes to your ears through the air. Listen to yourself eating celery and notice the difference. The sound of your own chewing reaches your ears through the air *and* through the bones in your jaw and skull. The transmission of sound within your head makes a difference in how the sounds are heard by you. The same effect occurs with a tape recording of your voice. Your tape-recorded voice is how everyone else hears you, but is not the voice you are accustomed to hearing.

You can hear sounds underwater where there is no air to transmit the compressional wave. The water alone is responsible for carrying the sound energy from one place to another. Try clapping rocks together underwater the next time you go swimming. You can easily hear distinct clicks. If you try talking underwater, the sound is very muffled and distorted. It is difficult to speak clearly underwater, because your lips and vocal cords are not strong enough to vibrate the water with articulate sounds of speech. (Also, one usually chokes on the water filling the mouth and throat.) Some marine life, however, communicate by sound waves underwater.

The speeds of compressional sound waves in some common materials are shown in Table 5-1. Notice that sound travels faster in solids and liquids than it does in gases. The speed of sound is greatest in solids because the forces that one particle exerts on its neighbor are larger in a solid than in a liquid or gas. Simply stated, solids are the most rigid state of matter and so they transmit sound faster than the less rigid liquids or gases.

The speed of sound in various materials depends not only on the rigidity of the material, but also on the density. An increase in density causes a decrease in the speed of sound. Notice in Table 5-1 that air, which is denser

Table 5-1
Approximate speed of sound in different materials at room temperature

Material	Speed, m/sec
Gases	
Air	340
Helium	970
Liquids	
Alcohol	1,230
Water	1,440
Solids	
Gold	3,240
Aluminum	5,100
Glass	5,640
Iron	5,950

than helium (see Table 3-1), has the slower speed of sound. Similarly gold, which has a much higher density than aluminum, has the slower speed of sound.

5.7 LISTENING TO SOUND

Although the speed of sound depends on several variables, the speed of sound does not depend on the frequency. High-frequency sounds travel at the same speed as low-frequency sounds in a given material. If sound traveled at different speeds for different frequencies, there could be no speech communication, let alone music. If, for example, the high-frequency sounds from a soprano were to move faster in air than the low-frequency sounds from a tenor, the voice of the soprano would arrive at your ears before the voice of the tenor. Instead of music, you would hear a jumbled, unintelligible mixture of confusing sounds if the speed of sound were not the same at all frequencies.

So far we have described sound in terms of physical properties, such as frequency and amplitude. These characteristics can be carefully measured by instruments and are thus agreed upon by everyone. As measured qualities, these properties are called *objective* properties. The word objective is related to the word *object.* An objective characteristic is one that is determined by using a measuring instrument.

Alternatively, we can describe sound waves by the way we hear them—by the way they interact with the human ear and the human brain. Then, of course, opinions vary. Not everyone has the same sense of beauty and not everyone has the same sensory response to sounds. Such descriptions are called *subjective* descriptions. Subjective means that a human response is involved. And, in general, different humans have different responses to the same experience.

FREQUENCY AND PITCH

One objective property of waves is frequency. When we listen to a sound wave, the frequency measured in hertz is the number of waves that impinge on our eardrum each second. The frequency can be measured by electronic instruments. We can produce sound waves of many different frequencies—and each in turn can be objectively measured. The measured ranges of frequencies of various voices and musical instruments are compared to the notes of a piano in Fig. 5-14.

But what happens when we listen to such sounds of different frequencies? The ear and brain sense the variations in frequency as variations along a musical scale. Some sounds are higher notes than others, as for example, the notes played by the keys on the right side of a piano keyboard compared to the notes played by the keys on the left side of a piano. This is a subjective characteristic of sound called *pitch.*

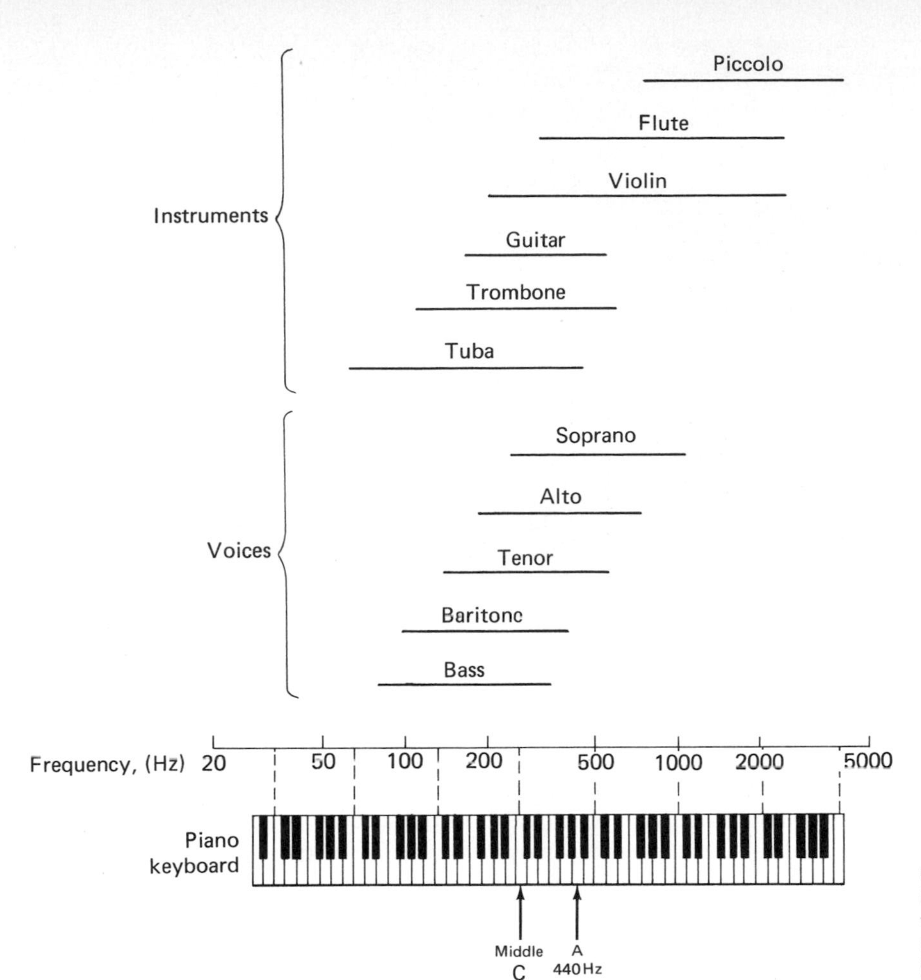

Fig. 5-14
Frequency ranges of various voices and musical instruments are compared with the notes of a piano.

Pitch *Pitch is that psychological property of sound characterized by highness or lowness, depending primarily upon frequency of the sound stimulus.*

Pitch is our subjective perception of high and low notes. Pitch is affected by several factors and is not related to frequency alone. Pitch also depends somewhat on the wave amplitude and the shape of the wave. Nevertheless the effects of these other factors are not very important. Changes in pitch depend primarily on changes in frequency.

Most of the sounds you hear have a low pitch or frequency. The rumbling sounds of cars, trucks, trains, thunder, and activities in other rooms are all low-pitched. Higher-pitched sounds exist, but do not pass through walls as easily as low-pitched sounds. Nevertheless you can easily produce high-frequency sounds in several ways. One way to produce high-pitched sound is by whispering. Even if you sing tenor or alto parts in a chorus, your

Fig. 5-15
Sizzling water generates high-pitched sound. (*Photograph by Korrine Nusbaum.*)

whisper is still high-pitched. Rubbing your fingertips together also produces sounds with high frequencies. Another source of high-pitched sound is water spattering on a griddle (see Fig. 5-15). Try flicking a few drops of water onto the hot pan the next time you fry eggs or pancakes. Listen to the pitch.

Human hearing varies considerably, but from 20 to 20,000 hertz is a reasonable range of frequencies for normal hearing. The hearing range is much larger than the range of human speech. The frequency range of human voices is from about 85 to 1,100 hertz. Yet a person who could hear only in the range from 85 to 1,100 hertz would have a severe hearing loss. The need to hear much higher frequency sounds than are found in the fundamental range of the human voice is based on an analysis of the sound of speech.

All voices have a fundamental frequency range, a range in which it is

easy to produce the sound of speech or song. But in addition to this comfortable frequency range everyone with normal speech also generates high-frequency sounds. The sounds of the letters S, Z, and sometimes C are naturally high-frequency sounds. Recite the sentence "She sells seashells by the seashore" and notice the high-pitched, hissing sounds. The human ear must be able to hear these high-pitched sounds if there is to be speech communication.

Sounds with frequencies below 20 hertz and above 20,000 hertz exist, but people cannot hear these compressional waves. Sound with frequencies below the range of human hearing is called *infrasonic* sound. Infrasonic waves are found inside moving cars and trucks. The low-frequency vibration of the windows and bodies of the vehicles causes the enclosed air to transmit infrasonic waves. Although infrasonic sound cannot be heard by human ears, it causes fatigue after exposure for several hours.

Sounds with frequencies above 20,000 hertz are called *ultrasonic* sound. Although humans cannot hear ultrasonic sound, many animals hear it. A dog whistle emits an ultrasonic wave with a frequency of typically 30,000 hertz. A dog hears the sound, but a human hears nothing. Bats navigate around obstacles and hunt insects by sending out ultrasonic sound and listening for the echo. Common bats found in the United States emit sound in a frequency range of 45,000 to 90,000 hertz.

INTENSITY AND LOUDNESS

We now turn our attention to descriptions of the strength of sound waves. Again there are two quantities—one objective and one subjective. How can the objective property related to the strength of a sound wave be described? The objective property could be the amplitude of the sound wave, that is, the maximum value of the displacement. For a very loud sound, one that is painful to hear, the amplitude of the compressional wave is about $\frac{1}{100}$ of a millimeter. For a very quiet sound, just barely audible, the amplitude is a million times smaller—about the size of a single air molecule.

Fig. 5-16
A dog can hear an ultrasonic whistle. A human cannot hear such high-pitched sounds.

A scale relating sound strengths to wave amplitudes might be useful for some purposes. However, the common scale of sound strengths is one that uses *intensities*. The intensity of a sound wave depends upon the square of the amplitude of the wave, which is related to the power carried by the sound.

Intensity *The intensity of a sound wave is defined to be the power crossing a small area divided by that area.*

Intensity is an objective property of sound. From the definition it can be seen that intensity is measured in units of watts per square meter. Table 5-2 lists the intensities of several common sounds.

For each of the sounds in Table 5-2 the intensity is also listed with another common unit of sound strength—the sound *level* measured in *decibels*. The decibel unit is based on a ratio of intensities and is abbreviated with the symbol dB. For example, a typical motorcycle produces up to 100 dB of sound.

The decibel scale is unlike most common scales in that small changes on the decibel scale represent large changes in intensity. The decibel scale is not a linear scale but a logarithmic scale. Each time the number of decibels is increased by *adding* 10, the intensity is increased by *multiplying* by 10. As shown in Table 5-2, the intensity of the sound of clashing shopping carts is 100 times larger than that of a sewing machine. But the difference on the decibel scale is only 20 dB, the difference between 80 dB and 60 dB.

Just as intensity is an objective measure of the strength of a sound wave, there is also a subjective measure of its strength. This subjective characteristic is *loudness*.

Table 5-2
Some common sound intensities

Sound source	Intensity, W/m^2	Level, dB
Rock band, near speakers	1	120
Jet plane, near wing tip	10^{-1}	110
Motorcycle, muffler removed	10^{-2}	100
Electric razor	10^{-3}	90
Clashing shopping carts	10^{-4}	80
Conversation in crowded room	10^{-5}	70
Sewing machine	10^{-6}	60
Hospital corridor	10^{-7}	50
Softly playing radio	10^{-8}	40
House in countryside	10^{-9}	30
Whisper	10^{-10}	20
Rustle of leaves	10^{-11}	10
Hearing threshold	10^{-12}	0

Loudness *Loudness is the magnitude of the psychological sensation produced by a sound, depending primarily on the intensity of that sound.*

Loudness is our own human response to the intensity of sound. Based on an average of many different hearing tests, scales and charts of loudness levels have been made. Such charts are not common, being used mainly by professional psychologists and audiologists. Hearing tests have shown that loudness depends primarily on the intensity of sound as measured in W/m^2. But very precise testing indicates that loudness depends to a small extent on frequency and the shape of the wave. Loudness, like the subjective quantity pitch, is not simply related to one objective characteristic. Since the effect of other characteristics is small, it is best to emphasize that *loudness depends primarily on intensity.*

Decibel readings such as those in Table 5-2 can be convenient indicators of sound levels. However, sound spreads out as it moves away from its source. So the intensity will decrease as the sound gets farther away from the source. For example, sound levels around the roaring engines of jet airplanes are dangerously high—so people working near them are required to wear ear protectors. But you need no ear protectors when watching the same planes take off half a mile away. At that distance the loudness is much less.

Loudness, the psychological perception of sound intensity, can be unpleasant. Noise is a form of pollution (sound pollution) that many people are trying to control. A sudden loud noise can be very startling and disturbing. Even when noise increases gradually to a very loud level, the sound is irritating if you are trying to sleep. Some cities have already set limits on the noise produced by traffic (see Fig. 5-17).

Whenever sounds are generated that have a level above 85 dB, special precautions must be taken. Sounds louder than this can make your ears "ring" or cause noticeable pain. Occasional short exposures to loud sounds will not harm your hearing. But long or repeated exposure, especially to the higher frequencies, may cause permanent loss of hearing.

WAVE SHAPE AND TONE QUALITY

The third objective property of a sound deals with the complexity of the shape of the sound wave. The shape of the sound wave is the record of its pressure variations. For example, Fig. 5-18 shows the shapes of sound waves produced by a tuning fork and a violin playing notes of the same pitch.

A tuning fork and a violin can produce notes of the same pitch with equal loudness. Yet these sounds from each instrument can easily be identified by the listener. This subjective property that distinguishes notes of the same pitch produced by different instruments is called *tone quality.*

Tone quality *Tone quality is the psychological property of sound that distinguishes between notes of the same loudness and pitch.*

The difference in tone quality of the sound from different instruments is the result of the different wave shapes of the same note.

Fig. 5-17
City officials of Torrance, California, post signs to limit the noise from traffic. (*Photograph by Tim Holton.*)

The note produced by a tuning fork has the wave shape of a very simple curve, such as the one in Fig. 5-18. The sound wave from a tuning fork has a single frequency, which is known as the *fundamental frequency.* If the tuning fork were built to produce middle C, the frequency of the wave would be 262 Hz.

The richer tone quality of a violin compared to a tuning fork is due to the greater complexity of the sound wave from the violin (see Fig. 5-18 once again). An analysis of the sound from any musical instrument playing middle

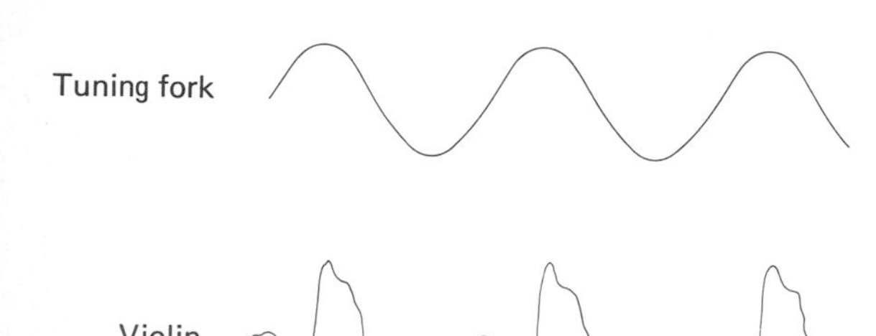

Fig. 5-18
The wave shapes representing the pressure variations of sound from a tuning fork and a violin.

C shows that waves of higher frequencies are being produced in addition to the fundamental frequency. The higher frequencies are known as *harmonics.* The first harmonic is the fundamental frequency of the instrument. The second harmonic is twice the fundamental frequency. The third harmonic is three times the fundamental frequency, and so on. In other words, a harmonic is a multiple of the fundamental frequency. For middle C the second harmonic has a frequency of 524 Hz (or twice 262 Hz), and the third harmonic has a frequency of 786 Hz.

A note played by a musical instrument or sung by a person can be analyzed in terms of the harmonic frequencies. Figure 5-19 shows the wave shape of a musical note separated into its fundamental frequency (first harmonic), second harmonic, and third harmonic. Combining the harmonics would reproduce the musical note shown at the top of Fig. 5-19.

The difference between the sound produced by various musical instruments playing the same note is a matter of the number and intensity of the harmonics. Middle C played on a piano is made up of the fundamental frequency and the first few harmonics. The same middle C when played on a violin contains over a dozen harmonics in addition to the fundamental frequency.

Remembering that the correlations are not exact, we can summarize the objective and subjective properties of sound:

Objective	Subjective
Frequency	Pitch
Intensity	Loudness
Wave shape	Tone quality

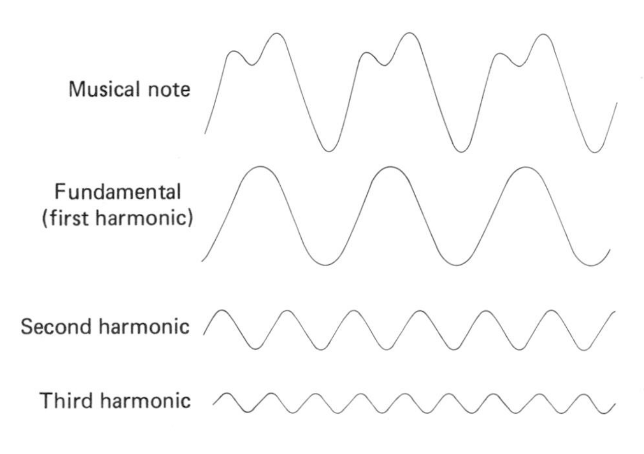

Fig. 5-19
The complicated wave shape of a musical note is made up of a series of harmonic waves with different frequencies (or correspondingly, different wavelengths).

One practical application which combines all these aspects of sound is the study of voiceprints. Voiceprints are an attempt to identify people by measuring (objectively) the frequency, intensity, and wave shape of a person's voice. A human then decides subjectively whether or not two voiceprints are the same. A few tests with volunteers have shown quite accurate results. But since some tests have been ambiguous, we cannot depend upon voiceprint comparisons as an accurate scientific technique.

5.8 NATURAL FREQUENCY AND RESONANCE

Vibrations can be set up in almost any physical object. Balloons, pencils, chairs, coins, and garbage cans all vibrate when they are struck with another object. If the vibrations are large enough, they are transferred to the surrounding air and we hear a sound. These sounds are often connected with the object itself. Indeed, it is hard to think of a garbage can without reflecting on the lid being loudly slammed down or thrown off to the side. One of the reasons for this is that a garbage can seems to possess a unique sound. Most of us would recognize that sound without opening our eyes!

Consider a few other sounds and try to imagine each without making any sound yourself: a dime dropped on a tabletop, a pencil dropped onto a hard kitchen counter top, a physician asking a patient to say "Ah."

Some of these sounds are quite complex. The ones with no definite pitch, which have a complicated wave pattern, are generally unpleasant. But other sounds are much simpler, with an easily recognized note or pitch. In such cases we speak of the *natural frequency* of the object.

Natural frequency

Natural frequency is defined as a frequency with which a system oscillates in the absence of external forces.

A chiming church bell can be used as an example. After the clapper strikes, the bell vibrates with a natural frequency. If a bell rubs even slightly against something, the clear, natural tone is completely changed. But if the bell is suspended so that it freely oscillates, it will pleasantly ring at its natural frequency. The triangle shown in Fig. 5-20 rings with a pleasant note.

A gentle tap on a wine glass produces an audible tone associated with the natural frequency of the glass. We also hear a natural frequency when we pluck a guitar string or jingle our car keys. (Throw a ring of car keys into the air. Then your fingers will not interfere with the natural oscillations of the keys.)

But not all natural frequencies lead to audible sounds. Dangle your telephone handset in the air, suspended by 3 or 4 feet of coiled cord. Jiggle it a bit and the handset will oscillate up and down at about 1 hertz. Thus, 1 hertz is the natural frequency for this system. But we do not hear a note with a frequency of 1 hertz because we cannot hear sound below about 20 hertz.

Many systems have such a low natural frequency that we cannot hear the associated sounds. And other systems oscillate with a natural frequency so high that we cannot hear these sounds either. But for frequencies that are

Fig. 5-20
A triangle vibrates at its natural frequency. (*Photograph from United Press International.*)

Fig. 5-21
The pitch changes when the bat is broken. (*Photograph by Ted Rozumalski. From Black Star.*)

audible we have all used the sound of the natural frequencies of vibration to test the condition of various objects. Have you ever dropped a baseball bat gently on its end? How does it sound? This is a common test for a broken bat, as shown in Fig. 5-21. A broken bat has a natural frequency which we recognize, and a bat with no cracks in it has a different natural frequency. Or have you ever seen someone test Ping-Pong balls by bouncing them? The same principle is at work.

All objects have more than one natural frequency. However, in many cases, only one of these (or perhaps several) is actually made to vibrate. A balloon stretched tight with air has many natural frequencies. But if the balloon sits quietly on your desk, none of these natural frequencies are excited. That is, the balloon is not vibrating with any of these frequencies. It is not vibrating at all. But send it into the air with a sharp swat, and you will momentarily hear a clear note. You have induced the balloon to vibrate with one (or more) of its natural frequencies.

You may have noticed another result of vibration at a natural frequency. It occurs when a hi-fi music system is turned up very loud. As the music plays, many different frequencies of sound are emitted by the speakers. In some music there is occasionally a prolonged note of unchanging frequency. When such a sound is made by the speakers, some object in the room may start to vibrate at one of its natural frequencies. You might notice a small vase rattling on a shelf in the room. Or more often, a loose sliver of wood or a piece of molding on the speaker cabinet is set into vibration. Because these vibrations interfere with our enjoyment of music, the offending rattle is eliminated. We move a quivering vase to another shelf or glue down a loose piece of wood. What causes these objects to vibrate?

The vibration of the vase or sliver of wood is caused by sound. Each sound wave that strikes "in time" with the vase or sliver gives a small push to the vibrating object. The compressions and expansions of the sound wave on the vase or sliver are similar to the gentle pushes given to a child on a swing. With each gentle push the child swings higher and higher. The height of the swing's arc increases with gentle pushes having the same frequency as the natural frequency of the swing. So too the amplitude of the vibration of the vase or sliver of wood increases when the natural frequency is the same as the sound frequency. This effect is known as *resonance.*

Resonance *Resonance is an increase in the vibration amplitude of an object caused by an external repetitive force having a frequency equal to the natural frequency of the object.*

Notice that neither the swing nor the vase is pushed very hard to increase its amplitude of vibration. Instead, many repeated pushes do the job. In the case of the swing we could see the momentary pushes, each one timed correctly and each slightly increasing the swing's amplitude. In the case of the vase we could not see the repetitive pushes and pulls of the passing sound, but they were present nonetheless.

The different pitched sounds of water being poured into a bottle are the result of resonance. The splashing water shown in Fig. 5-22 generates sounds of many different pitches. However, the natural frequency of the air space above the water enhances only frequencies of splashing water near the natural frequency of the air space. The air space gets smaller as the bottle fills. As this happens, the natural frequency increases. As a result, you hear the pitch of splashing water in a bottle getting higher as the bottle fills.

Resonance can be used to tune a guitar or other string instrument, even if the player does not have a good sense of pitch. A lower note string is played to produce a higher note, the same note that the next higher *open* string should produce. If the two strings are "in tune," playing the note on the lower string will cause the open string to resonate. If the open string is "out

Fig. 5-22
As the water level rises, the splashing water is heard as higher and higher pitched sound. (*Photograph by Ted Rozumalski. From Black Star.*)

of tune," it can be adjusted to resonate with the same note on the lower string. All the strings can be tuned by repeating the process. The strings may not be tuned to the absolute musical scale, but at least they will be consistent and won't sound too bad.

REVIEW You should be able to define the following terms: wave, wavelength, amplitude, frequency, speed of a wave, sound waves, reflection, pitch, intensity, loudness, tone quality, natural frequency, and resonance.

You should be able to write the wave formula and give examples of its application.

PROBLEMS

1. An explosion produces a sound wave with a frequency of 85 hertz. If the wavelength of the sound is 4 meters in air, calculate the speed of sound in air.
2. A sonar depth gauge used by a fisherman operates at a frequency of 14,000 hertz. If the wavelength of the sound wave is 10.25 centimeters in water, calculate the speed of sound in water.
3. The note known as middle C has a frequency of 262 hertz. If the speed of sound in air is 340 meters per second, what is the wavelength of middle C?
4. The brown bat emits an ultrasonic sound at about 90,000 hertz. If the speed of sound is 340 meters per second, what is the wavelength of the sound?
5. While watching a lightning storm, you observe a bolt of lightning. If it takes 12 seconds before you hear the thunder, how far away was the lightning?
6. At Echo Canyon you shout and listen for your echo, which returns in 10 seconds. How far away is the canyon wall that reflects the sound of your voice back to you?
7. The G string on a violin, which has a fundamental frequency of 196 hertz, is able to sound its twenty-first harmonic. Calculate the frequency of the twenty-first harmonic.
8. A French horn plays the note D above middle C. This note has a fundamental frequency of 294 hertz, but only four harmonics. What is the frequency of the fourth harmonic?

DISCUSSION QUESTIONS

1. Give an example of a common visible wave that can be generated in the home. What is the approximate amplitude of the wave you chose?
2. Estimate the frequency of (*a*) your heartbeat and (*b*) a Ping-Pong ball bouncing on a hard table. What factors might affect your estimates in each case?
3. Estimate the speed of a wave on the ocean as it nears the shore.
4. Waves on water can be observed from above the wave, from in front of the wave, and from various angles. What particular viewpoint should be used to examine the shape of the wave?
5. Describe a sound wave in air from both the macroscopic and microscopic points of view.
6. Does sound travel through a vacuum?
7. Ask a friend to eat some celery or potato chips with her mouth closed. Then do the same yourself. How does the sound reach your ears in each case?
8. How does sound get from one room to another in an apartment building or a motel?
9. What is an echo? Can there be multiple echoes?
10. Could an echo ever be louder than the sound that produced it?
11. If a sound wave were to reflect back and forth between two large buildings with flat, parallel walls, would the wave exhibit perpetual motion?
12. Why are echoes commonly heard from walls of large buildings and not from lakes?
13. What is the approximate frequency range of human hearing?
14. What is the difference between ultrasonic and supersonic?
15. What is the difference between infrasonic and subsonic?
16. Normally we do not hear the sounds of ants walk-

ing or moving grains of sand. Are the frequencies outside the range of human hearing?

17. What is the difference between pitch and frequency?

18. With continued space exploration we may eventually communicate with other intelligent beings who can hear. What would become of the ideas of frequency and pitch?

19. How does a Ping-Pong player check for a cracked ball?

20. How do baseball players check for a cracked bat? Could a xylophone or marimba player check for a cracked key in the same way?

21. Describe what is meant by resonance.

ANNOTATED BIBLIOGRAPHY

Of the many small books on sound which you might explore, two good ones are I. M. Freeman, *Sound and Ultrasonics,* Random House, Inc., New York, 1968, and W. E. Kock, *Seeing Sound,* Interscience Publishers, a division of John Wiley & Sons, Inc., New York, 1971. Freeman has an unusually clear style and broad coverage. *Seeing Sound* specializes in methods of recording sounds in patterns and diagrams.

If you are interested in the musical aspects of sound, do not miss G. Anfilov, *Physics and Music,* Mir Publishers, Moscow, 1966. This Russian author has chapters on Stradivarius violins, voice analysis, and computer music. The entire book seems to tell a story. It reads like a fascinating novel.

The story of human hearing is given in J. S. Wilentz, *Senses of Man,* Thomas Y. Crowell Company, New York, 1968. A chapter called "The Auditory Code" weaves together psychological tidbits with anatomical and physiological detail. Also on human hearing read "I Am Joc's Ear," *Reader's Digest,* October 1971, p. 131. For some fascinating experiments that show that left and right ears perceive differently, try "Musical Illusions," *Scientific American,* October 1975, p. 92.

Two master lecturers have had their lectures on sound preserved in paperback format. Six lectures by Sir William Bragg are in *The World of Sound,* Dover Publications, Inc., New York, 1968. A single lecture entitled "On the Physiological Causes of Harmony in Music" by H. von Helmholtz is in *Popular Scientific Lectures,* Dover Publications, Inc., New York, 1962.

6 ELECTRICITY AND MAGNETISM

The widespread use of electricity in everyday life is the basis of much of modern technology. In this chapter we will investigate electric charge, electricity, magnetism, and the laws governing their behavior.

6.1 ELECTRIC CHARGE

Benjamin Franklin, famous for his part in the American Revolution, is also famous for his experiments with kites and lightning. He discovered the electrical nature of lightning by observing electric sparks at the end of a kite string he was holding. (Do not try this experiment yourself. If lightning were to strike the kite, it could cause death. Several nineteenth-century experimenters were electrocuted repeating Franklin's experiment. Franklin was lucky.) On the basis of this and other studies, Franklin introduced the terms *positive* and *negative* for electric charge. These terms are still in common use.

In order to describe electric charge it is necessary to expand the brief discussion of atoms provided in Chap. 3. We saw that an atom is the smallest part of matter that keeps the identity of a chemical element. Regardless of which chemical element is being considered, an atom of any element may be described as a nucleus surrounded by orbiting particles.* The nucleus has a

*The structure of the atom is often described as similar to that of the sun and its system of orbiting planets. A more refined description of the nature of electrons and their motion in the atom is given in Chap. 9.

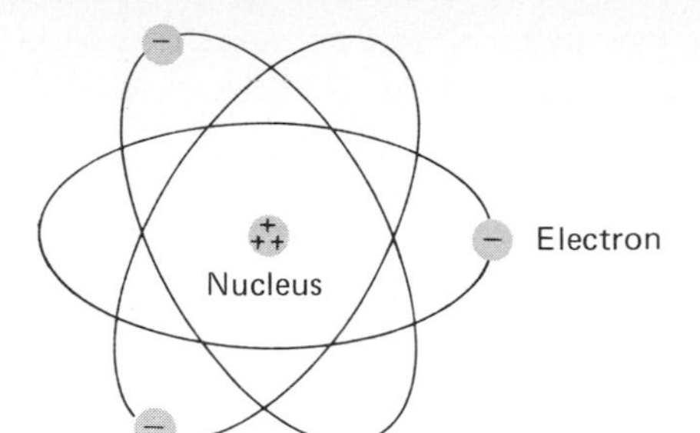

Fig. 6-1
Electric forces exist within an atom between the negatively charged electrons and the positively charged nucleus, which are not drawn to scale.

certain amount of positive charge (see Fig. 6-1). This positive charge is due to particles known as *protons.* Protons are particles with the type of electrical characteristic that has come to be known as *positive charge.* All protons have an equal amount of positive charge.

The particles orbiting the nucleus are *electrons,* particles with a different type of electrical characteristic known as *negative charge.* All electrons possess the same amount of negative charge. The charge of an electron is equal in magnitude but opposite in type (and thus sign) from the charge of a proton. The magnitude of this charge is 1.6×10^{-19} coulomb, where the unit of charge is called the coulomb (abbreviated C) in honor of Charles Coulomb, who studied the forces between electric charges.

An atom in its normal state is electrically neutral. Charge neutrality occurs in an atom because the number of protons in the nucleus equals the number of electrons surrounding the nucleus. Therefore, an atom contains equal amounts of positive and negative charge. Thus an object made up of neutral atoms is also uncharged. The coins in your pocket are not charged. They are made up of atoms and molecules which are electrically neutral. You might think of the coins as having both positive and negative charge within them. But overall there is a balance between the positive and negative charge. The coins have no *net* electric charge.

There are occasions, however, when an object is not electrically neutral. Your body becomes charged when you slide across the seat cover of your car. Then you experience a shock as you touch the door handle. It is possible to understand some aspects of the role of charge in this example. At first, both you and the car seat are electrically neutral. Then, as you slide on the seat, a separation of charge occurs. Some net charge is left on you, and some net charge is left on the car seat. These two charges are equal in amount. If the two charges were measured, we would find that the number of coulombs of each is exactly the same. But the two charges are of opposite sign. One is positive charge, and one is negative charge.

Small amounts of electric charge are relatively easy to isolate. Not only can sliding across car seats produce shocks (and often a visible spark), but walking across a thick carpet can often cause the same effects. Combing dry hair will also produce a crackling sound as sparks jump back and forth between the hair and the comb. Removing nylon blouses and shirts in a dark room often produces dozens of miniature visible sparks as an electric charge jumps between one part of the garment and another.

Charge separation, followed by recombination, is very common. When the recombination is strong enough to produce a spark, we can even see the effect. Remember, though, that we do not see charges. A car seat may be highly charged, and yet it *looks* exactly the same as when it is electrically neutral.

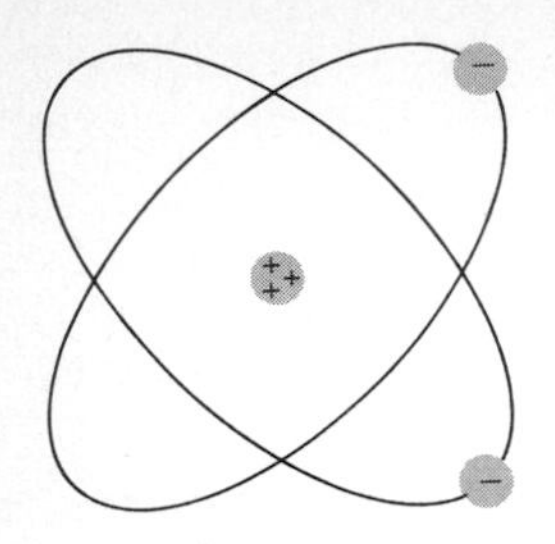

Positive ion

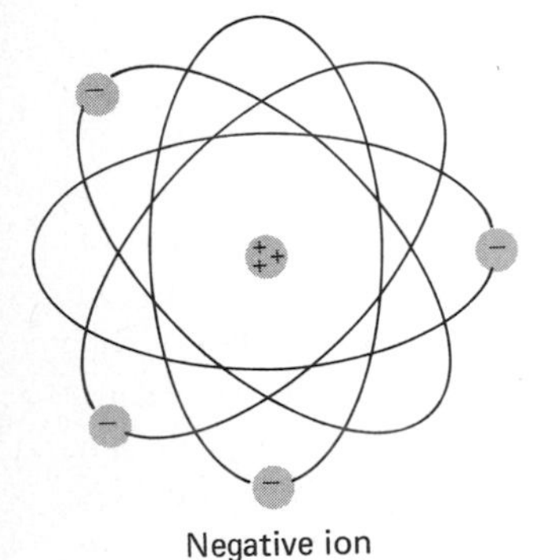

Negative ion

Fig. 6-2
Ions have a net charge.

Objects become charged when they are rubbed or touched against another object, by means of a process known as *ion formation.* An ion is formed when a neutral atom loses or gains one or more electrons. If an atom gains electrons, it possesses an excess of negative charge. Then the charged atom is called a *negative ion.* A *positive ion* is formed by the loss of one or more electrons from a neutral atom. These two cases are shown in Fig. 6-2.

In some ways, ions are well understood. Chemists have determined much about the behavior of ions in liquid solutions. And physiologists know many details concerning certain ions in our body cells. In some other aspects, ion behavior is a mystery. For example, return to the car seat for a moment. When we slide, and charge ourselves and the seat, which way does the charge flow? Does the seat become positive and leave us with a negative charge? Or is it the other way around?

Experiments have shown that sometimes the seat becomes positively charged and other times the seat becomes negatively charged. We do not yet know all the factors which control ion formation. Certainly our clothing has some effect. The behavior of cotton and wool is different in this regard. It is also known that a little surface moisture or surface dirt has an effect. But no one is sure of the details.

6.2 ELECTRIC FORCES AND FIELDS

One of the simplest games you can play with a comb is to comb your hair and then use the comb to lift small bits of shredded paper. This lifting force is electrical in nature and is greater in strength than the gravitational force. (If the electric force were not larger than the gravitational force, the pieces of paper could not be lifted.)

More systematic studies of the electric force bring out the fact that whether two charged objects are alike (both positive or both negative) or unlike (one positive and one negative) makes a profound difference in the direction of the electric force. If the charged objects are unlike, the objects experience a force of attraction. If the charged objects are alike, the objects experience a force of repulsion. These facts are often summarized by the statement "opposites attract and likes repel." Figure 6-3 shows two charged Styrofoam cups.

The distance between the charged objects affects the magnitude of the electric force. Similar to the gravitational force, the magnitude of the electric force decreases as the separation increases. More precisely, the electric force varies inversely as the square of the distance between the charged objects.

The overall behavior of the electric force between charged objects is summarized in *Coulomb's law.*

Coulomb's law

The force acting between two charged objects is directly proportional to the product of the two charges and inversely proportional to the square of the distance between them.

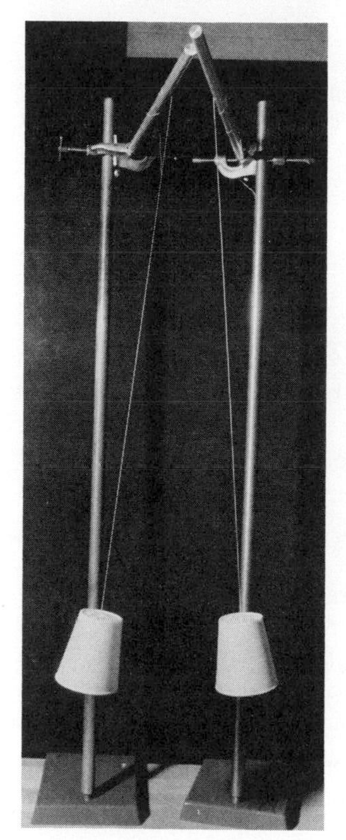

Fig. 6-3
Unlike charges attract. Are these coffee cups charged alike? (*Photograph by Ted Rozumalski. From Black Star.*)

This law can be expressed in a formula, namely,

$$F = k\frac{q_1 q_2}{d^2}$$

where F = electric force, measured in newtons
q_1 and q_2 = charges of the objects, measured in coulombs
d = distance separating the charges, measured in meters

Coulomb's constant k is equal to $9 \times 10^9\ \mathrm{N \cdot m^2/C^2}$. See Fig. 6-4.

A study of matter at the atomic level reveals that the influence of the electric force is overwhelming. Gravitational forces between electrons, for example, are extremely small when compared to the electric forces between them. If two electrons could be isolated in a vacuum, the electric force of repulsion would be 4×10^{42} times (that is, 4 million, trillion, trillion, trillion times) larger than the gravitational force of attraction. Electrons repel each other!

At the opposite extreme (at the level of planets), the influence of the gravitational force is overwhelming. There is no detectable electric force between celestial bodies (for example, the earth and its moon). Apparently on the earth and all other planets, every electron is paired with a proton. Some of these pairs may be separated to form ions. But overall the net charge is zero. If there were any electric force between the earth and its moon, these two bodies would have different orbits that could be detected by astronomers.

The idea of a *field* in physics is that a force is acting in a region of space. For a gravitational field, which we encountered in Chap. 1, one mass attracts another. Apparently all masses attract other masses. Although theoretical speculation exists concerning gravitational repulsion, no such repulsion has ever been measured.

Electric charges either attract or repel each other by means of an *electric field.*

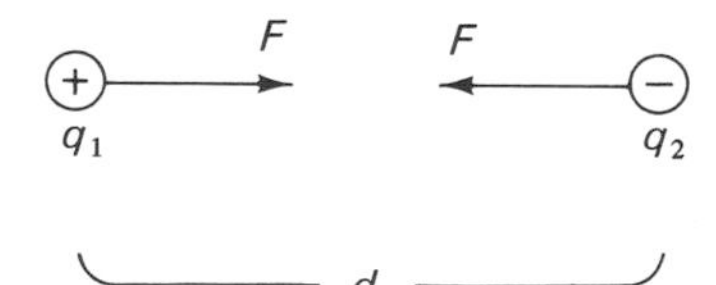

Fig. 6-4
Coulomb's law describes the force between two charges.

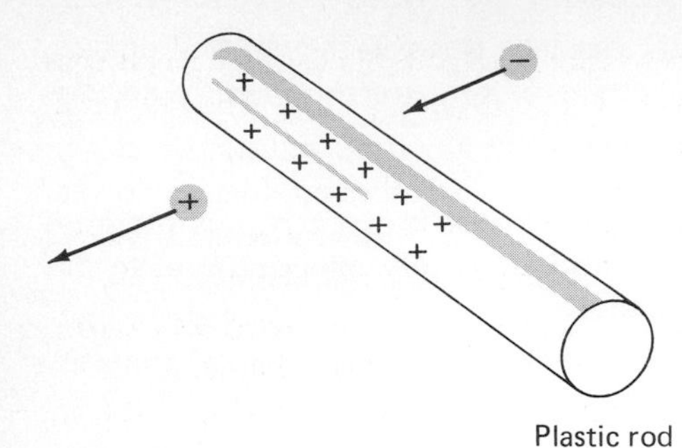

Fig. 6-5
The positive charge on the plastic rod sets up an electric field in the surrounding space. Positive and negative charges in the field experience a force and accelerate according to Newton's second law of motion.

Electric field *The strength of an electric field is defined as the force experienced by an electric charge divided by the value of that electric charge.*

An electric field is set up in a region of space by a charge. Because of the presence of the electric field, a second charge is either attracted or repelled. The forces of attraction and repulsion follow the rule that electric charges of opposite sign attract and electric charges of the same sign repel. Fig. 6-5 shows, above, the effect of the electric field of a plastic rod on two other charges.

Electric forces of attraction and repulsion are utilized in many industrial processes. One example is in the manufacturing of sandpaper. Before the sand grains are sprayed toward a glue-coated sheet of paper, the sand grains are charged negatively. Then as the sand grains are sprayed into the air, they repel each other. This results in a more even distribution of sand. Thus there are fewer clumps. The paper, which is charged positively, attracts the grains of sand (see Fig. 6-6). There is better coverage and sticking as the sand grains are driven into the glue by electric forces.

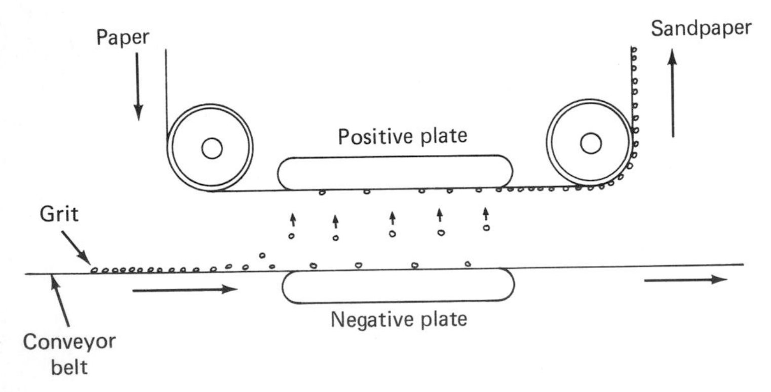

Fig. 6-6
Sandpaper can be manufactured using electric forces.

6.3 ELECTRICAL CONDUCTION

In a vacuum there is nothing to impede the movement of charges. If a charge is placed in an electric field, the charge will experience a force. And the force will cause the charge to accelerate under such vacuum conditions.

In a material the motion of charges is more complicated. Electric forces can still act on charges within the material. But now there may be other forces acting as well. When an electric field is applied to a material, there is a motion of electrons or ions. This motion, or drift of charges, is called an *electric current,* and the material is said to be conducting electric charge.

Just as in the case of thermal conduction (discussed in Chap. 4), all objects are capable of conducting charges to some extent. However, there is a great variation in electrical conductivities among different materials.

Conductors

Those materials which readily conduct charge are called conductors.

Those materials which are *not* good conductors of charge (although there can be *some* charge flow in any material) are called *electrical insulators.*

There is a great similarity in the ability of a material to conduct both heat and electricity. The ability to conduct heat and electricity is based on the ability of electrons to migrate throughout the interior of the material. In all metals there are "free" electrons that can be easily moved by the influence of an electric field. Thus all metals are good conductors of both electric charge and heat. There is of course some variation in both electrical and thermal conductivity from one metal to another. An aluminum lawn chair easily conducts electric charge, as shown in Fig. 6-7.

There are many common electrical insulators. Glass, rubber, dry wood, and most plastics have a very small electrical conductivity. In these materials the electrons are strongly bound to molecules and are not free to participate in the conduction process. Electric charge does not flow easily in insulators. Table 6-1 shows the electrical conductivities (relative ease in conducting charge) of several typical materials.

Water, especially impure water, is an electrical conductor. Unlike metals, water does not conduct electricity by means of the movement of electrons. Typically some impurity such as salt dissolves in water to form positive and negative ions. These ions, rather than free electrons, are the charges that move in the liquid. Most water outside a laboratory contains some dissolved salt or minerals. Such chemically impure water may be excellent drinking water, but it does contain ions in solution. These ions make ordinary water an electrical conductor.

Dry wood, which is normally an insulator, becomes a conductor if wet. Also, humidity that condenses on electrical equipment can cause the surfaces of insulators to become conducting. Moist surfaces are dangerous around electrical equipment. The equipment often malfunctions because of the unwanted conduction.

Electricity is usually understood to be due to the motion of electric charge. The motion of charge is called *current.*

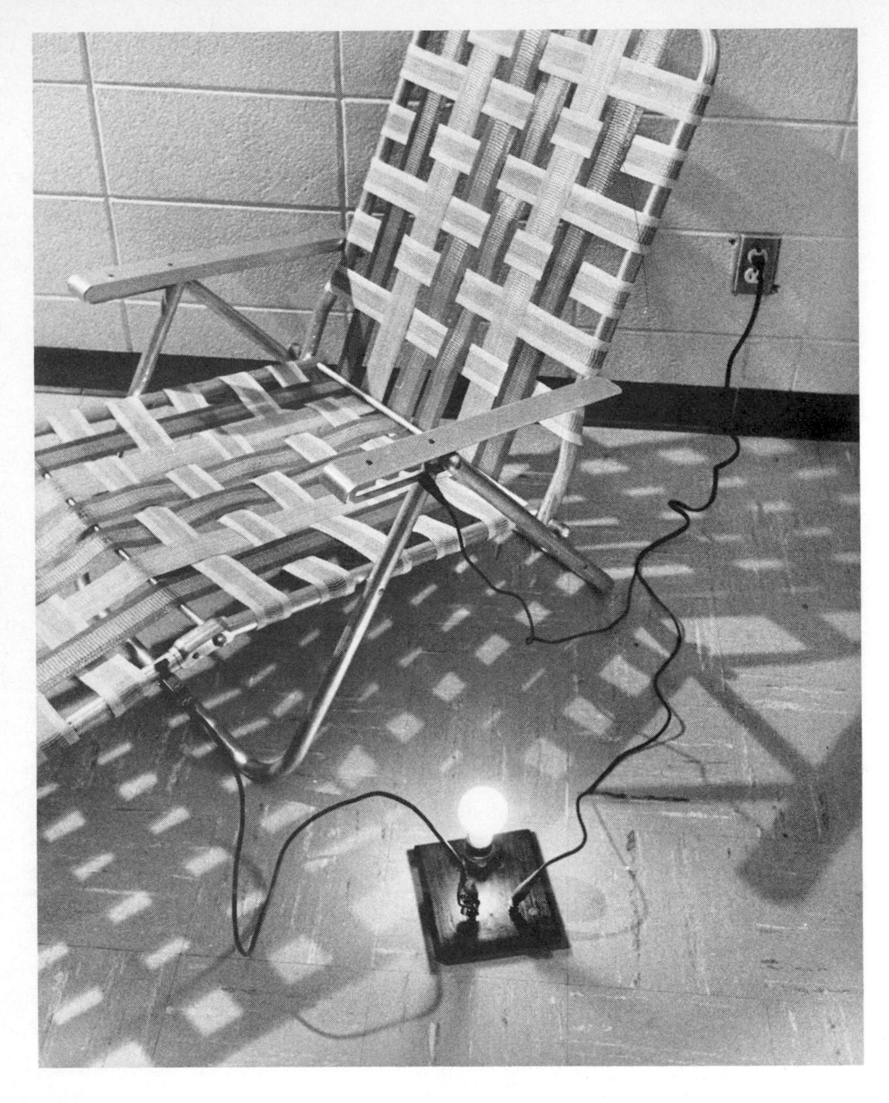

Fig. 6-7
Aluminum is a good electrical conductor. (*Photograph by Ted Rozumalski. From Black Star.*)

Current *The magnitude of current is defined as the amount of charge that flows past a point divided by the time it takes for the charge to flow past.*

In the metric system, current is measured in units of amperes (sometimes simply called "amps," and abbreviated as A). One ampere is one coulomb of charge flowing each second. The ampere is named after André Ampère, who established the relationship between electricity and magnetism.

Picture a long copper wire in which there is a current of 10 amperes. If you could see the moving charges, you would see 10 coulombs of charge per second pass any point along the wire (see Fig. 6-8). You would see 20 coulombs pass if you watched for 2 seconds. And you would see 30 coulombs

Table 6-1
Electrical conductivities

Material	Electrical conductivity at room temperature, $C^2/J \cdot sec \cdot m$	
Rubber	10^{-15}	↑
Glass	10^{-12}	better insulators
Fresh water	2×10^{-4}	
Ocean water	4	
Brass	1.59×10^{7}	better conductors
Aluminum	3.54×10^{7}	↓
Silver	6.14×10^{7}	

pass in 3 seconds, 40 coulombs in 4 seconds, and so on. This is the meaning of 10 amperes of current. Ten amperes is a typical current for an iron, which is listed in Table 6-2.

As charges move through a wire or a liquid, the charges encounter more or less difficulty in moving among atoms and ions. A metallic wire conducts electricity well, but some appliances offer more *resistance* to the flow of electric current.

Resistance *The resistance of a material is defined as its opposition to the flow of electric current.*

In the metric system, resistance is measured in units of ohms, in honor of Georg Ohm. One ohm is a small resistance and is typical of ordinary metallic

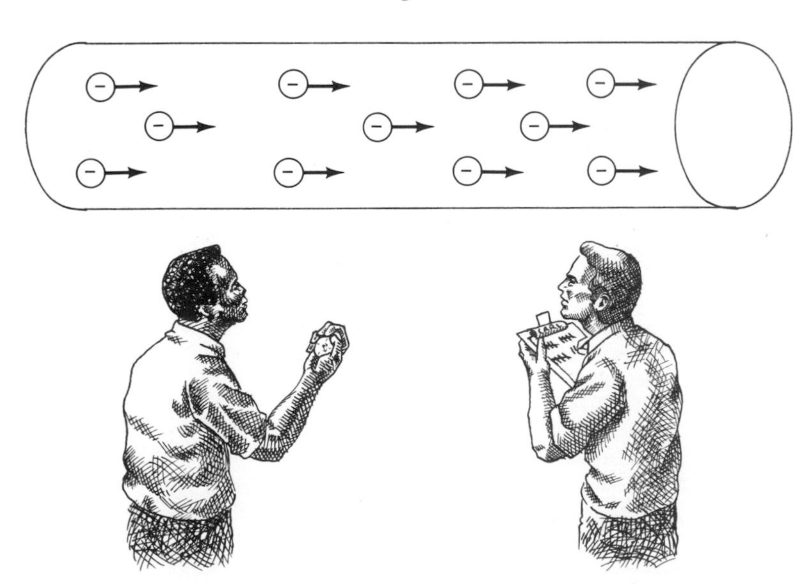

Fig. 6-8
Current is the amount of charge that flows past a point divided by the time. *If* you could actually see charges in a wire, the current could be calculated with the help of a stopwatch.

Table 6-2
Representative values of electric current

Item	Current, A
Spark plug	0.001
Clock	0.025
Table radio	0.2
60-watt bulb	0.5
Heating pad	0.7
Television set	2
Iron	10
Electric clothes dryer	16
Arc welder	50
Lightning bolt	20,000

wires a few meters long. Household appliances are intermediate in resistance, having values between about eight ohms and several thousand ohms. At the upper end of resistance values we find that small pieces of some insulators have a resistance of several million or even a billion ohms.

Besides current and resistance, there is a third fundamental electrical quantity. Electric charges do not move in a conductor just because they are free to do so. Electric fields exert a force on the charges, which move through a conductor. The work done by this force is related to the *voltage.*

Voltage *The magnitude of the voltage is defined as the work done by the electric field divided by the electric charge.*

In the metric system voltage is measured in volts. As an example of voltage, consider the 120-volt electrical outlets in buildings. The electric field does 120 joules of work on each coulomb of charge which passes through the wires. Table 6-3 lists some common voltage values. Figure 6-9 shows two of the most frequently used batteries.

Table 6-3
Representative values of voltage

Item	Voltage, V
Human nerve impulse	0.1
Flashlight battery	1.5
Car battery	12
Arc welder	30
Home wall outlet	120
Spark plug	25,000
Storm cloud	100,000,000

Fig. 6-9
Most flashlight batteries are rated at $1\frac{1}{2}$ volts, and car batteries are rated at 12 volts. (*Photograph by Ted Rozumalski. From Black Star.*)

6.4 OHM'S LAW

The three concepts of current, resistance, and voltage are related by Ohm's law for electricity. Ohm's law is very similar in both form and idea to Poiseuille's law for fluid flow. Recall that in Chap. 3 Poiseuille's law dealt with the volume flow rate for a fluid. The "driving force" for fluid flow in a pipe is the pressure drop. Opposition to the flow occurs because of the resistance of the pipe.

In identical fashion the same concepts can be applied to electricity. The "driving force" for current flow in a wire is the voltage. Opposition to the current occurs because of the resistance of the wire.

Ohm's law **The current depends directly on the voltage and inversely on the resistance.**

Ohm's law can be expressed in the simple formula

$$\text{Current} = \frac{\text{voltage}}{\text{resistance}}$$

or

$$I = \frac{V}{R}$$

where I = current measured in amperes
V = voltage measured in volts
R = resistance measured in ohms

The units of amperes, volts, and ohms are metric and are universally used.

Table 6-4
Comparison of fluid flow and charge flow

Fluids		Electricity	
Volume flow rate:	$\mathfrak{F}$	Current:	I
Pressure drop:	Δp	Voltage:	V
Resistance:	R	Resistance:	R
Poiseuille's law: $\mathfrak{F} = \frac{\Delta p}{R}$		Ohm's law: $I = \frac{V}{R}$	

Ohm's law can be used to make simple calculations. A television set which uses 120 volts can have, for example, a resistance of 60 ohms. What is the current through such a television set? Using Ohm's law, we write

$$\text{Current} = \frac{\text{voltage}}{\text{resistance}}$$

or

$$I = \frac{120 \text{ volts}}{60 \text{ ohms}}$$

$$= 2 \text{ A}$$

The concept of voltage involves the concept of work (energy) as one of its parts. The energy idea can be brought out more clearly by calculating the electric *power* used, for example, in an appliance. The formula for electric power can be written as

Electric power

$$\text{Electric power} = (\text{current})(\text{voltage})$$

or

$$P = IV$$

where P stands for the power.

The power is given in watts if the current is measured in amperes and the voltage in volts. As an example of the use of this equation, let us calculate the power used by the television set of the previous example. The set utilized 120 volts and had a current of 2 A. Thus,

$$\text{Electric power} = (\text{current})(\text{voltage})$$

or

$$P = (2 \text{ amperes})(120 \text{ volts})$$

$$= 240 \text{ watts}$$

In an example such as this one, the power is the rate at which energy is used. We can find the total energy used by multiplying power by the time of use.

For example, how much energy is used by a 240-watt television set in 10 hours?

$$\text{Energy} = (\text{power})(\text{time})$$

or

$$= (240 \text{ watts})(10 \text{ hours})$$

$$= 2{,}400 \text{ watthours}$$

The 2,400 watthours of energy can also be expressed using the larger energy unit of kilowatthour. Since 1,000 watthours equals 1 kilowatthour, 2,400 watthours equals 2.4 kilowatthours. In your home, the electric bill shows the number of kilowatthours measured on an "electricity" meter (see Fig. 6-10). The bill is a charge for *energy,* and so you are charged for kilowatthours. One kilowatthour typically costs about 5 to 10 cents. At this rate, it costs about 12 to 24 cents to run a television set for 10 hours.

Although the kilowatthour is the accepted unit of energy for selling electricity, the more common unit in physics is the joule. In order to convert 1 kilowatthour to the equivalent number of joules, recall that 1 watt is equal to 1 joule per second (see the discussion of power in Sec. 2-4). And 1 kilowatt is equal to 1000 joules per second. Thus 1 kilowatthour is equal to 3,600,000 joules (the product of 1000 joules per second and the 3,600 seconds in an hour).

Lightning is an electrical event that delivers tremendous power in a

Fig. 6-10
Electricity bills are based upon the total *energy* used, as measured by an "electricity" meter. (*Photograph by Korrine Nusbaum.*)

SUPERCONDUCTIVITY

There are many physical properties which are temperature-dependent. Change the temperature, and something else happens. Warm some water in a pan, and it will evaporate faster than if left at room temperature. Heat an iron poker in the fireplace, and the metal changes color. But these changes are commonplace when compared with a remarkable discovery which took place in the early twentieth century. The property being investigated was electrical conductivity—the ability of a material to carry an electrical current. The whole story begins with some investigations at very low temperatures.

During the late 1800s several experimenters were producing lower and lower temperatures in their laboratories in order to turn gases into liquids. Oxygen becomes a liquid at 90K, and nitrogen becomes a liquid at 77K. In 1898 James Dewar liquefied hydrogen at 20K. One gas, helium, resisted all efforts until 1908 when it was liquefied at the very low temperature of 4K. Liquid helium is shown in the photograph below.

(Photograph by Solid State Research Group, University of Wisconsin—Milwaukee.)

Three years later a Dutch physicist, Kamerlingh-Onnes, used liquid helium to study the electrical conductivity of various elements at low temperatures. A big surprise came when he studied the element mercury. As he lowered the temperature, the conductivity of mercury slowly changed. Suddenly, at a temperature between 4 and 5K, the mercury became a perfect conductor. It had no resistance at all!

Careful measurements confirmed the idea. Mercury becomes a superconductor at a temperature of 4.2K. A superconductor has no resistance at all to the passage of an electrical current. In the years that followed closely, many other superconducting materials were discovered. Over 100 elements, compounds, and alloys are now known to become superconducting at low temperatures.

There are many practical uses for superconducting materials. Since they do not suffer power losses due to resistance, any electrical device can be made more efficient with the use of superconductors. Both motors and electromagnets have benefited by being constructed of superconducting wire. Such a superconducting electromagnet is shown in the photograph below. But there is a drawback. All known superconductors regain resistance if they are warmed above their transition temperature. The transition temperatures of all known superconductors lie below 30K. This creates a special problem. If we want a material to be superconducting, we must keep it cold. A refrigeration system must be used which can maintain temperatures below 30K. At present, because of this, superconducting materials are restricted to rather special applications. Widespread application of superconductors requires the discovery of new compounds with higher transition temperatures. Maybe someday we will see a room-temperature superconductor!

(Photograph by Calvin Campbell, courtesy of MIT News Office.)

single lightning bolt. A typical current in a lightning bolt is 20,000 amperes. The voltage between a cloud and the ground during an electrical storm may be as high as 100 million volts. The power in a lightning bolt is tremendous. Our formula shows

$$\begin{aligned} P &= IV \\ &= (20{,}000 \text{ amperes})(100{,}000{,}000 \text{ volts}) \\ &= 2{,}000{,}000{,}000{,}000 \text{ watts} \end{aligned}$$

Although the power in a single lightning bolt may reach 2 trillion watts, we should not expect to be able to use this power in any practical way. First of all, the lightning only lasts for a fraction of a second. Second, we never know where it will strike!

6.5 ELECTRICAL CIRCUITS

Unlike water that may flow out the end of a pipe, electric charge does *not* flow from the end of a wire and collect in a puddle. Air is an insulator, and so electric current must be guided through a wire or some other electrical conductor. This pathway for electricity is known as a *circuit.*

Circuit *A circuit is simply a complete electrical pathway for electricity.*

A light bulb is an example of a circuit. The filament of the light completes the electrical pathway between the two connections on the base of the light (see Fig. 6-11).

In order for a circuit to sustain a current, there must be a complete loop, that is, a complete pathway for the current. A single piece of straight wire (not connected to any other conductors) cannot have a current through it. Such an isolated piece of wire can have both a resistance and a voltage, but no current can flow unless we complete the circuit.

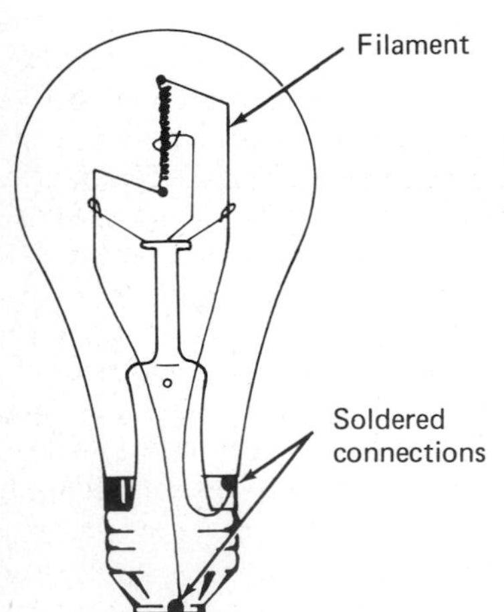

Fig. 6-11
The filament provides a complete electrical path through the bulb. (*Photograph courtesy of General Electric.*)

In any practical problem, such as wiring a string of Christmas tree lights or an entire house, each device must be part of the circuit. There are basically two different circuits: a *series circuit* and a *parallel circuit.*

In a series circuit each bulb or appliance is connected with one wire to the preceding one. Thus the same current flows through each bulb or appliance in the circuit, as can be seen in Fig. 6-12. Unfortunately, if any appliance or bulb in such a circuit is turned off or removed, then everything else in the series circuit stops working. In a series circuit the current is the same in each part. If the current is zero for one part, it is zero for all others.

Perhaps you have seen a string of Christmas tree bulbs wired in series. If one bulb burns out, then the entire string is darkened. A series circuit is not satisfactory for wiring a house. You do not want the lights or appliances in your house all working or all turned off at once!

A more practical electrical circuit for homes is the parallel circuit. In a parallel circuit all the light bulbs or appliances have one of their two electrical connections wired together. The other electrical connections are also wired together. Then the two common connections are each connected to the source

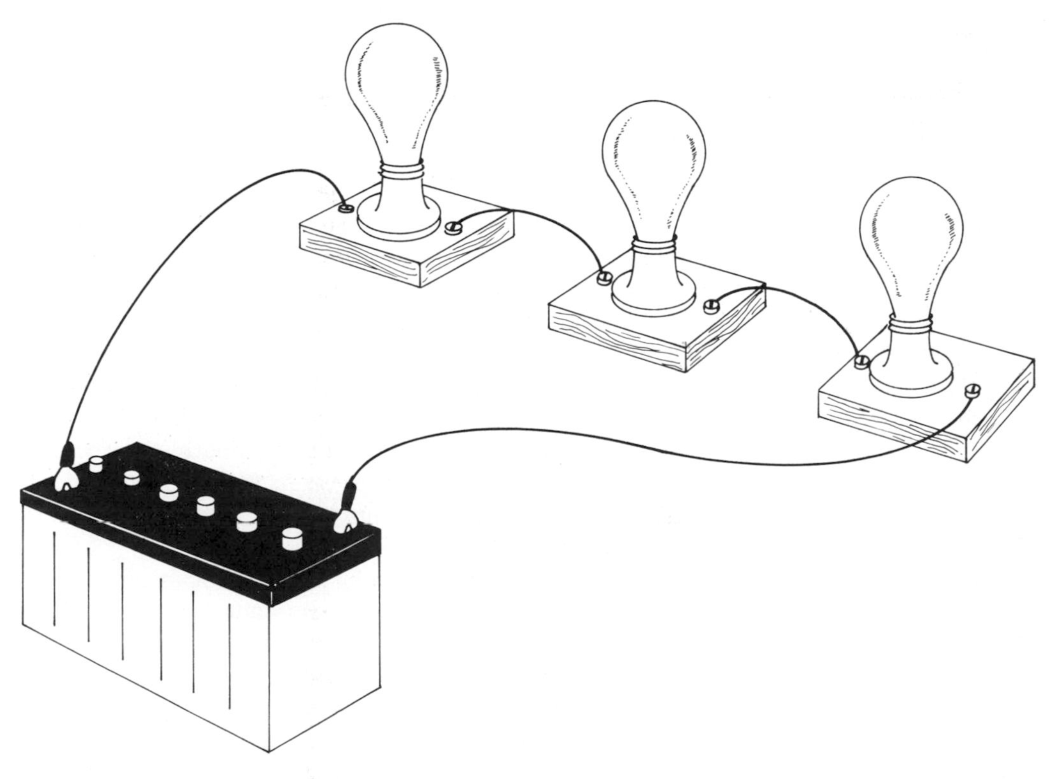

Fig. 6-12
In a series circuit each bulb has the same current.

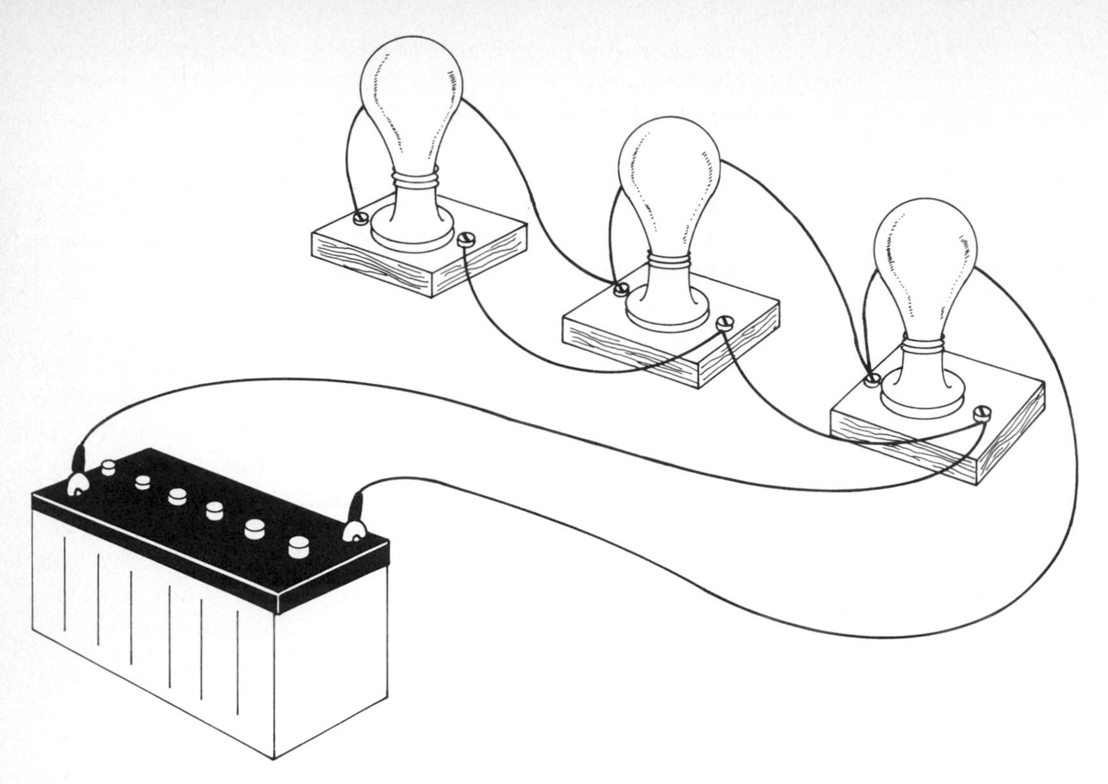

Fig. 6-13
In a parallel circuit the current from the battery splits into portions going through each bulb.

of voltage, as shown in Fig. 6-13. As a result, the current in a parallel circuit divides and a part of the total current passes through each electrical device. If one bulb or appliance is turned off or removed, the other devices will continue to function. Better-quality Christmas tree lights have the bulbs wired in parallel. If one bulb burns out, this does not darken the entire string.

One often hears about a "short circuit." A short circuit is usually an unwanted (and perhaps dangerous) circuit. If the insulation on an extension cord breaks, it is possible for the wires to touch. Since the wires have only a small resistance by themselves, Ohm's law shows that the current can become quite large. When the current in a home exceeds a fixed value (usually 15 or 20 amperes), a fuse or circuit breaker shuts off all the electricity in that circuit.

All industrial plants are also protected by fuses or circuit breakers. In some cases these allow currents as high as several hundred amperes to flow safely. The wires are specially designed for this purpose. Such large currents would not be safe in a home.

6.6 TYPES OF ELECTRICITY

Water can be made to flow from one place to another. It would seem that charges must also move from one place to another. Indeed that happens when the electricity is *direct current* (abbreviated dc). Curiously, electrical energy can also be transported if the charges vibrate back and forth as *alternating current* (abbreviated ac).

Electricity that flows as dc is caused by a voltage source in which the sign of the voltage does *not* change. For example, a flashlight battery is an electrochemical device with two terminals. Each of the terminals has a fixed voltage labeled "plus" or "minus." Chemical reactions inside the battery maintain these voltages. The voltage goes to zero when the chemical reactions stop. But the sign of the voltage at the terminals does *not* switch back and forth.

Typically the top terminal of a battery is positive and the bottom terminal is negative. When the battery is inserted into a flashlight, electrons leave its bottom terminal, pass through the filament of the bulb, enter the top terminal on the battery, pass through the battery, and repeat the process. The electricity flows steadily in one direction as direct current.

Electricity that flows as ac is caused by a voltage source in which the sign of the voltage changes. An electric power station generates voltage from a spinning dynamo. The dynamo has two terminals for the voltage output. As the dynamo spins, the output voltage oscillates back and forth between a maximum positive value and a maximum negative value. For the electricity sent to our homes, the frequency of the voltage change is 60 hertz. This means 60 *complete* voltage changes are made every second. The sign of the voltage at the electrical outlet on the wall switches back and forth from positive to negative.

When a desk lamp is connected to a 60-hertz voltage outlet, the electrons throughout the wiring and the light bulb move back and forth continuously. But the electrons make no progress along the wire. You might wonder how electrical energy can be transported from a power station to your home if the alternating current means electrons only vibrate. Although the *electrons* in ac do not flow from one terminal to another, *energy* can still flow through the circuit. The vibrating electrons act as carriers of electrical energy. The energy is passed along by the electrons. In similar fashion water waves act as carriers of storm energy. The individual water molecules do not scurry from an offshore hurricane to the beach. It is the waves that carry the energy and not the individual particles.

A mixture of both direct current and alternating current occurs whenever there is a spark or a bolt of lightning. In either electrical discharge the current first exists as dc until a pathway is made. Once the circuit has been established, there can be continuing direct current, pulses of current, alternating current, or complicated mixtures of all three.

One type of electrical discharge is a spark, which can occur when two "live" wires are brought close together. Usually air is a good insulator, but not always. Sometimes the voltage between two charged objects may become quite large (for example, between parts of television circuits). At other times the gap between two charged objects may become small. If either or both of these conditions occur, the air "breaks down." This means that the air becomes ionized and becomes a conductor. Electricity in the form of a spark then jumps from one charged object to the other.

How much voltage is needed to cause a spark to jump from one object to another? The answer depends on the curvature of the surfaces (flat or pointed) between which the voltage is applied. Typically 30,000 to 75,000 volts will produce a 1-inch spark. Or 12,000 to 30,000 volts will produce a 1-centimeter spark. Although there is some uncertainty about the precise value of the voltage, sparks of any appreciable length are not caused by a few tens or hundreds of volts. These sparks are caused by many thousands of volts.

It is very common during the winter to get "shocked" when touching a doorknob after walking across a rug. If you touch the doorknob slowly and estimate the length of the spark, you can calculate the voltage that was on your body. A half-centimeter spark means that the voltage was between about 6,000 and 15,000 volts. In summer, the higher humidity in the air moistens the surface of your shoes. The moisture provides a low-resistance circuit that prevents charge buildup.

Lightning is much more complex than the simple spark caused by the electrical breakdown of air. The lower part of a cloud (usually a "thunderhead") is charged with an excess of electrons, as shown in Fig. 6-14. The electrons are brought to the lower part of the cloud by the action of swirling, turbulent air masses. The negatively charged electrons on the cloud and the positively charged ions on the ground attract each other.

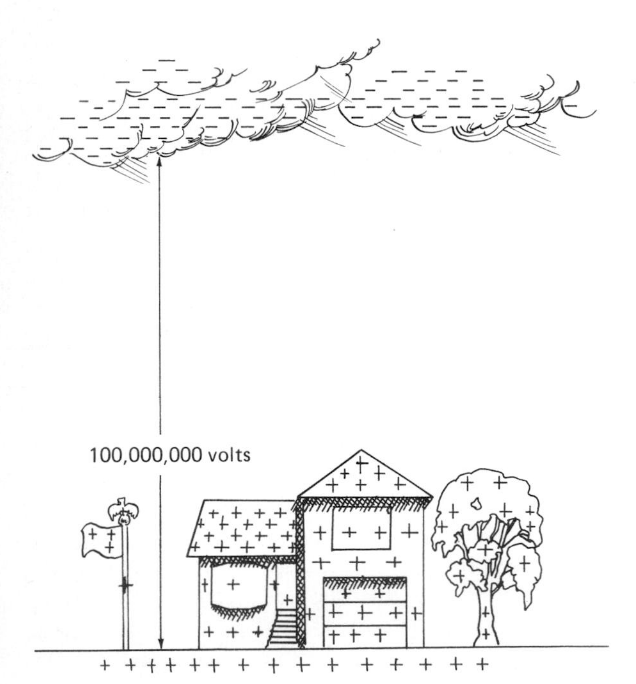

Fig. 6-14
The charge distribution just before a lightning bolt strikes.

Fig. 6-15
The return stroke of lightning is highly visible. (*Photograph by M. E. Warren. From Photo Researchers, Inc.*)

Electrical attraction causes electrons to leave the cloud and move through the air. The electrons follow a jagged course from the cloud as a low resistance pathway is established. This path, which is a channel filled with electrons, is known as a *step leader.* It is not bright enough to be seen with the naked eye. The step leader continues to grow in length, often reaching 2 or 3 miles.

As the step leader approaches the ground, a building, or a tree, the force of electrical attraction between the step leader and the positive charges increases greatly. The end of the step leader makes contact with the positive charge, and a tremendous current flows violently through the jagged pathway. The current, which is the movement of electrons from the cloud, neutralizes the positive charges. This violent flow of electrons is known as a *return stroke.* The return stroke is spectacularly bright and is called a lightning bolt in everyday conversation (see Fig. 6-15).

We have described lightning in its most common form. Usually the lower parts of the clouds are negative, and positive charges accumulate on the ground. But storms are capricious, and occasionally lightning is produced from positive charges on the clouds and negative charges on the ground. And quite often, lightning strikes within a single cloud. These bright bolts of lightning are usually hidden from our view.

6.7 MAGNETISM

In Chap. 1 we saw that a mass gives rise to a gravitational field. In this chapter it was seen that a charge sets up an electric field. Another important field is the *magnetic field.* We have all experienced magnetic fields when playing with magnets. Let us examine magnetism more closely.

The basic idea of any field is that a force is experienced by an object placed in the field. The source of the field is a similar object some distance away. The field transmits the force between the two similar objects. For masses, the gravitational field transmits the force. For electric charges, the electric force is transmitted by the electric field. For magnets, it is the magnetic field that transmits the magnetic force.

The specific kind of field that is generated depends upon two factors: (1) the type of objects and (2) the relative motion between the two objects. In the case of electric charges, we can identify three different fields, depending on the state of motion.

In the first case the charges may be at rest, that is, there is *no* relative motion between the charges. Then an electric field (or more precisely, an electrostatic field) exists in the space surrounding the stationary charges. If there are no other forces, then the charges will accelerate as described by Newton's second law of motion.

In the second case the electric charges are in motion with constant speed, as, for example, in a wire carrying direct current. A compass will show a curious behavior in the presence of the wire. Instead of pointing north, the compass needle will be affected by the direct current. The compass needle near the wire points at right angles to the wire, as shown in Fig. 6-16. As a result of this experiment, we conclude that a *magnetic field* exists in the space surrounding moving charges. The electric charges create a magnetic field simply by being in motion. The electric field of the charges does not disappear if the charges are put in motion. Rather, the electric and magnetic fields are both found in the region of moving electric charges.

When direct current exists in a wire, the electrons are in uniform motion. This motion of the electrons causes a magnetic field around the wire. The dc can be turned on or off, and so the magnetic field can also be turned on or off. A practical example of this kind of magnet is an electromagnet. Figure 6-17 shows an electromagnet being used in industry to load and unload scrap iron and steel.

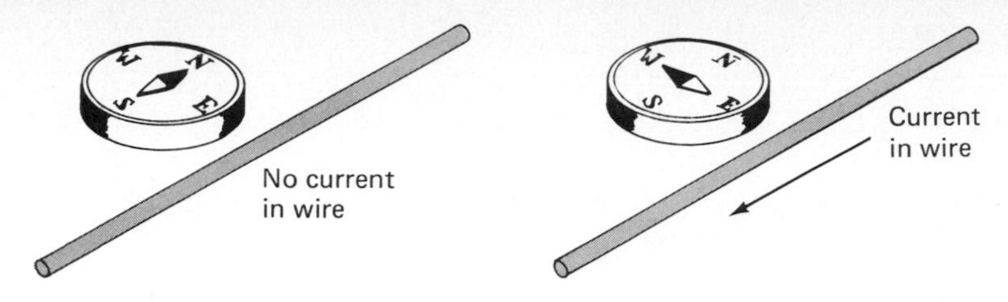

Fig. 6-16
The needle of a compass near a wire carrying a large direct current points at right angles to the wire instead of north.

The orbiting electrons in an atom or molecule can be viewed as tiny direct currents. Orbiting electrons cause their own magnetic fields. Ordinarily these separate microscopic magnetic fields are randomly oriented. The net result is that these random magnetic fields cancel each other out. Most materials thus show little or no overall magnetic field either in a single atom or in a number of atoms.

An exception to this kind of behavior is the group of elements known as *ferromagnetic* elements. These are the metals iron, nickel, and cobalt. Objects made from these metals or their mixtures show large and easily observed magnetic fields. The motion of the electrons in such substances is not completely random in its orientation. Instead, the separate magnetic fields of individual electrons combine to form a macroscopic magnetic field. In other words, ferromagnetic materials are used to make magnets. Figure 6-18 compares the random magnetic fields (schematically shown as arrows) of an unmagnetized ferromagnetic material with a magnetized ferromagnetic material.

How do the *micro*scopic magnetic fields arise in a ferromagnetic material? Ferromagnetic fields are caused by *electron spin.* You can imagine electron spin to be the result of the rotation of an electron about an axis through its center. In a simple picture the electron spin is similar to the spinning of a ballerina or ice skater. A spinning electron is a moving charge and thus produces a magnetic field. For most materials the magnetic fields caused by each electron spin cancel one another. In ferromagnetic materials, however, cancellation is not complete and a *macro*scopic magnetic field is produced.

Ferromagnetic materials can be made into permanent magnets in any shape. Traditionally many magnets have been made of iron in the shape of a rod or a horseshoe. But modern magnets are frequently made as discs, rings,

Fig. 6-17
Magnetic fields can be turned on and off by starting and stopping the current in an electromagnet. (*Photograph from Leo de Wys, Inc.*)

and many other special-purpose shapes. Magnetic materials can also be embedded in rubber to make flexible magnets. And some extra-strong magnets, called ceramic magnets, are brittle and break like china cups!

A magnet has two sides called *north* and *south.* Since magnets were first used as compasses to navigate, the two types of magnetism were named for the earth's poles, the north pole and the south pole. Convection of charges in the molten core of the earth sets up a magnetic field. The magnetic field aligns the small magnet in a compass on the surface of the earth. Early explorers and experimenters may have thought their compasses were pointing to magnetic objects at the poles. But the earth's south and north poles are

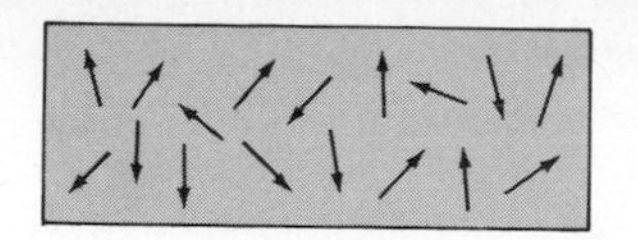

Unmagnetized

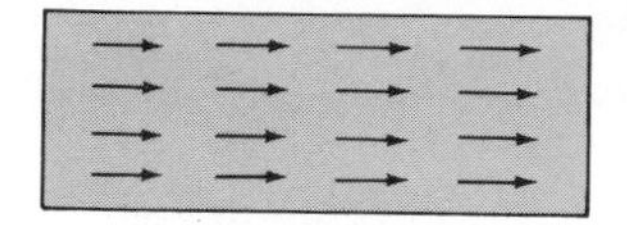

Magnetized

Fig. 6-18
In a ferromagnetic material the contributions of microscopic magnetic fields can be made to align, producing a bar magnet.

Table 6-5
Generation of fields for various states of motion of an electric charge

State of motion of charge	Type of field
At rest (motionless)	Electric (electrostatic) field
In steady motion (direct current)	Magnetic field
In oscillating motion (alternating current)	Electromagnetic waves

barren, wind-swept regions. The origin of the earth's magnetic field lies far beneath the surface of the earth in its swirling charged, molten core.

The force between magnetic poles is similar to that of charges: "Opposites attract, and likes repel." Even the mathematical form of the force law for magnets is similar to Coulomb's law for charges. But the similarity ends there. Charges can be isolated and separated from each other. Magnetic poles *cannot* be isolated and separated from each other. If you cut a bar magnet in two, you end up with two smaller magnets each with a north and south pole. It is impossible to cut a magnet into one north pole and one south pole.

So far we have seen that a motionless electric charge generates an electric field and that direct current generates a magnetic field. Now consider a third case. What does alternating current generate? While the charge undergoes vibration, there is a *varying* magnetic field and a *varying* electric field. Altogether it is called a *varying electromagnetic field* or *electromagnetic wave.* Electromagnetic waves emanate from an oscillating or vibrating electric charge. They travel outward in space. The behavior of these electromagnetic waves will be studied in detail in Chap. 7. Table 6-5 summarizes the origins of the three types of fields caused by various states of motion of electric charge.

6.8 ELECTRICAL SAFETY

It was inevitable that the increasing use of electricity would present increasing electrical dangers. The problem is not one of bare wires. Instead, the insulation on a wire often ages and breaks. Breaks in the insulation usually occur out of sight, for example, inside the housing of an electric mixer, a desk lamp, a hand drill, a hair dryer, or a lawn mower.

Once the insulation is broken, it is possible for the current to pass throughout the metal housing, as well as through the wiring. If you were to hold a defective appliance and touch a water pipe or furnace vent, you would "ground" the appliance and probably electrocute yourself. A *ground* is a short circuit to the ground, which is capable of handling all the current in the circuit. Grounding an appliance by allowing current to go through your body sets up the condition needed for your electrocution!

In order to provide some protection against unwanted ground connections, modern household, commercial, and industrial electrical wiring has a built-in grounding circuit. Older buildings have electrical outlets that will accept two-prong connections. The danger with two-prong wiring is the lack of a separate grounding circuit. Modern homes and buildings have a third hole in the electrical outlet. The third hole connects an appliance to a wire that goes to the service entrance panel or fuse box. At this point the ground wire is either fastened to a pipe driven into the ground or fastened to a cold-water pipe. This connection provides a low-resistance electrical path for any unwanted current.

If for some reason you have a short circuit, there will be an immediate increase in the current through the wiring. The voltage will remain at the rated 120 volts, but the current can easily surge from a few amperes to dozens of amperes. Left alone, this high current will heat the wiring and can cause a fire. Your electrical protection, however, is a *fuse* or *circuit breaker*, both of which are shown in electrical boxes in Fig. 6-19. These devices are designed to limit the current to a safe level, typically, 15 or 20 amperes. If a fuse is burned out, it may be that too many appliances are plugged into the same circuit. Unplug some of the appliances and replace the fuse. *Never* replace a fuse with a coin. A coin will *not* limit the current to a safe level. Houses have burned down because of this foolish action.

In addition to the danger of fires, a short circuit with you as part of the circuit can electrocute you. The important factor is the amount of current flowing through the body. Table 6-6 lists the physiological effects of various currents flowing between your arms.

Low values of current produce a tingling sensation. If the current should be larger, the electrical jolt is painful and sometimes paralyzes the muscles. When your muscles become paralyzed, you cannot open your hands and break the circuit.

Table 6-6
Physiological effects of electric current

Current (in amperes) passing through body	Physiological effect
0.001	Threshold of feeling
0.005	Accepted maximum harmless current
0.007 to 0.015	"Can't let go" current (paralysis)
0.050	Pain, possible fainting, and exhaustion
0.1 to 0.3	Ventricular fibrillation and death
Above 0.3	Ventricular paralysis and burns

Fig. 6-19
Fuses or circuit breakers provide homes with protection against overloaded circuits. (*Photographs by Korrine Nusbaum.*)

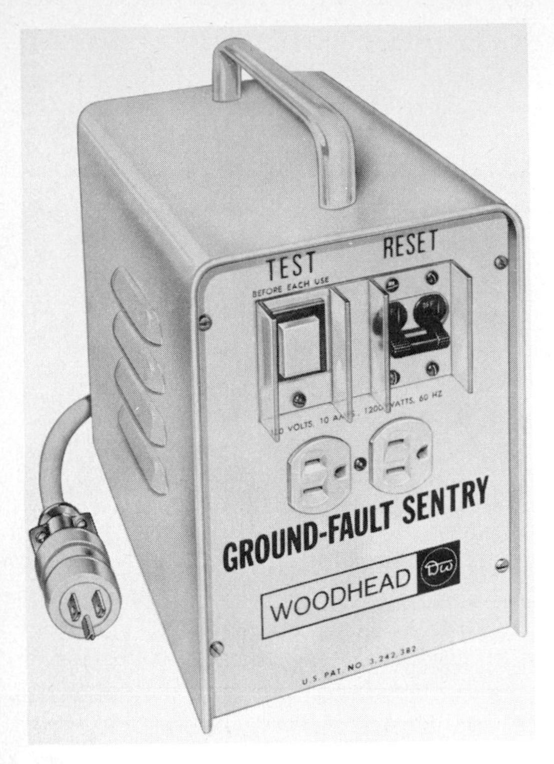

Fig. 6-20
A ground fault interrupter provides people with protection against electrocution. (*Photograph courtesy of Woodhead.*)

If the current should be even larger, the current flowing through your body can disrupt the beating of your heart. Your heart stops its rhythmic pumping and flutters uselessly (so-called ventricular fibrillation). During ventricular fibrillation, blood stops flowing and you die from lack of oxygen in the brain after a few minutes. Surprisingly, currents larger than about 0.3 amperes paralyze your heart instead of causing ventricular fibrillation. Occasionally a person may survive a large brief current, while a smaller sustained current will cause death.

Since the current needed for electrocution is much less than the limit set by a fuse or circuit breaker, is there any protection available? The resistance of *dry* skin is normally quite high, and this provides some protection. The high resistance of dry skin prevents much current from passing through your body. However, skin resistance is *not* a dependable protection, because the resistance depends greatly upon perspiration and skin oils.

Some modern circuits are protected by a special device known as a *ground fault interrupter* (shown in Fig. 6-20). A ground fault interrupter detects small leakage currents that go through a person in a short circuit to ground instead of through the normal circuit. If as little as 0.005 ampere leaks out of the normal circuit, presumably through an unwanted short circuit, the ground fault interrupter breaks the circuit and stops all the current. These devices have already saved many lives.

REVIEW You should be able to define the following terms: electric field, conductors, current, resistance, voltage, and circuit.

You should be able to state and give examples of Coulomb's law, Ohm's law, and the formula for electric power.

PROBLEMS

1. A 3-cell flashlight including the light bulb has a resistance of 20 ohms. Calculate the current when the flashlight is turned on.
2. A 2-cell flashlight has a resistance of 15 ohms. Calculate the current when the flashlight is turned on?
3. A 12-volt car battery provides a current of 48 amperes when the car is being started. Calculate the resistance of the starter.
4. A hot plate is connected to the 120-volt outlet in a house. If 10 amperes of current exist in the heating coils, what is the resistance of the hot plate?
5. An automobile spark plug typically operates at a voltage of 25,000 volts and a current of 0.001 ampere. What is the power used in the spark plug?
6. An electric clothes dryer operates on a voltage of 240 volts and a current of 16 amperes. Calculate the power to run the clothes dryer.
7. A 100-watt light bulb is used in a desk lamp plugged into a 120-volt wall outlet. What current exists in the light bulb?
8. A refrigerator uses 330 watts of power when connected to a wall outlet. What is the current used by the refrigerator?

DISCUSSION QUESTIONS

1. Describe the electric forces that exist between positive and negative charges.
2. How can electric forces be used to collect dust?
3. What is an electric field?
4. Do all materials conduct electricity?
5. What voltages are usually provided by a flashlight battery? A wall outlet in the home? An automobile battery?
6. If a battery were available that did not "run down" (that is, never used up all its available electrical energy), could such a battery be used in a perpetual motion machine?
7. How much current is usually available from a wall outlet in the home?
8. How is the operation of a fuse related to current?
9. Since most household appliances require the same voltage, why do these appliances require different currents?
10. What information is needed to calculate the electric power used by an appliance?
11. What is the difference between kilowatt and kilowatthour?
12. Do high-power appliances always drastically increase your electricity bill?
13. What is meant by dc electricity? By ac electricity?
14. How much voltage is required to cause sparks in air?
15. How are magnetic fields related to electric current?
16. Does an extension cord get warm if it is being properly used?
17. Does the current in a refrigerator motor warm the refrigerator?
18. What electrical factor is most important in determining electric shock hazard?
19. What amount of electricity is usually considered to cause death?
20. Does lightning often strike from one cloud to another?
21. Distinguish between the leader stroke and return stroke for lightning.

ANNOTATED BIBLIOGRAPHY

One of the best introductions to static electricity is in A. D. Moore, *Electrostatics,* Anchor Books, Doubleday & Company, Inc., Garden City, N.Y., 1968. This small paperback has some fascinating reading. We'll bet if you try pages 7 to 14 that you'll try more! Another paperback in the same series (Science Study Series), *Magnets* by Francis Bitter, has a discussion of the relationships between electricity and magnetism. It begins on page 33 and is written very clearly.

T. Bernstein gives additional details on safe and unsafe amounts of electricity in "Effects of Electricity and Lightning on Man and Animals," *Journal of Forensic Sciences,* vol. 18, 1973, pp. 3–11.

An excellent summary of lightning can be found in a few pages in *The Encyclopedia Britannica,* Chicago, 1974. See "Lightning" by R. E. Orville. For some special sounds to hear with lightning see the article "Thunder" by A. A. Few, *Scientific American,* July 1975, p. 80.

Electrical wiring in the home is subject to many rules and regulations, mostly for safety reasons. If you ever do your own wiring, consult *Wiring Simplified* by H. P. Richter. The book has been published in many paperback editions by Park Publishing, Inc. Look for the most recent edition.

7 ELECTROMAGNETIC WAVES

An oscillating electromagnetic field travels through space as an electromagnetic wave. The range of wavelengths of electromagnetic waves extends from distances longer than a city block to distances smaller than individual atoms. These waves with different wavelengths exhibit different behavior and have different applications.

7.1 THE SUN'S VISIBLE RADIATION

Sunbathers know that the sun is a tremendous source of heat. But what exactly *is* heat from the sun? In line with the discussion in Chap. 4, a partial answer is that solar heat is radiation. A better understanding of radiation comes from an examination of the electromagnetic wave nature of sunlight.

The sunlight that you see is actually electromagnetic waves with wavelengths shorter than the thickness of the finest thread. These small waves are capable of stimulating the human eye to produce the sense of sight. Visible light is nothing more than very short electromagnetic waves that stimulate the retina of the human eye. The waves striking the retina can be slightly different in length. This difference will be seen as different *colors* by the eye. When you see different colors, electromagnetic waves of different wavelengths are entering your eyes.

Of course, the light need not come from the sun in order to be seen. An ordinary light bulb emits visible electromagnetic waves. And a neon advertising sign glowing brightly over a shopping center is also sending out visible electromagnetic waves.

Our eyes can see many different colors of light. In order to make some sense out of all the different colors surrounding us, we think about changing the wavelength of electromagnetic waves in a systematic way. The electromagnetic wave of longest visible wavelength is seen as the color red. As the wavelength is shortened, the color of the light changes, as shown in Fig. 7-1. From red light, the color changes to orange, then yellow, green, blue, and finally violet. Each color of light blends smoothly and continuously into the next color. There are not any sudden changes in the color as the wavelength is gradually changed.

For the sake of discussion, six colors with common names have been isolated from their neighboring colors.

Spectrum of colors *The sequence of six colors ranging from red to violet is called the spectrum of colors.*

The spectrum of colors is produced by changing the wavelength of visible electromagnetic waves, starting with red as the longest visible electromagnetic wave and ending with violet as the shortest visible electromagnetic wave.

All colors, tones, tints, and hues are made up from the basic spectrum of colors. In most cases the combination of electromagnetic waves is very complicated. Of the infinite number of possibilities, each combination is made up of various amounts of different parts of the full spectrum.

Two cases of color identification, however, are especially simple. One is the color white. When equal amounts of each part of the spectrum of colors enter the eye, we do not see a jumble of distinct colors. Instead, the brain combines all the visual sensations to produce the color white.

White *The entire spectrum of colors is mixed in equal amounts to produce white light.*

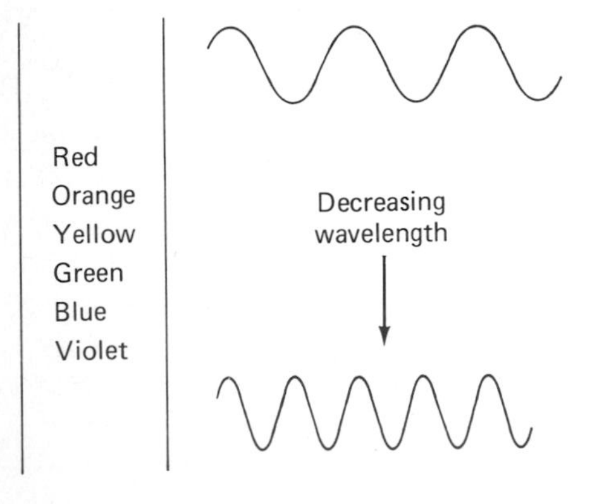

Fig. 7-1
The wavelength of red light is about 0.63 millionths of a meter and the wavelength of violet light is about 0.42 millionths of a meter.

A color identified as "off-white" simply has a little more of one part of the visible spectrum. For example, the color pink has an extra amount of the red part of the spectrum. And the color called antique white usually contains an extra amount of yellow.

Another color that is an exception to the usual complicated mixture of electromagnetic waves is the color black.

Black *Black as a color is the total absence of electromagnetic waves.*

When you see a pure, flat black color, then no electromagnetic waves are entering your eyes. The printing on this page is such an example. You see the white of the page surrounding the printing, so you know the shape of the black letters. However, if the page, the book, and everything in the room were pure, flat black, then you would see nothing. You are able to see black objects only because of their surroundings. Sometimes a black object can be seen because of an uneven surface texture, dust on the surface, or a combination of these effects. A totally darkened room or cave would be described as "pitch black" if no electromagnetic waves enter the eye.

7.2 THE SUN'S INVISIBLE RADIATION

The spectrum of colors is unique insofar as these electromagnetic waves can be seen as visible light. What about the electromagnetic waves outside the range of wavelengths making up the spectrum of colors? Such waves are not visible, but, nevertheless, are emitted from the sun and accompany visible sunlight.

Infrared radiation *If invisible electromagnetic waves have wavelengths longer than red light, they are known as infrared radiation.*

Ultraviolet radiation *If invisible electromagnetic waves have wavelengths shorter than violet light, they are known as ultraviolet radiation.*

In this sense, infrared means "below the red" (in frequency) and ultraviolet means "beyond the violet" (in frequency). Both infrared and ultraviolet electromagnetic waves are invisible to human eyes, although these forms of radiation can cause other sensations and effects.

One effect of infrared radiation on the skin is to produce a warming sensation. The sun is one of the most common sources of infrared radiation. In this case, the infrared radiation is accompanied by visible light. So we get both heat and light from the sun. But pure infrared radiation entering your eyes would only make them warmer. Infrared radiation cannot be seen.

Another source of infrared radiation is a piece of hot metal. If metal is taken from the glowing charcoal of a barbecue, the metal will often be red-hot, which would be at a temperature of about 750°C. At this temperature, visible red light is being emitted as well as invisible infrared radiation. As the metal cools, the reddish color vanishes and the natural color of the metal, usually black, returns. However, the black metal can still have a temperature of several hundred degrees and be too hot to handle with bare hands. The hot black metal is still emitting electromagnetic waves in the form of infrared radiation (see Fig. 7-2). The infrared radiation is absorbed by the skin of your hands, which are warmed up. Human beings have nerve endings that respond to conducted heat rather than radiated heat. These nerve endings in the skin directly sense the increase in temperature and therefore indirectly the infrared radiation.

Although human beings are unable to see infrared, snakes have special receptors that directly sense infrared radiation. A snake can hunt warm-blooded animals, such as mice, on completely dark nights by sensing the infrared electromagnetic waves emitted by the small mammals, as shown in Fig. 7-3.

Infrared radiation is not unique in its ability to raise the temperature of objects. The visible component of sunlight can also warm an object that absorbs light (called *opaque*). Sunglasses not only cut down on the glare on a

Fig. 7-2
Increasing the temperature of the horseshoe increases the emission of infrared radiation. (*Photograph from United Press International.*)

Fig. 7-3
Infrared receptors are located in small depressions between the snake's eyes and nostrils.

sunny day but also cool your eyes by absorbing some of the solar radiation. When infrared radiation is filtered out of sunlight, the remaining visible light is quite capable of warming an object. For example, a glass-enclosed porch can be comfortably warm even in the winter. Long-wavelength infrared from the sun is unable to pass through the glass, but visible light is transmitted. This visible sunlight can be easily felt on your bare arm inside a glass-enclosed porch.

The objects inside a glass enclosure absorb some visible light, become warmer, and so emit infrared radiation. But the infrared cannot escape through the glass, because glass is generally opaque to much of the infrared.* The infrared trapped in the glass enclosure causes the temperature to rise.

*Some materials are just the opposite of glass. The elements silicon and germanium are opaque to visible light (you cannot see through these materials) but are transparent to infrared.

We understand you meant to be kind in taking your dog with you today, but you could be risking his life.

On a hot summer day the inside of a car heats very quickly. On an **85** degree day, for example, the temperature inside your car--with the windows slightly opened--will reach **102** degrees in **10** minutes. In **30** minutes it will go up to **120** degrees. On warmer days it will go even higher.

A dog's normal body temperature is 101.5 to 102.2 degrees Fahrenheit. A dog can withstand a body temperature of 107-108 degrees Fahrenheit for only a very short time before suffering irreparable brain damage--or even death. The closed car interferes with the dog's normal cooling process, that is, evaporation through panting.

IF YOUR DOG IS OVERCOME BY HEAT EXHAUSTION, YOU CAN GIVE IMMEDIATE FIRST AID BY IMMERSING HIM OR HER IN COLD WATER UNTIL BODY TEMPERATURE IS LOWERED.

Published by **Animal Protection Institute**
P. O. Box 22505, Sacramento, Ca. 95822

Fig. 7-4
A warning to pet owners.
(Photograph courtesy of Animal Protection Institute.)

Such heating occurs in a greenhouse and can be used to extend the growing season of plants and flowers.

Greenhouse effect *The trapping of infrared radiation inside a glass enclosure is known as the greenhouse effect.*

The greenhouse effect also occurs in automobiles with closed windows, especially in the summer. Opening a window not only lets cool air into a hot automobile but also lets infrared radiation out of the automobile (see Fig. 7-4).

The third main electromagnetic component of sunlight is ultraviolet radiation (see Fig. 7-5). Ultraviolet is invisible, just as infrared. But unlike infrared, the effect of ultraviolet on the skin is not immediately noticed and is more complicated. If pure ultraviolet radiation were to shine on your skin for several minutes, there would be little noticeable heating as in the case with infrared or visible light. Some time later, depending on the exposure time and the condition of your skin, there would be a reddening of the skin. Ultraviolet radiation does not physically warm the skin. Instead, ultraviolet radiation causes a chemical reaction (bonding changes) in the skin. The most obvious effect of exposure to ultraviolet radiation is a sunburn. Some days later there may be a darkening of the skin, that is, a suntan.

Occasional exposure to ultraviolet radiation can be beneficial. The tanning of one's skin is thought by some to be a sign of good health. So having a suntan can increase one's psychological well-being. A more impor-

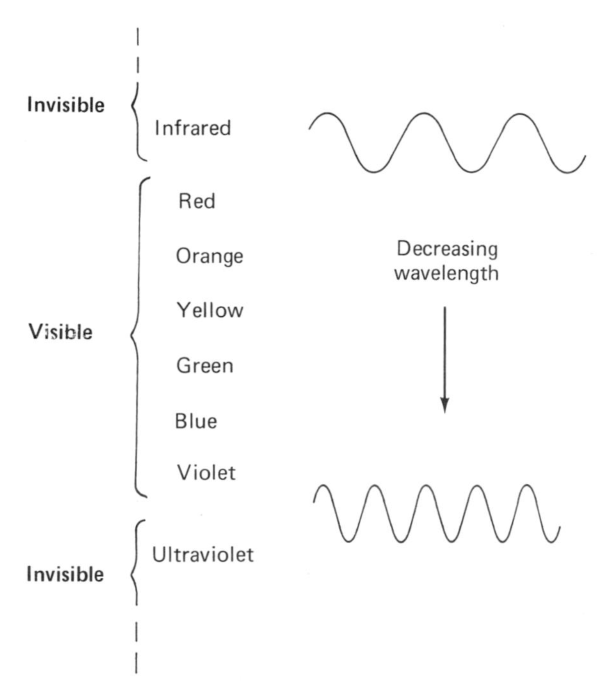

Fig. 7-5
The wavelength of the radiation decreases from the infrared to the ultraviolet.

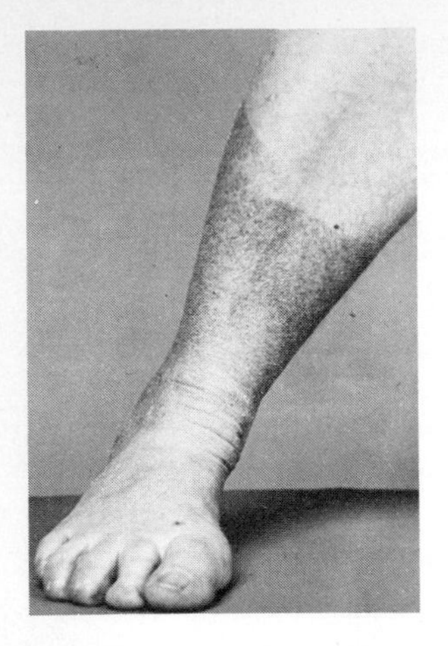

Fig. 7-6
Which component of solar radiation causes a sunburn?
(Photograph from Lester V. Bergman and Associates, Inc.)

tant effect of exposure to ultraviolet radiation is to produce vitamin D naturally in your body. But there are other ways of obtaining your required vitamin D. You probably obtain D from that which has been artificially added to milk.

Like many good ideas, suntanning is overdone. Young people (and older people trying to look young) expose their skin to too much ultraviolet radiation (see Fig. 7-6). Many people with light complexions have experienced the pain of overexposure to ultraviolet radiation. Some have also known the peeling of skin that follows an overexposure to ultraviolet radiation. What is not so well known is that repeated overexposure to ultraviolet radiation does increase the chance of developing skin cancer.

Skin cancer depends not only on overexposure to ultraviolet radiation but also on several other variables, including the hereditary background of the individual. So not everyone with a light complexion develops skin cancer. Nevertheless, it is known that many farmers and sailors develop skin cancer because of their outdoor occupations. Nowadays, the incidence of skin cancer is rapidly increasing in all occupational groups, principally due to increasing exposure to ultraviolet radiation from the sun in order to obtain a deep tan.

7.3 FEATURES OF ELECTROMAGNETIC WAVES

Table 6-5 in Chap. 6 summarized the three types of fields generated by an electric charge in different states of motion. In this chapter our attention is centered on the electromagnetic waves that radiate from an oscillating or vibrating charge. The electric and magnetic field components of an electromagnetic wave both oscillate perpendicular to the direction of travel of the wave. In other words, an electromagnetic wave is a *transverse* wave. If we could see the fields at some instant, they would appear as in Fig. 7-7.

Waves on a rope are also transverse waves. These waves are disturbances transmitted by the rope. But light and other electromagnetic waves can be transmitted across the empty space between the sun and the earth. What then is the medium, and what is "waving" for electromagnetic waves?

Empty space, either in a vacuum chamber on the earth or in outer space, is capable of possessing and transmitting any kind of a field. A perfect vacuum may have an electric, magnetic, or gravitational field. There is no need for a substance (material medium) to transmit a field from one place to another. For example, throughout the universe, there are gravitational fields. These fields extend to regions of outer space that are empty of atoms and molecules. These same regions of space can also contain electric fields, magnetic fields, and electromagnetic waves. The oscillating, moving, transverse electric and magnetic fields are *themselves* the electromagnetic wave.

All electromagnetic waves, whether visible light, infrared, ultraviolet, or waves yet to be discussed, move at the *same* very high speed in outer space. This speed common to all electromagnetic waves was first measured for light,

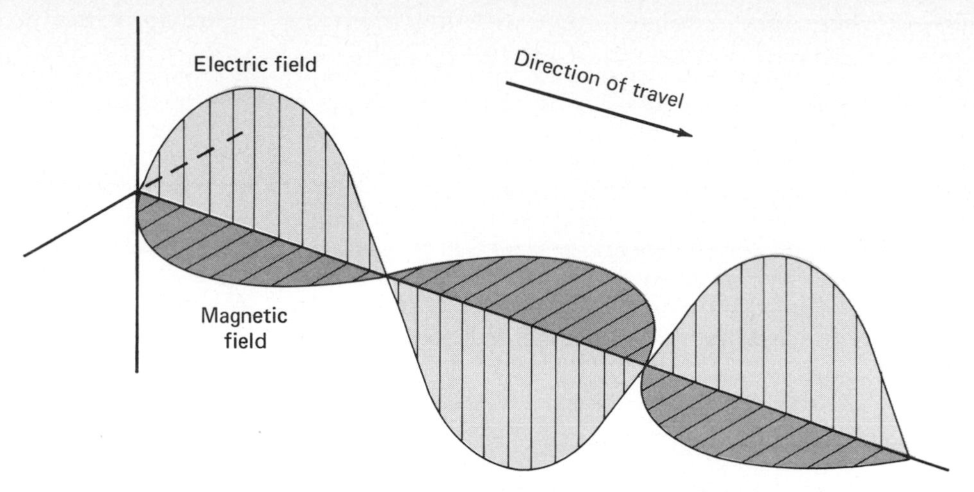

Fig. 7-7
The electric and magnetic fields oscillate perpendicular to each other and to the direction of travel of an electromagnetic wave.

and so it was named the *speed of light.* The speed of light is now known to be the speed for *all* electromagnetic waves in vacuum. The numerical value of this speed is about 186,000 miles per second, which is equal to about 300,000 kilometers per second. At this speed, an electromagnetic wave is able to travel the 150 million kilometers from the sun to the earth in about 8 minutes 20 seconds. The infrared, visible light, and ultraviolet electromagnetic waves all travel together at the same speed in the journey from the sun to the earth.

Electromagnetic waves can be described in terms of the wave properties introduced in Chap. 5, namely, the speed, frequency, and wavelength. The speed has already been given and is the same for all electromagnetic waves. But the frequency and the wavelength are different for each type of electromagnetic wave. All electromagnetic waves obey the wave formula given in Sec. 5-2, which can be rewritten here as

Wave formula for electromagnetic waves

$$c = f\lambda$$

where c = speed of light (about 300,000,000 meters per second in vacuum)
f = frequency measured in hertz (that is, cycles per second)
λ = wavelength measured in meters

The product of the frequency and the wavelength of any particular electromagnetic wave always equals the speed of light. Since the product of the frequency and the wavelength is always a constant number, when the frequency increases, the wavelength decreases. The higher the frequency of an oscillating electric charge, the shorter the wavelength of the electromagnetic wave that is generated. The converse is also true. When the frequency decreases, the wavelength increases. As an example, consider red and violet light. Since red light has a longer wavelength than violet light, it has a lower frequency than violet light.

COMMUNICATION WITH EXTRATERRESTRIAL LIFE

Do intelligent beings exist far away from the earth in the remote regions of space? Such a question is bound to provoke far more questions than answers. If such beings exist, can we communicate with them? Will they respond? Might they be more intelligent than we? It seems that such questions will remain unanswered in the twentieth century. Yet because of human curiosity, attempts are being made to find out.

Perhaps the most important part of "communication" is the message or information. To communicate with another human being we send letters, use the telephone, or talk in person. And we express our love, anxieties, happiness, or other feelings. But these kinds of messages may have limited usefulness. Perhaps not all intelligent beings would understand. We need a more universal message.

Some physical objects may also be useless. A blade of grass and a hammer seem common enough to us. But they tell nothing about our location in the universe. What is needed is something which any advanced civilization understands. Most people agree that this is the subject of astronomy.

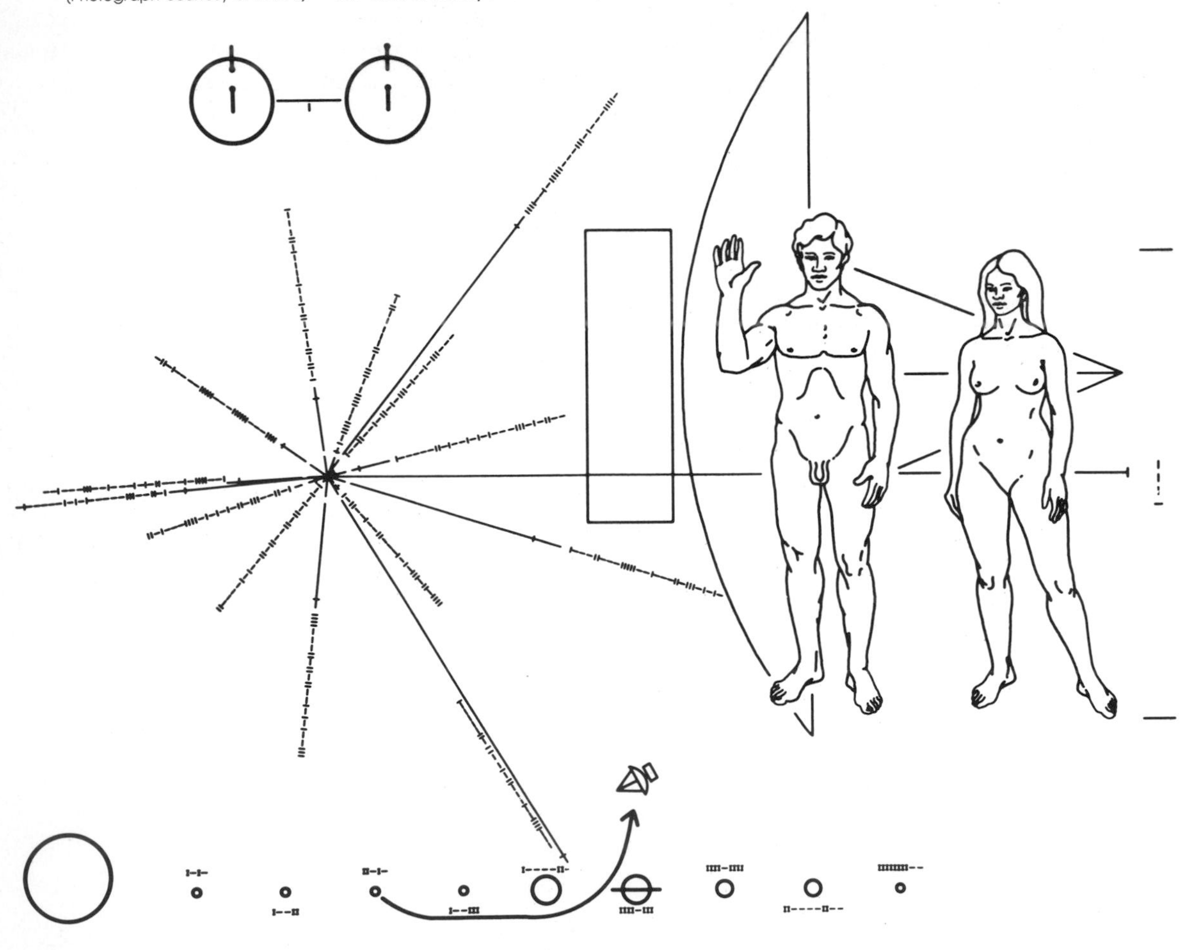

(Photograph courtesy of NASA.)

An attempt to communicate by the use of astronomical information was started by the United States in 1972. In that year the spacecraft Pioneer 10 was launched on a journey that would lead out of our solar system and toward the constellation Taurus. Pioneer 10 will pass the orbit of Pluto in 1987 and then take about 8 million years to reach Taurus!

On board Pioneer 10 is a copy of the plaque shown on the opposite page. The sunburst design on the left side contains information about pulsars. Pulsars are astronomical objects which send out very strong radio waves in pulses. The pulsations repeat with very accurate periods. It is very much like a lighthouse sending out flashes of light at regular intervals. In this case, however, the pulsar sends out radio waves, not light. The bursts of radio waves come at intervals which vary from one pulsar to another. One pulsar has been found which sends out 30 pulses each second. And others send out waves each second or each half second. Pulsars are thought to be very small when compared with ordinary stars. Typical sizes are shown in the diagram below. It is not drawn to scale!

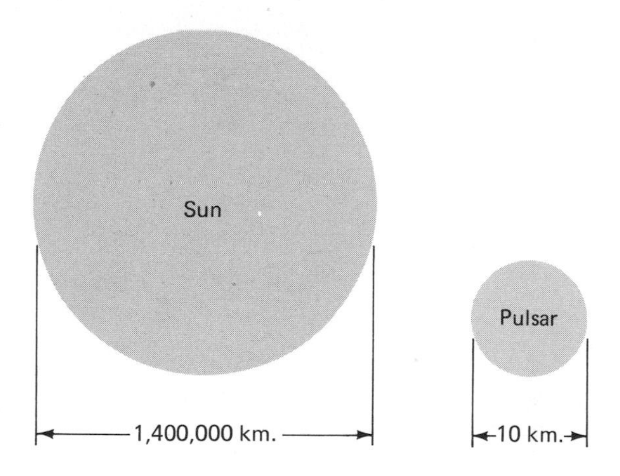

The sunburst design on the plaque of Pioneer 10 records data on 14 pulsars. The length of each line shows the relative distance from us. And the direction of each line shows the bearing relative to our solar system. It is believed that any advanced civilization will have knowledge of pulsars. And that information should let them decode the message.

In the future we can expect many more attempts to communicate with intelligent beings. Surely the most exciting aspect is hoping for a return message.

Table 7-1.
The electromagnetic spectrum

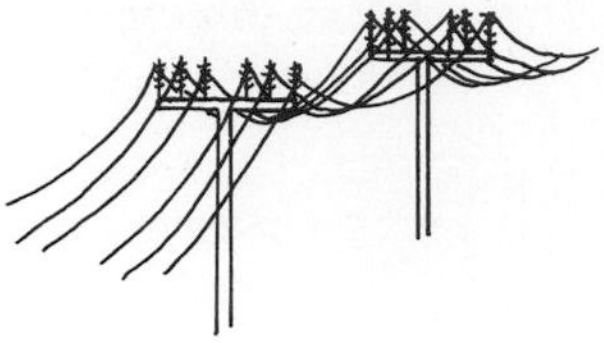

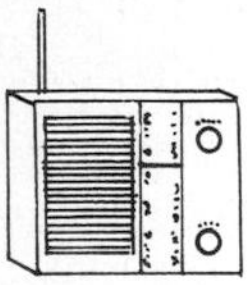

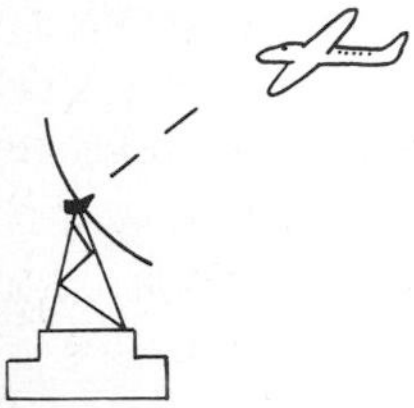

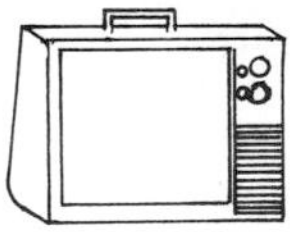

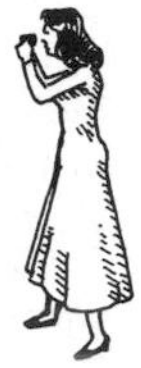

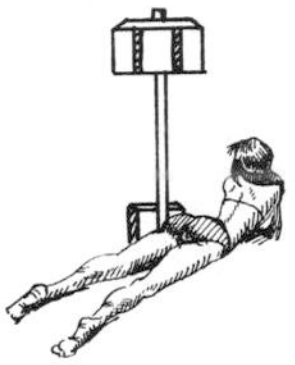

	Wavelength in meters	
	10,000,000	AC electric waves
	1,000,000	
	100,000	
	10,000	
	1,000	
am radio		
cb radio	100	
fm radio	10	television
	1	
microwaves radar	$\frac{1}{10}$	
	$\frac{1}{100}$	
	$\frac{1}{1,000}$	
	$\frac{1}{10,000}$	infrared
	$\frac{1}{100,000}$	
	$\frac{1}{1,000,000}$	
visible light		
	$\frac{1}{10,000,000}$	
	$\frac{1}{100,000,000}$	ultraviolet
	$\frac{1}{1,000,000,000}$	
x rays	$\frac{1}{10,000,000,000}$	
	$\frac{1}{100,000,000,000}$	
	$\frac{1}{1,000,000,000,000}$	gamma rays
	$\frac{1}{10,000,000,000,000}$	

7.4 ELECTROMAGNETIC WAVE SPECTRUM

Just as there is a spectrum of colors ranging from red to violet, there is a larger spectrum of electromagnetic waves that includes the spectrum of colors.

Electromagnetic wave spectrum

The totality of all electromagnetic waves is known as the electromagnetic wave spectrum.

All electromagnetic waves are identical insofar as they are transverse electric and magnetic fields which travel at the speed of light. But electromagnetic waves in various regions of the spectrum have different frequencies, which means the corresponding wavelengths are also different. The common names for different regions of the electromagnetic spectrum are shown along with their frequencies and wavelengths in Table 7-1. This spectrum is smooth and continuous just like the spectrum of colors. There are no sharp dividing lines between the various regions.

Theoretically, all the various phenomena shown in Table 7-1 could be generated simply by oscillating an electron at higher and higher frequencies in a wire. The wire would be a transmitter of electromagnetic waves known successively as ac electric waves, AM radio, police radio, FM radio, TV, microwaves, infrared waves, visible light, ultraviolet waves, x-rays, and gamma rays. As a practical matter, quite different techniques are used to generate the various kinds of electromagnetic waves.

7.5 FOOD PROCESSING

Electromagnetic radiation is used beneficially in many applications. Two of the more interesting ones are the use of microwaves for cooking in the home and the use of ultraviolet radiation to kill bacteria. Both of these uses will be discussed in this section.

COOKING WITH MICROWAVES

Figure 7-8, which previously appeared as Fig. 4-14, illustrates the use of heat transfer by means of radiation to keep foods warm. The type of radiation used is infrared radiation, sometimes known as *heat rays.* Infrared is not the only electromagnetic radiation that will raise the temperature of food. Visible light could be used to heat food, but the bright light would be annoying. So most heat lamps emit infrared radiation. These lamps also emit some visible red light together with the invisible infrared. The red light serves as a warning that the infrared heat lamp is on and working. An infrared heat lamp cannot be used to cook food, since the radiation does not penetrate deeply into the food.

Food can, of course, be cooked using a gas or electric stove. These stoves raise the temperature of food by conduction and convection of heat. The temperature of food can also be raised by radiation. Cooking by radiation is being done more and more frequently using the microwave region of the electromagnetic wave spectrum. The practical device used for this purpose is called a *microwave oven,* which is shown in Fig. 7-9. A microwave oven cooks electronically using *microwaves.*

Fig. 7-8
Heat radiation from the lamps keeps the food warm. *(Photograph by John Harmon.)*

Fig. 7-9
Microwave ovens are fast and efficient. *(Photograph courtesy of Amana Refrigeration, N.Y.)*

Microwaves *Microwaves are electromagnetic waves with wavelengths longer than those of infrared radiation.*

When microwaves strike the food in the oven, they actually penetrate the food. The microwaves cause the molecules of the food to vibrate more rapidly, thus creating the high temperature that does the cooking. The high temperature is in the food. The oven itself remains cool.

The principal advantage of a microwave oven is speed—most foods are cooked 2 to 4 times as quickly as in a gas or electric oven. This is due to the penetration of the microwaves into the interior of the food. In an ordinary oven, heat energy reaches the interior of food by conduction, which is a much slower process. A microwave oven saves electrical energy also, both because the cooking time is shorter and because its power rating is lower than that of an electric oven.

Table 7-1 shows that microwaves and radar occupy the same region of the spectrum of electromagnetic waves. There is no physical difference between microwaves and radar. A microwave oven is simply a small radar transmitter enclosed in a metal housing. The radar or microwaves are not allowed to escape from the oven but are confined to the interior of the oven where the food is cooked.

FIGHTING BACTERIA WITH ULTRAVIOLET

We are constantly surrounded by bacteria in great numbers. Some bacteria and microorganisms are beneficial to us. Others are harmful. If we are healthy, our bodies can handle both kinds. The organisms which help us are utilized to make essential vitamins in our bodies. The harmful bacteria are killed within the body.

Ultraviolet radiation is an effective agent for killing bacteria, a bactericide. When this radiation is projected onto bacteria, the slaughter begins immediately. At first the numbers of bacteria are decreased. If the radiation is stopped at this point, the bacteria will simply reproduce and multiply again. But if the exposure to ultraviolet radiation is continued, all the bacteria will be killed. Ultraviolet rays are sometimes used in specially designed cabinets (see Fig. 7-10) to sterilize instruments.

Ultraviolet radiation is also used to prevent mold in storage bins. In one commercial use, ultraviolet lamps are installed inside tanks containing sugar dissolved in water. The growth of bacteria in the sugar solution is retarded by the ultraviolet radiation. The bacteria-free sugar solution can be blended later with other baking ingredients.

Dairy products are well known for their bacteria content. Butter, milk, yogurt, and cheese are all rich in bacteria. When these foods are processed, ultraviolet radiation is used extensively to keep the bacteria under control. But complete sterilization of these foods would be undesirable. The taste would be changed drastically.

Fig. 7-10
Ultraviolet radiation kills bacteria in a sanitizing storage cabinet. (*Photograph courtesy of Applied Ultra Violet Technology.*)

7.6 COMMUNICATIONS

One obvious way to communicate with another person is by talking. But the range of the sound of a voice is limited to a small distance unless some form of technology is used. The telephone is a common device for extending the range of the human voice. When you speak into a telephone, the pressure variations in the sound of your voice drive a small microphone. The purpose of the microphone is to generate an electrical pattern that reproduces the pressure variations in the sound of your voice. The electrical pattern is transmitted through a wire to another telephone, which contains a small speaker in the earpiece. The electrical pattern drives the speaker, and the sound of your voice is reproduced.

Communication by telephone is limited to those persons who are connected to the telephone wire. Sometimes there is a need to communicate with someone who does not have telephone service or to communicate with many different people in different places. Then "wireless" communication can be used. Wireless communication is simply a form of radio communication. In radio communication, an electromagnetic wave replaces the wire as the carrier of the signal representing the sound of a voice. The citizen's band (CB) radio, shown in Fig. 7-11, is an example of wireless communication of conversations. The transmission of voices is only one use of radio communication. Electromagnetic waves are also used to carry music and pictures, for example, in AM radio, FM radio, and television.

All forms of communication using electromagnetic waves are similar. The information to be communicated, whether a voice, music, or a picture, is changed into a complicated voltage pattern. The voltage pattern is then used to vibrate the electrons in a transmitting antenna. The resulting electromag-

Fig. 7-11
The citizen's band radio utilizes electromagnetic waves for communication. (*Photograph by David Strickler. From Monkmeyer.*)

netic wave radiates in all directions through the atmosphere and sometimes even into outer space. The electrons of the receiving antenna are then driven by the electromagnetic wave to reproduce the original voltage pattern. Some sort of electronic device converts the complicated voltage pattern back into a voice, music, or a picture. Let us look at the more common types of communication using electromagnetic waves.

AM RADIO

Transmission of information on an electromagnetic wave requires that some feature of the electromagnetic wave be changed. One feature of an electromagnetic wave that can be varied is the *amplitude*.

AM radio

When the amplitude of an electromagnetic wave is modulated, the radio communication is called AM radio. The letters AM stand for amplitude modulation, or in other words, the changing of the wave amplitude.

An AM radio station broadcasts an electromagnetic carrier wave at one assigned frequency. Before the audio signal representing conversation or

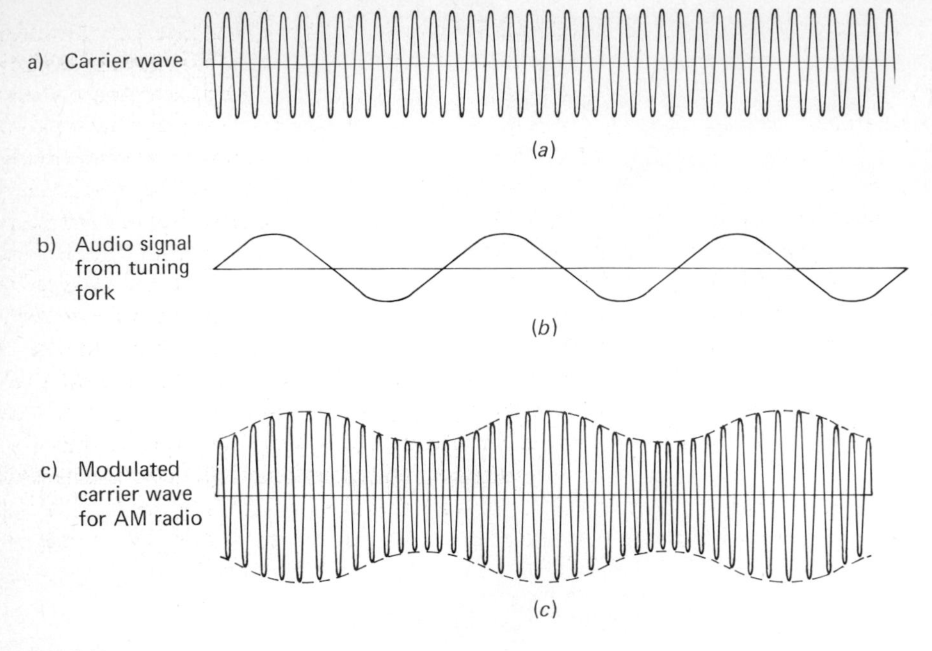

Fig. 7-12
Carrier wave, audio signal, and modulated carrier wave for AM radio.

music is added to the carrier wave, the electromagnetic wave can be imagined to look like the wave shown in Fig. 7-12*a*. Notice that the pure carrier wave has a constant amplitude and frequency. The audio signal of a vibrating tuning fork has the form shown in Fig. 7-12*b*. This audio signal modulates the amplitude of the pure electromagnetic carrier wave. Combining these two waves produces an altered electromagnetic wave (shown in Fig. 7-12*c*) that is broadcast from the radio station. The amplitude of the electromagnetic wave has been modulated by the audio signal, but the frequency remains unchanged.

The dial on your AM radio is typically marked with numbers between 5.4 and 16. These numbers show the range of carrier frequencies of the electromagnetic waves to be between 540 and 1,600 kilohertz. Every AM radio station is assigned a frequency within this range. The assignments are 10 kilohertz (10 kHz) apart. There are stations, for example, that broadcast at 910 kHz, 920 kHz, and 930 kHz. But no station operates at 925 kHz. Having unassigned frequencies between station broadcast frequencies prevents overlap in the reception of broadcasts.

The citizen's band radio used in trucks, cars, and vans is also a form of AM communication. The voice signal modulates the amplitude of the carrier wave. The frequencies, or so-called channels, for CB radio are much higher frequencies than the frequencies for AM radio. The 40 CB channels are located in the range between 26,965 and 27,405 kilohertz, or equivalently, 26.965 and 27.405 megahertz.

FM RADIO Another feature of an electromagnetic wave that can be varied in order to carry communications is its *frequency.*

FM radio *When the frequency of an electromagnetic wave is modulated, the radio communication is called FM radio. FM stands for frequency modulation, or the changing of the wave frequency.*

An FM radio station is assigned one fixed frequency for broadcasting. The pure electromagnetic carrier wave can be imagined to have the shape shown in Fig. 7-13*a*, which shows a wave with constant amplitude and frequency. An audio signal from a vibrating tuning fork is shown in Fig. 7-13*b*. In FM broadcasting it is the frequency of the electromagnetic carrier wave that is modulated by the audio signal. These two waves are combined to produce an electromagnetic wave as shown in Fig. 7-13*c*. The frequency of the electromagnetic wave has been modulated by the audio signal, but the amplitude remains unchanged.

FM radio dials usually show the stations numbered between 88 and 108. These numbers represent the range of carrier frequencies of the electromagnetic waves, which vary between 88 and 108 megahertz. From the wave formula the wavelengths of FM waves are found to range from 11.2 to 9.1 feet in length, or equivalently, from 3.41 to 2.78 meters. (Can you calculate these wavelengths? Remember that FM electromagnetic waves travel at a speed of 300,000,000 meters per second.)

Fig. 7-13
Carrier wave, audio signal, and modulated carrier wave for FM radio.

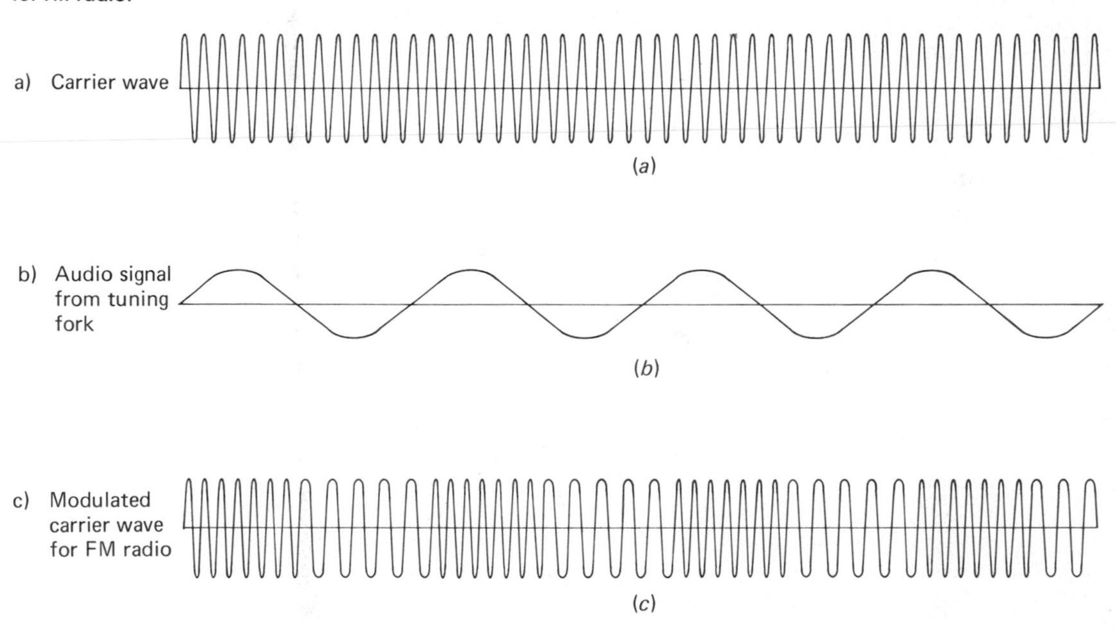

TELEVISION

Television is more complicated than radio communications, because a picture (video signal) is transmitted along with the sound (audio signal). The video signal is broadcast on the electromagnetic carrier wave as an amplitude modulation. This corresponds to the brightness of various portions of the image. The television camera electronically divides an image into a series of horizontal lines. Each line is then reduced to a variation in brightness, which is used to vary the amplitude of the carrier wave. At the same time the station must broadcast the audio signal, that is, voice or music. The audio signal is transmitted by means of frequency modulation, just like FM radio.

The various TV channels are assigned frequency intervals (or ranges) for their broadcast signals, unlike AM and FM radio, which have fixed frequencies. Each TV channel broadcasts both its audio and video signal using its assigned range of carrier frequencies. All TV channels taken together broadcast in the range between 54 and 890 megahertz. The wavelengths of the carrier waves corresponding to these frequencies are 18.2 feet and 13.3 inches (5.56 meters and 0.34 meters).

Not all electromagnetic waves in the frequency range of 54 to 890 megahertz are assigned for TV broadcasts. Certain parts of this range are reserved for FM radio broadcasts, police calls, and aircraft communication. As a matter of fact, all the FM stations are tucked in between TV channels 6 and 7!

7.7 ANALYSIS USING ELECTROMAGNETIC WAVES

Electromagnetic waves with different frequencies provide clues to the identity of the source of the waves. Using a suitable detector, a researcher can gain information by collecting and analyzing the specific kind of electromagnetic wave an object emits. A radio station emits an electromagnetic wave which has a different behavior than an x-ray from a dentist's x-ray machine. We shall conclude this chapter by sketching several uses of electromagnetic waves in astronomy, engineering, navigation, and forgery detection.

RADIOASTRONOMY

People have always been able to observe the sun, moon, planets, and stars by the visible light coming from them. Observation of light was, throughout most of recorded history, the only means of doing research in astronomy. All the astronomy known prior to the twentieth century (a large amount of information) was based on the very small part of the electromagnetic spectrum that the human eye can see.

During the first half of the twentieth century sensitive detectors and receivers of low-frequency electromagnetic waves were developed as part of the growth of wireless communication. It was discovered that AM radio and microwaves were greatly affected by sunspots. Sunspots are associated with turbulence in the atmosphere of the sun. The radio and microwave signals arise primarily from the vibrating charges in the turbulent gas. Within our solar system Jupiter emits radio waves because of its turbulent atmosphere.

These radio waves coming from the sun and planets are received by giant antennas known as *radio telescopes.* A radio telescope is shown in Fig. 7-14.

Radio signals have also been received from galaxies far beyond our own Milky Way. Dust and gas clouds are stirred into vast turbulence by either collisions or explosions of galaxies. Microwaves have been detected from galaxies at far greater distances than can be seen using an optical telescope. An example of the information that can be obtained from a radio telescope is shown in Fig. 7-15. The microwave intensity was measured in many different directions in a small region of the sky. Each measurement was carefully plotted on a graph, and then lines were drawn connecting points of equal radio intensity.

During the second half of the twentieth century research in astronomy has expanded to use the entire electromagnetic spectrum. Infrared astronomy has discovered an object at the center of our galaxy which is emitting 0.1 percent of the total radiation from the galaxy. This amount of energy may not seem like very much energy until you recall that there are billions of stars radiating energy in our galaxy.

Electromagnetic waves with frequencies higher than that of visible light are also being detected as part of the research in astronomy. Ultraviolet radiation, x-rays, and gamma rays are all detected by devices carried aloft by rockets or satellites. The detectors must be placed above the earth's atmosphere, since the atmosphere absorbs much of the ultraviolet radiation, x-rays, and gamma rays. Data from these detectors are used to investigate our galaxy.

Fig. 7-14
You do not see visible light through a radio telescope. Instead, this giant antenna receives extraterrestrial radio waves from various points in the sky. *(Photograph courtesy of National Radio Astronomy Observatory, Greenbank, West Virginia.)*

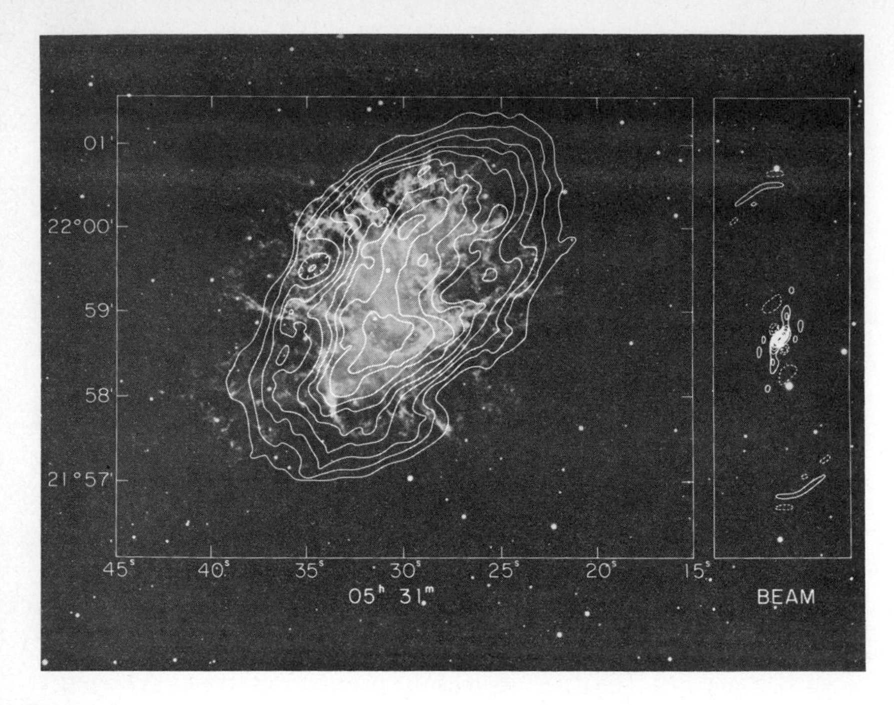

Fig. 7-15
The data from a radio telescope are shown as white lines connecting points of equal radio intensity. The background is a photograph taken through an optical telescope which uses visible light to show the same region of the sky. (*Photograph courtesy of National Radio Astronomy Observatory, Charlottesville, Virginia.*)

It has been discovered that the center of our galaxy is also an x-ray source, as well as a source of infrared radiation and radio waves.

TESTING FOR FLAWS *As the wavelength of ultraviolet rays is made shorter and shorter* (*or equivalently, the frequency becomes higher*), *the electromagnetic radiation merges into that part of the spectrum known as x-rays.*

X-rays All electromagnetic waves from the long-wavelength AM radio signals to the short-wavelength ultraviolet radiation are completely stopped and absorbed by metals, as well as by many other solids. But x-rays of the appropriate wavelength can pass through all metals and solids with only slight loss in amplitude or intensity.

Engineers can use x-rays to check the inside of metals (see Fig. 7-16). Before the development of x-ray technology it was impossible to examine the seams where metal parts were welded together. Today x-ray photographs are taken of metal parts before they are put into complicated pieces of machinery. A giant generator of electricity should not be allowed to stand idle because of breakage along weak seams connecting its metal parts.

Metal parts that were flawless when built can be checked by x-rays as the parts age. The metal wings of airplanes are regularly checked using x-rays. Flaws can develop because of the flexing and bending that normally occurs in the wings. In a similar fashion, a paper clip, which is bent back and forth, weakens and eventually breaks. Airplane wings obviously are not bent so severely, but the wings are nevertheless checked on a regular basis.

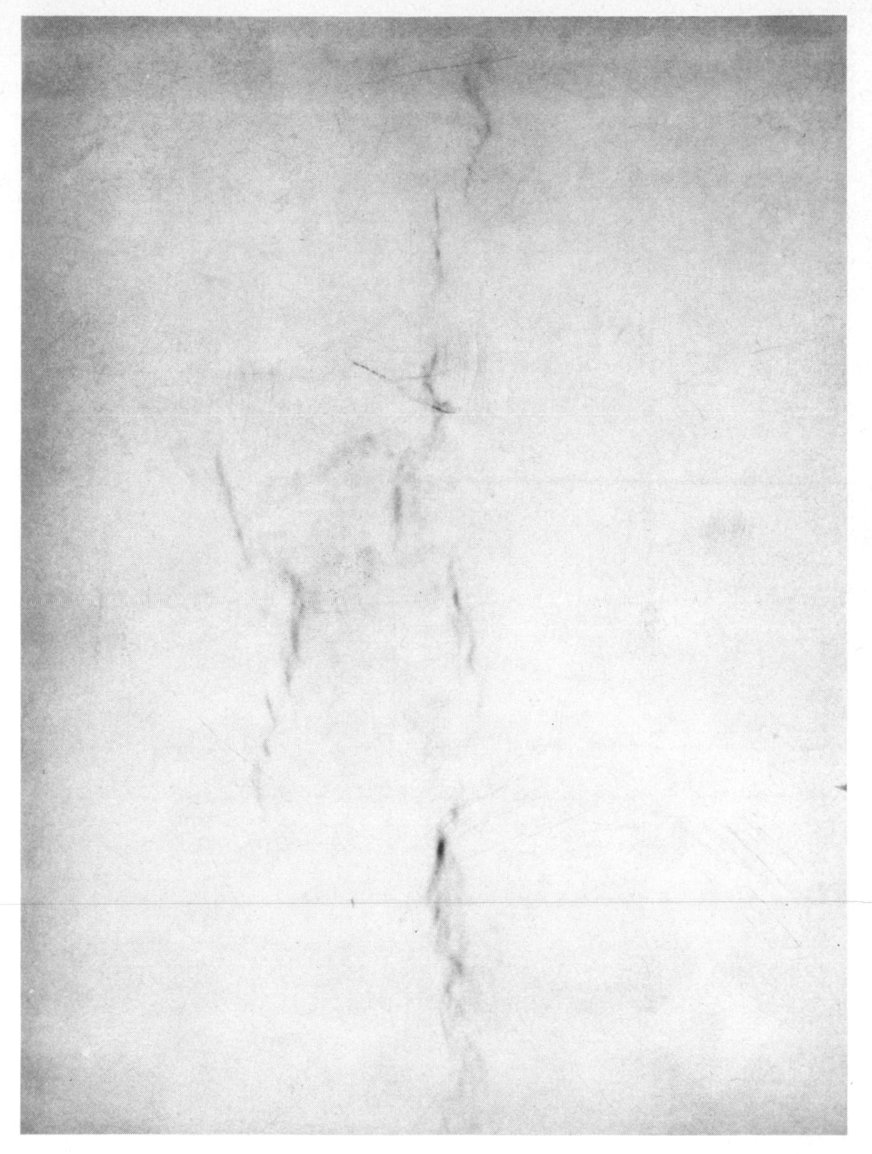

Fig. 7-16
Short-wavelength x-rays can penetrate thick layers of steel. Above, a modern x-ray machine is used to examine metal castings. On the right, an x-ray photograph shows flaws in a casting. (*Photographs courtesy of Peabody Testing Service.*)

As the wavelength of x-rays is made shorter and shorter, a new region of the electromagnetic spectrum is identified: the gamma rays. Gamma rays are the shortest wavelength (or equivalently, the highest frequency) of all the electromagnetic waves. Gamma rays penetrate materials even more easily than x-rays.

Some very thick metal parts are impossible to check for flaws using x-rays. For such objects it is possible to take photographs with the more penetrating gamma rays. Gamma rays, unlike other electromagnetic waves in the spectrum, are not easily generated. Only by using "atom smashers" or particle accelerators can scientists and engineers turn gamma rays on or off. However, some materials spontaneously and continuously emit gamma rays. For example, the nucleus of one kind of cobalt atom emits gamma rays. If this type of nucleus is isolated from other cobalt nuclei, it is possible to have a portable source of gamma rays. Because the rays are continuously being emitted, one must use special precautions when making gamma ray photographs. Nevertheless, safe procedures have been established, and engineers routinely use gamma rays for testing metal parts.

RADAR

Prior to World War II, experiments with high-frequency radio communication showed that electromagnetic waves in the form of microwaves were interrupted by ships and aircraft. Instead of being used for communication, microwaves were developed to *detect* ships and airplanes.

Radar *Radar is a name for a technique to detect objects and comes from radio detection and ranging.*

Although radar was developed under wartime conditions, it has many peaceful applications. Present-day aircraft use on-board radar to determine altitude, especially at night and under cloudy conditions. Radar is also used by the control towers at airports to accurately locate arriving and departing airplanes (see Fig. 7-17). Air traffic controllers continually monitor airplanes to prevent collisions.

Ships at sea need radar to detect obstacles, especially at night and in fog.

Fig. 7-17
An air traffic controller uses radar to prevent midair collisions. (*Photograph courtesy of Federal Aviation Administration.*)

Fig. 7-18
Rainstorms, hurricanes, and tornadoes can be tracked using radar. (*Photograph courtesy of National Center for Atmospheric Research.*)

Although ships may be in voice communication with other ships and with the shore, it is also necessary to know the distance to other ships and the shoreline. Far out in the ocean, even in the absence of other ships, the presence of drifting icebergs must be accurately known, especially in the north Atlantic Ocean.

Another use of radar is weather surveillance. Rainfall is easily detected by radar. Weather fronts and hurricanes are usually accompanied by heavy rainfall. Radar can map out these weather patterns over large areas. An example of a radar weather map is shown in Fig. 7-18. This is particularly useful over the oceans, where it is difficult to maintain many weather stations.

DETECTING FORGERIES

There are many kinds of forgeries. Some are quite innocent, such as a note sent to a friend, but with someone else's signature on the note. Or perhaps the letters with "invisible" ink that we all tried as children. Others are more serious and range from forged wills and counterfeit currency to forged paintings.

A technique is available for detecting all these frauds. It involves ultraviolet radiation. When these electromagnetic waves are used to illuminate a particular ink or paint, the invisible ultraviolet is absorbed and visible light is given off. This emitted light is *not* simply reflected light. The original radiation is invisible to humans, but when ultraviolet rays interact with the molecules of the pigment, visible light is emitted.

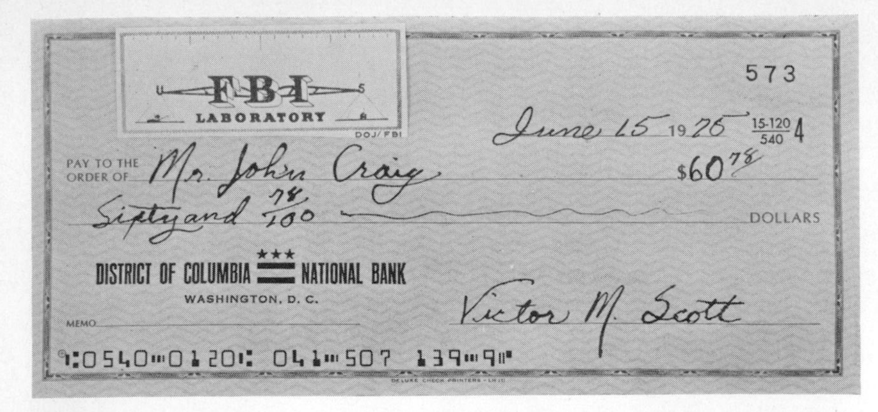

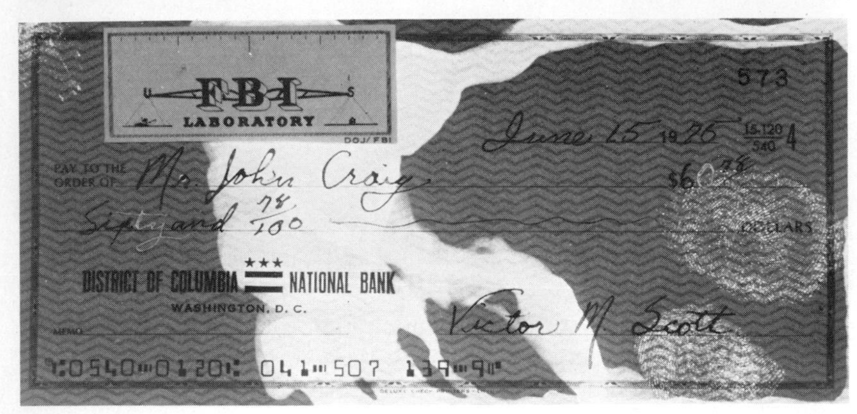

Fig. 7-19
Hidden details of a forged document are revealed using electromagnetic radiation with different wavelengths. The document to the left is illuminated with visible light, and the same document below is illuminated with ultraviolet radiation. (*Photographs courtesy of the FBI.*)

This technique is useful for forgery detection because the visible light given off is characteristic of the absorbing pigment. An ink made by one manufacturer glows differently under ultraviolet radiation than an ink made by another manufacturer (see Fig. 7-19), even when the two kinds of ink look identical under white light. It is the ultraviolet radiation which brings out the difference in inks.

Numerous ancient historical documents have been inspected using these techniques during recent years. Some papers have been discovered with changes made years after the original was written. A few documents have turned out to be complete fakes. Certain chemicals used in making the ink were not even discovered until centuries *after* the documents were supposed to have been written!

REVIEW

You should be able to define or describe the following terms: spectrum of colors, white, black, infrared radiation, ultraviolet radiation, electromagnetic wave spectrum, microwaves, AM radio, FM radio, x-rays, and radar.

You should be able to state the following effect and formula and give examples of their application: the greenhouse effect and the wave formula for electromagnetic waves.

PROBLEMS

1. The dial on your AM radio is set at 11 to tune in the local rock-and-roll station. What is the frequency of the carrier wave?
2. What is the wavelength of an AM radio station broadcasting at an assigned carrier frequency of 620 kilohertz?
3. The wavelength of red light is approximately 650 nanometers (650×10^{-9} meters). What is the frequency of this electromagnetic wave?
4. Violet light has a wavelength of approximately 400 nanometers. What is the frequency of this electromagnetic wave?

DISCUSSION QUESTIONS

1. Why does the spectrum of colors not include black or white?
2. Does green light come from the sun?
3. What is meant by infrared radiation?
4. How is ultraviolet radiation used to detect document forgery?
5. Can human beings see ultraviolet radiation?
6. Do both infrared radiation and ultraviolet radiation cause a suntan?
7. List some practical uses of ultraviolet radiation.
8. Does ultraviolet radiation from the sun travel through ordinary window glass?
9. What is an electromagnetic wave?
10. How fast do microwaves travel?
11. How does a microwave oven cook food?
12. Describe amplitude modulation as applied to AM radio.
13. Describe frequency modulation as applied to FM radio.
14. How is the signal for television carried by electromagnetic waves?
15. How is it known that some stars emit very little light but emit large quantities of energy?
16. What is a radio telescope?
17. What similarities and differences exist for UV rays and x-rays?
18. What similarities and differences exist for x-rays and gamma rays?

ANNOTATED BIBLIOGRAPHY

The British Broadcasting Corporation is known for quality programming. One series of seven broadcasts on electromagnetic radiation was later turned into a book with many good examples. See A. Hewish, *Seeing beyond the Visible,* American Elsevier Publishing Company, Inc., New York, 1970. Another book with a very large range of topics is J. Dogigli, *The Magic of Rays,* Alfred A. Knopf, Inc., New York, 1960. Look here for some delightful descriptions of art forgery and its detection.

Solar radiation effects are discussed by R. J. Wurtman in "The Effects of Light on the Human Body," *Scientific American,* July 1975. Sunlight is also compared with radiation from incandescent and fluorescent lamps.

A paperback devoted to radar alone has been written by R. M. Page. Holder of 40 patents involving radar, he tells an interesting story in *The*

Origins of Radar, Doubleday & Company, Inc., Garden City, N.Y., 1962.

Microwave ovens become more popular each year. A discussion about the economics and possible cost savings using a microwave oven can be found in *House Beautiful,* April 1977, p. 104. The article presents sample calculations showing how to estimate your savings.

The use of microwaves to beam solar power from a satellite to the earth is discussed in *Physics Today,* February 1977, p. 30.

A good source for information on both television and AM and FM radio is P. Davidovits, *Communication,* Holt, Rinehart and Winston, Inc., New York, 1972.

The question of infrared detection by human beings, insects, and reptiles is discussed in J. Wilentz's book *Senses of Man,* Thomas Y. Crowell Company, New York, 1968. The hazards of ultraviolet radiation on the skin of human beings are explained in the article "Some Thoughts on Teaching about Ultraviolet Radiation" by W. Thumm in *The Physics Teacher,* March 1975, p. 135.

8 LIGHT

Visible light, though only a very small part of the spectrum of electromagnetic waves, is very important to humans. The way in which we use light is controlled by the underlying principles of optics. The laws of reflection, refraction, and dispersion will be applied to several everyday phenomena and optical instruments.

8.1 REPRESENTATION OF LIGHT RAYS

In Chap. 7 the discussion of electromagnetic waves relied heavily on the concepts of frequency and wavelength. Although crests and troughs of electromagnetic waves cannot actually be seen, it is useful to draw the waves as if these features were visible.

In discussing the visible portion of the electromagnetic spectrum, namely light, the "wave picture" has some shortcomings. Much of the study of light involves changes in its direction of travel. It is therefore useful to represent the direction of travel of a light beam. For this purpose, we often speak of a *ray* of light.

Ray *A ray is defined as a line representing the pathway of a light beam.*

The definition of a ray agrees with our experience of seeing sunlight shining between rain clouds or in a dusty room (see Fig. 8-1).

Fig. 8-1
Narrow beams clearly show the rays of sunlight. (*Photograph by Myron Wood. From Photo Researchers, Inc.*)

When we see rays, or sunbeams, we are not looking directly at the sun. A beam of sunlight is invisible if the light passes through clean air. But if there are dust particles or small water droplets in the air, we can see the beam. Some of the sunlight is scattered off the small particles. This scattered light then enters our eyes. The pinpoints of light sketch out a sunbeam against a darker background.

A misleading idea about the role of light in vision is expressed in the

Fig. 8-2
Vision, whether normal or x-ray, involves electromagnetic waves entering the eyes—not leaving the eyes. (*Drawing from 1962 D.C. Comics, Inc.*)

statement, "Look out the window." It is wrong to think that vision involves "looking out" rather than "light rays coming in." Vision is possible because light enters our eyes. This misunderstanding is expressed in Superman comics. When Superman is shown using his x-ray vision, he appears to be sending out x-rays (see Fig. 8-2). If anyone really could have x-ray vision, only objects illuminated by x-ray machines would be visible. X-rays do not usually illuminate objects around us except occasionally in a dentist's office or a hospital laboratory.

8.2 REFLECTION

When light falls on a rough opaque surface, some light is absorbed, but most of the light rays are scattered in all directions (see Fig. 8-3). The roughness of the surface, although not noticeable to the touch of fingers, is still larger than the wavelength of light. The aluminum surface of a saucepan, for example,

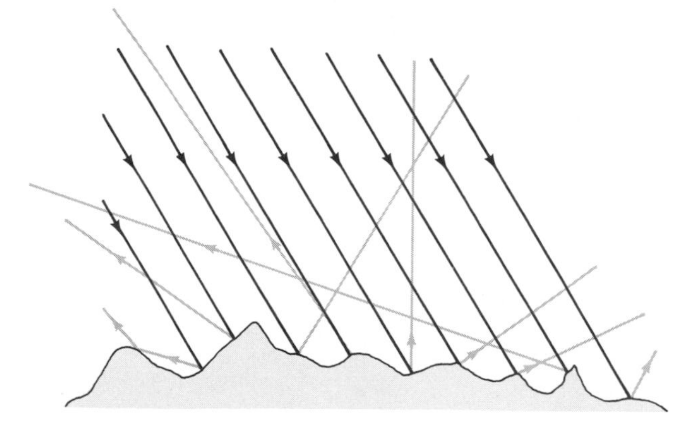

Fig. 8-3
Rays of light being scattered from a rough surface.

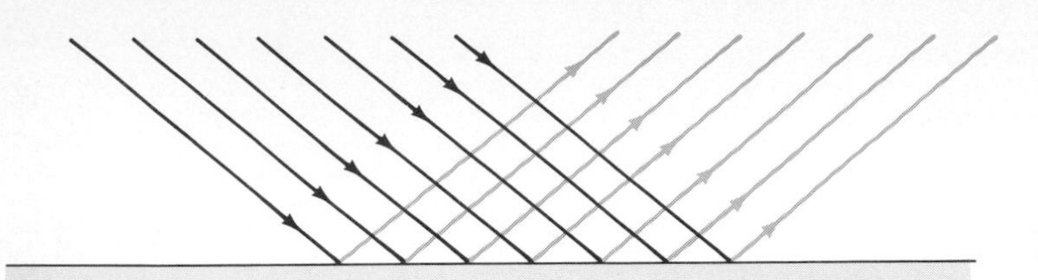

Fig. 8-4
Rays of light being reflected from a smooth surface.

scatters light in all directions. Light rays striking all parts of the surface are scattered to the observer's eyes, and so the sheet of aluminum is visible. The image of the source of light is not seen, however.

If the surface is polished to decrease the roughness to less than the wavelength of light, the light rays are no longer scattered. Instead, the light rays are reflected together, as shown in Fig. 8-4. Aluminum foil is an example of a surface that collectively reflects rays of light. The aluminum foil reflects an image of the source of light.

Mirror surface *A mirror surface is defined as a smooth reflecting surface.*

A mirror could be made by putting aluminum foil behind a protective glass plate. Generally, however, mirrors are made by depositing aluminum onto a smooth glass surface. The glass, which is in front of the reflecting aluminum surface, can be washed. The smooth aluminum surface reflects the image of the illuminated object. Figure 8-5 shows an aluminum saucepan, aluminum foil, and a mirror being illuminated.

The effect of light striking a mirror can be understood by considering a single ray of light. Light reflection behaves like the perfect bounce of a Ping-Pong ball. That is, the angle of the incident ray striking the mirror

Fig. 8-5
The rough surface of an aluminum saucepan scatters light rays. Aluminum foil and a mirror reflect images. (*Photograph by Ted Rozumalski. From Black Star.*)

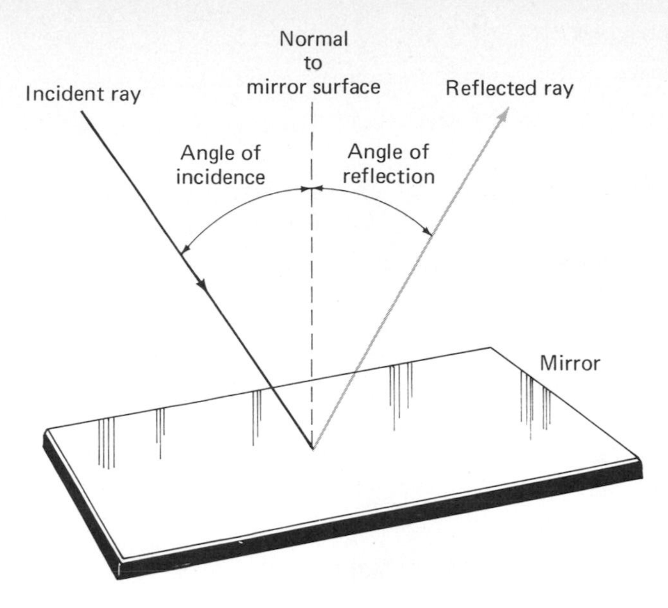

Fig. 8-6
Reflection from a plane mirror surface. The angle of incidence equals the angle of reflection.

surface equals the angle of the reflected ray bouncing from the mirror. It is customary to measure both the incident ray and the reflected ray from a line perpendicular to the mirror surface. This perpendicular line is called the *normal* (see Fig. 8-6).

The *law of reflection* for light can be stated.

Law of reflection **Whenever a ray of light strikes a mirror surface, the angle formed by the incident ray and normal is equal to the angle formed by the reflected ray and normal. The incident ray, normal, and reflected ray all lie in a plane.**

Most mirrors have *plane* reflecting surfaces—that is, the mirror is flat. A dentist puts a small plane mirror in your mouth to look behind your teeth. The rear-view mirror in a car is simply a flat mirror. The mirrors in a fun house, however, are wavy, which has the effect of stretching or shrinking your image, as shown in Fig. 8-7.

Mirrors that are not flat have many useful purposes. A *convex mirror* is a mirror that has the center bulging outward (see Fig. 8-8). The effect of a convex mirror is to increase the extent to which a scene is visible to the viewer. Small convex mirrors are often built into the flat rear-view mirrors of trucks and vans, as shown in Fig. 8-9. Convex mirrors are also often found mounted in corners of stores so that clerks can watch for shoplifters.

The opposite of a convex shape is a concave shape. A *concave mirror* has its center bulging inward. When such a mirror is placed in sunlight, all the rays are brought together at a point called the *focus* as shown in Fig. 8-10. An

Fig. 8-7
The "trick mirror" at the fair or amusement park produces a distorted image because of the wavy surface of the mirror. (*Photograph by Robert Houser. From Photo Researchers, Inc.*)

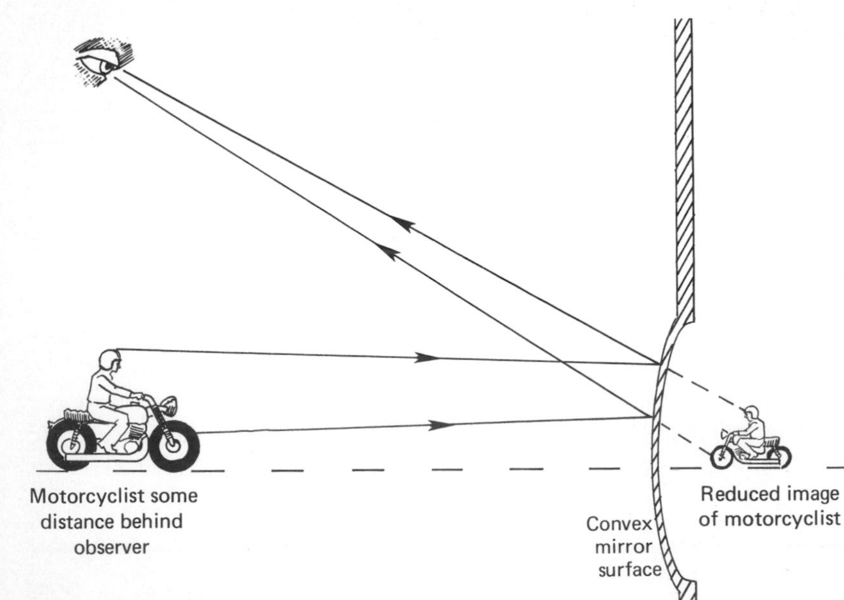

Fig. 8-8
A convex mirror provides a larger field of view than a plane mirror by reducing the size of the image.

Fig. 8-9
The view seen in a rear-view mirror of a truck or van.
(Photograph by Ted Rozumalski. From Black Star.)

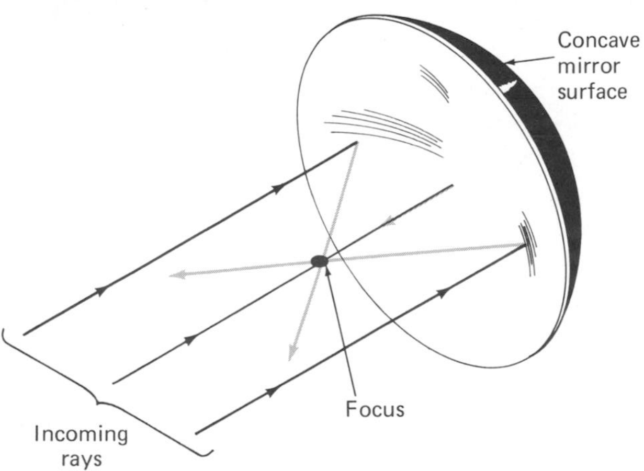

Fig. 8-10
A concave mirror can be used to bring light rays together at a focus.

Fig. 8-11
Food can be cooked by focusing the sun's rays with a concave mirror. (*Photograph furnished by Dr. James Silverberg.*)

object placed at the focus of a large concave mirror can be raised several thousand degrees in temperature by the heat from the sun.

Most mirrors have a relatively thick metal coating that forms the mirror surface. If the coating is thin enough, part of the incident light passes through the metal and part of the light is reflected. Then the mirror is called a *half-silvered mirror.* The use of a half-silvered mirror in normal illumination produces both a reflection and a view of the scene behind the mirror. If the half-silvered mirror is used as a window between a brightly lit room and a dimly lit room, then the mirror becomes a "one-way" mirror. A person inside the dimly lit room can see through the mirror into the brightly lit room. But a person in the brightly lit room sees only a reflection in the mirror.

One-way mirrors are often found above the meat racks in large supermarkets (see Fig. 8-12). The meatcutters easily see the shoppers (and perhaps someone stealing meat), but the customers see only their reflections. Gambling casinos in Nevada and New Jersey have numerous one-way mirrors

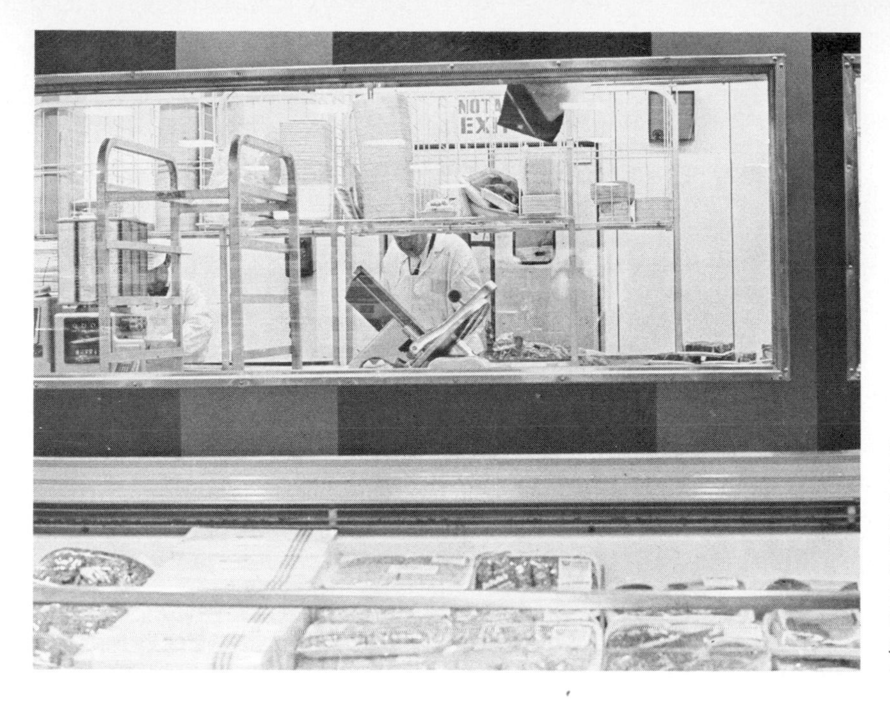

Fig. 8-12
The light has been adjusted on both sides of the one-way mirror so that you can see the meatcutter. Normally he would be working in dimmer light and not be visible to the customers. (*Photograph from Beckwith Studios, Brooklyn, N.Y.*)

located above the roulette, card, and dice tables. Guards behind the one-way mirrors watch for cheaters.

8.3 REFRACTION

The study of reflection involves light striking a mirror surface. Another common occurrence is for light to strike a transparent object in air. When light falls at an angle upon the smooth surface of a transparent material such as glass or water, a small part of the incident light is reflected back into the air according to the law of reflection. But most of the light passes from the air into the transparent material. Surprisingly, the light inside the transparent material does *not* proceed in the direction of the incident ray. Instead, the light in the transparent material moves off at an angle to the original direction, as shown in Fig. 8-13. The ray of light has changed its direction of travel at the interface. The new ray is called the *refracted ray.*

Refraction

Refraction is defined as the deflection of a ray of light from a straight line as the ray passes from one transparent medium to another.

To understand refraction it is necessary to learn more about the speed of light. In Chap. 7 the speed of light was given as 186,000 miles per second or 300,000 kilometers per second. This is the speed of light in air, in a vacuum, or in outer space.* However, in a material such as water or glass, the typical

*The speed of light in air is slightly less than the speed of light in a vacuum. We will ignore that difference here.

FIBER OPTICS

What internal organs of the human body have you seen? On television we are occasionally shown an operation on the heart, stomach, or lungs. And some books have beautiful color pictures of almost every anatomical detail that you can imagine. Such pictures are made in two ways. The body can be surgically opened, or we can insert viewers into the body openings. Special instruments have been designed to look into the nose, throat, urethra, and rectum. These instruments have the advantage of not requiring any surgery on, and subsequent healing of, the patient.

One type of instrument employs a hollow tube. A physician simply looks through the tube as you would look through the cardboard tube in the center of a roll of paper towels. Such medical tubes are often several feet long and about the diameter of a pencil, as shown in the photograph. Sometimes the inside of the tube is plated with a thin layer of gold. In this way light travels more easily down the tube by reflections off the sides.

Another type of instrument uses bundles of solid glass rods. The method of light travel is shown in the diagram below. Light entering the end of a curved rod undergoes reflections from the inner surface of the rod. The rod can be twisted or bent, but the light still emerges at the other end. Light that reflects from the inside of the rod is said to undergo total internal reflection. Light travels around the bends of a glass rod by total internal reflection unless the corners are made quite sharp. Then some light emerges from the surface.

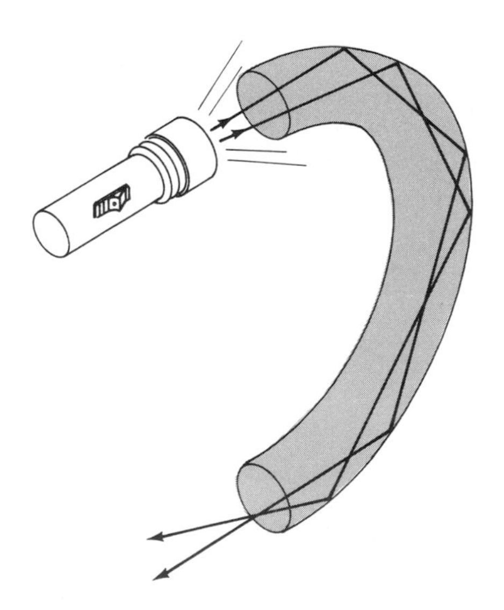

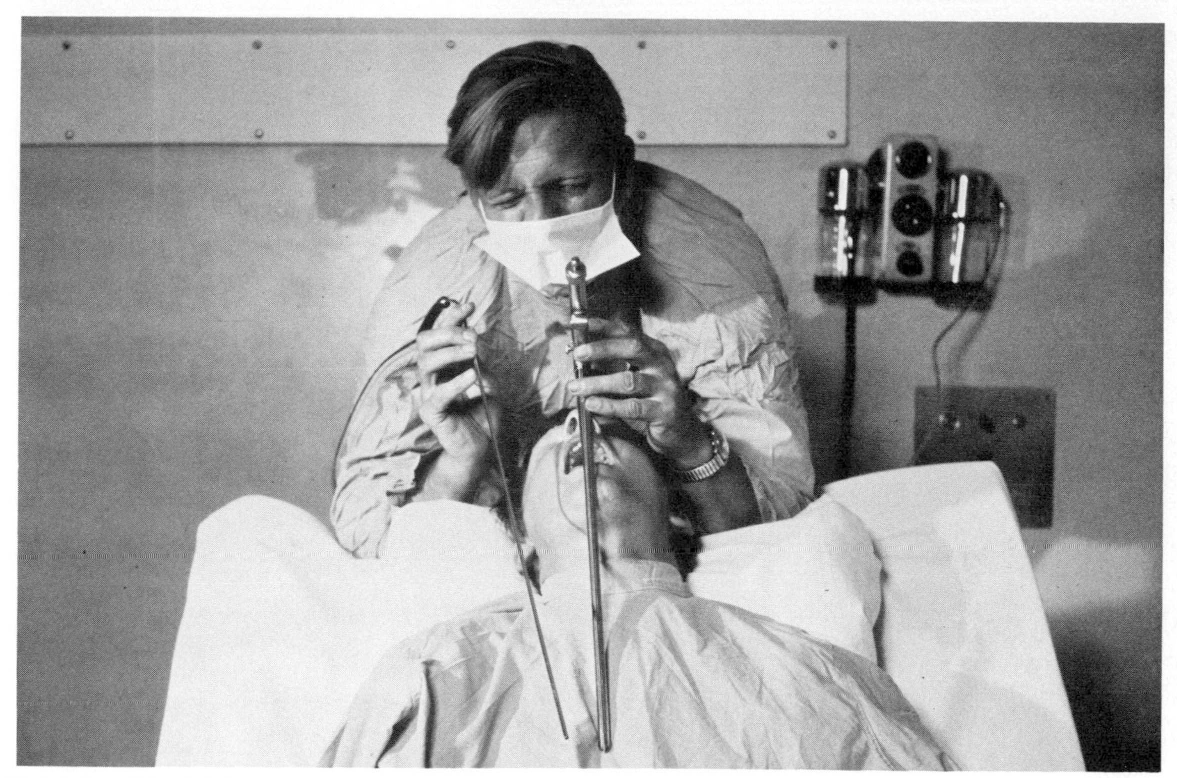

(Photo by Barry Richards © Aldus Books.)

The endoscope, a medical instrument for peering into the stomach, employs a bundle of thin glass fibers (see photograph above). These fibers are about $\frac{1}{4}$ of a millimeter in diameter. Each individual fiber transmits light by total internal reflection. Some of the fibers are used to carry light into the patient, to illuminate the stomach or other organ. Other fibers are used to return light (and the view of the interior) to the observer. Endoscopes are often connected to a camera or TV monitor. Color TV endoscopy provides a full color view of the interior without surgery!

Because of the flexibility of thin glass fibers, many organs can be easily reached. The rectum can be viewed. And the bronchial tubes can be examined. Even the bladder can be seen. But some organs like the kidneys and small intestines are still beyond reach. For those organs we still must use surgery.

Applications of glass fibers for transmitting light have spread into many areas. The whole subject has been given the name fiber optics. Magnifying glasses can now be made using fiber optics. And the interior of jet plane engines can be examined using fiber optics (without turning off the engine!). Many futuristic suggestions have been made for the use of fiber optics. Maybe someday fiber optics will be used to carry telephone messages, or even a television picture into your home.

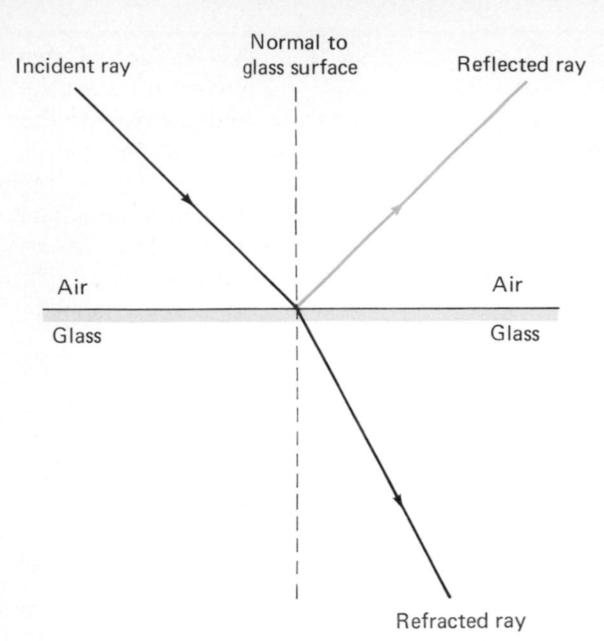

Fig. 8-13
At a boundary between two transparent materials both reflection and refraction occur.

speed of light is significantly slower. In water the speed of light is only three-fourths that of the speed of light in a vacuum, that is, approximately 225,000 kilometers per second. In an ordinary piece of glass the speed of light is even less, only two-thirds that of the speed of light in a vacuum, that is, approximately 200,000 kilometers per second. Table 8-1 lists the speed of light in various materials.

The apparent decrease in the speed of light in various transparent materials is caused by the positive and negative charges of the atoms and molecules. The electric and magnetic fields of the light beam moving through the transparent material affect the positive and negative charges. New electric and magnetic fields are produced, which act counter to the movement of the light. The effect is that the resulting waves appear to move more slowly through the material.

With this background the *law of refraction* can now be stated.

Table 8-1
Typical speed of light in various transparent materials

Material	Typical speed, mi/sec	Typical speed, km/sec
Vacuum	186,000	300,000
Ice	143,000	230,000
Water	140,000	225,000
Glass	125,000	200,000
Diamond	75,000	120,000

Law of refraction **Whenever the speed of light is decreased at the surface of two transparent materials, the refracted ray is bent toward the normal. When the speed is increased, the refracted ray is bent away from the normal. The incident ray, normal, and refracted ray all lie in a plane.**

The refraction of light toward the normal (as the speed of light decreases) and the refraction of light away from the normal (as the speed of light increases) are both shown in Fig. 8-14. Refraction toward the normal occurs as light passes from air to glass. Refraction away from the normal occurs as light passes from glass back into air.

The fact that light is refracted upon passing from one transparent material to another can be understood by playing with a sheet of sandpaper and two wheels mounted on an axle (see Fig. 8-15). The wheels and axle represent a ray of light as it moves along. The sheet of sandpaper will affect the speed at which the wheels roll. Suppose the wheels roll on a smooth table and strike the sandpaper at an angle. The sandpaper slows down the motion of the wheels, which is analogous to the slower speed of light in a transparent substance. The wheel that strikes the sandpaper first is slowed down first, and so the direction of motion is changed to bring the path of the wheels closer to the normal. At the other edge of the sandpaper the same wheel that struck first is the first one off the sandpaper. Again the direction of motion is changed, except that the direction of motion is bent away from the normal.

Although the speed of light in air can often be treated as equal to the

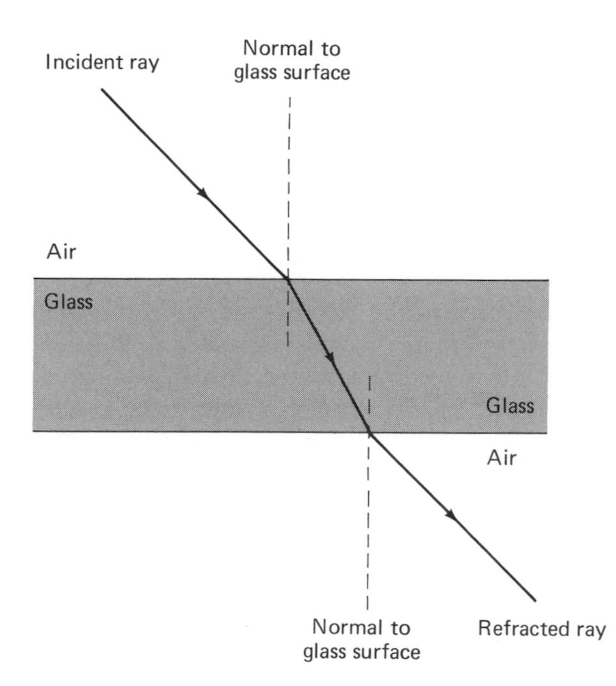

Fig. 8-14
Refraction at the surfaces of two transparent materials. The direction of the ray of light is changed at both the upper and lower glass surfaces.

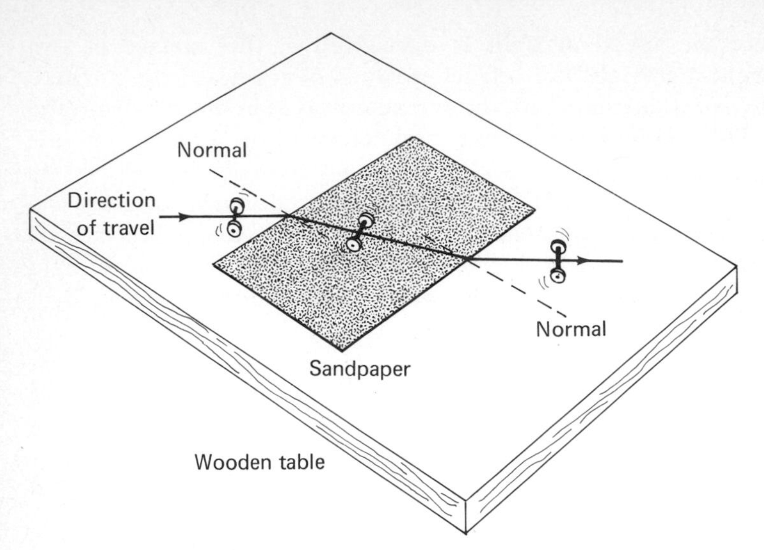

Fig. 8-15
Refraction of light is similar to the behavior of the rolling wheels as they travel from a wooden surface onto a sheet of sandpaper.

speed of light in a vacuum, there are circumstances where this approximation is not valid. One case deals with sunlight traveling hundreds of miles through the earth's atmosphere at sunset. As light first enters the upper atmosphere from outer space, there is a slight decrease in the speed of light. Thus, refraction occurs, and the light is bent toward the earth. This slight bending continues during the flight of the sunbeam, because the light is gently slowed as the density of the air increases. A person watching the setting sun near the horizon sees light from the sun actually *below* the horizon. Atmospheric refraction has bent the light to follow the curvature of the earth. Each day atmospheric refraction adds a few extra minutes of sunlight. A view of the bending of light rays in the atmosphere is shown in Fig. 8-16.

Have you ever seen the wavy thermal patterns in the air above the burners or heating elements on a stove? (Figure 8-17 reminds you of the effect.) We often speak of the objects behind the rising hot air as "shimmer-

Fig. 8-16
Refraction of sunlight by the earth's atmosphere makes the sun appear higher in the sky.

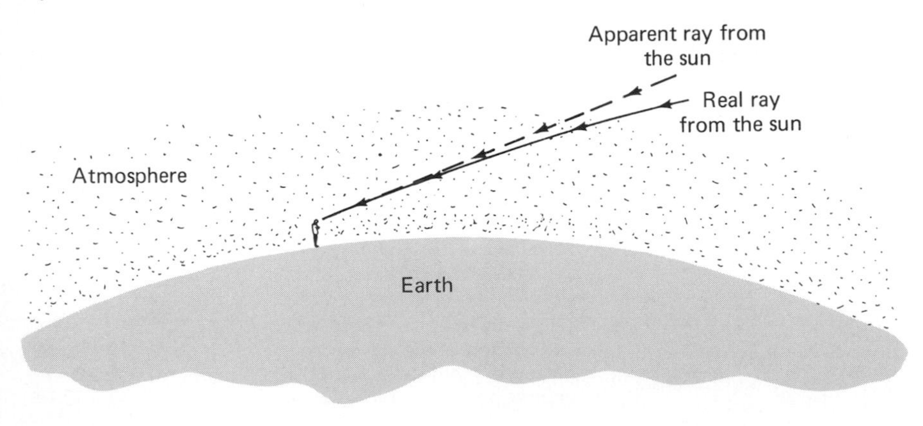

Fig. 8-17
Refraction of light by the hot air rising from a burner causes the clock face to appear distorted.

ing." Some people falsely think that the shimmering is heat itself rising. Heat is an abstract idea, but the effect of heat—that is, refraction in the air—is easily noticed. Refraction is caused here by the mixing layers of hot and cold air. The speed of light varies with density changes in the air. In warmer air, the density is less and the speed of light is larger. The same shimmering of objects can also be observed when looking above a campfire or any open flame.

On a hot summer day the refraction of light above a wide flat surface, such as a highway or desert, causes a *mirage* (see Fig. 8-18). In order to form a mirage, light comes from the sky near the horizon and skims along the surface of the ground. The layer of hot air near the ground bends the light

Fig. 8-18
A mirage in the desert.
(Photograph by Jerome Wyckoff.)

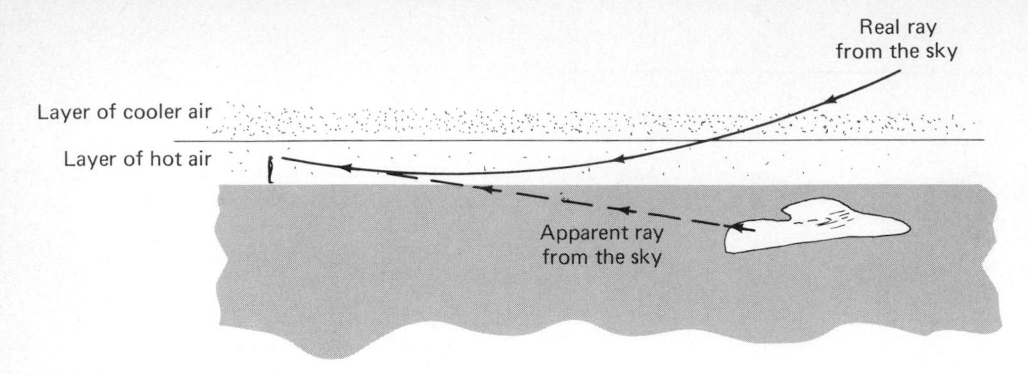

Fig. 8-19
A mirage is caused by the refraction of light rays moving through a layer of hot air lying beneath a layer of cooler air.

upward, as shown in Fig. 8-19. Refraction of the light rays makes light from the sky appear to be coming from the surface of the ground or highway. Blue light appearing to be coming from the ground tricks small children into thinking that "water" is on the highway. Of course the "water" is only a fleeting mirage. The mirage fades into the distance because the light rays need some distance for the air layers to refract light into the eyes of the observer.

8.4 PRISMS AND LENSES

A *prism* is a piece of material used for refracting light. Often prisms are made of glass. But they can be made of plastic, ice, or any clear material. Prisms have a triangular cross section. When light passes through a prism, the rays are refracted at two surfaces, once entering the glass from the air and once leaving the glass returning to the air. At the first surface the light is bent *toward* the normal due to the decrease in the speed of light upon entering glass. At the second surface the light is bent *away* from the normal due to the increase in the speed of the light upon leaving the glass. Figure 8-20 illustrates the refraction occurring in a prism. (Light of only one color is used in this discussion. The effect of a prism on white light will be discussed in Sec. 8-5.)

An understanding of the refraction of light in a prism can be the basis of understanding the action of a *lens.*

Lens *A lens is a piece of transparent material used to change the direction of rays of light.*

Lenses are classified by their cross sections. A lens that is thicker in the middle than on the edges is a *convex* lens. A lens that is thicker on the edges than in the middle is a *concave* lens. Figure 8-21 shows cross sections of several standard forms of convex and concave lenses.

The effect of a lens on rays of light can be understood by imagining the lens to be built of prisms and blocks of glass stacked on one another (see Fig. 8-22). As a result of stacking the prisms and blocks to imitate the gross

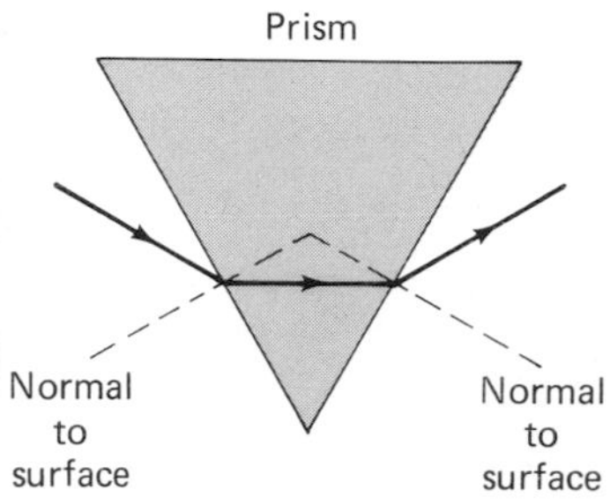

Fig. 8-20
On the left, laser light is refracted twice in passing through a prism. On the right, a schematic diagram shows the refracted ray and the two normals of a prism, as viewed from directly overhead. (*Photograph by Ted Rozumalski. From Black Star.*)

behavior of the two types of lens, the rays of light are refracted differently. A convex lens converges parallel rays of light to the focus labeled f. Appropriately enough a convex lens is also known as a *converging* lens.

The prisms and blocks stacked differently produce light refraction similar to that produced by a concave lens. A concave lens diverges parallel rays of light *as if* the rays came from the focus f'. The real rays only apparently originate at the focus of a concave lens. Since the concave lens diverges the rays of light, it is known as a *diverging* lens. The images seen through a convex and a concave lens are shown in Fig. 8-23.

The simplest optical instrument is the magnifying glass, or reading glass. A magnifying glass is nothing more than a single convex lens. The object to be viewed is placed close to the lens at a distance less than the distance to the focus. Light coming from the object will be refracted as it goes through the lens. After passing through the lens, the light will diverge less, as shown in Fig. 8-24. The light will enter the eye in such a way that the object appears

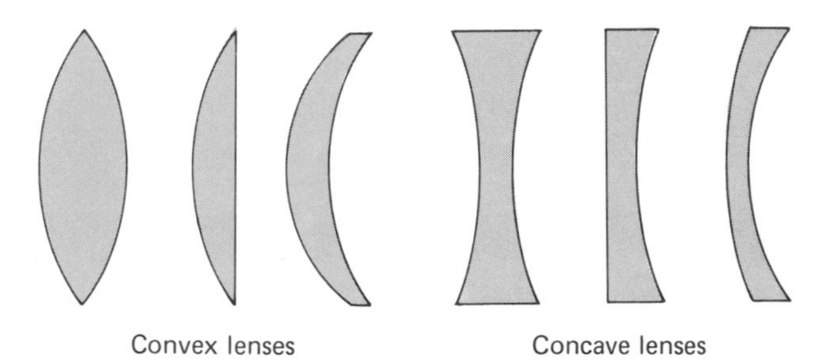

Fig. 8-21
Cross sections of standard forms of convex and concave lenses.

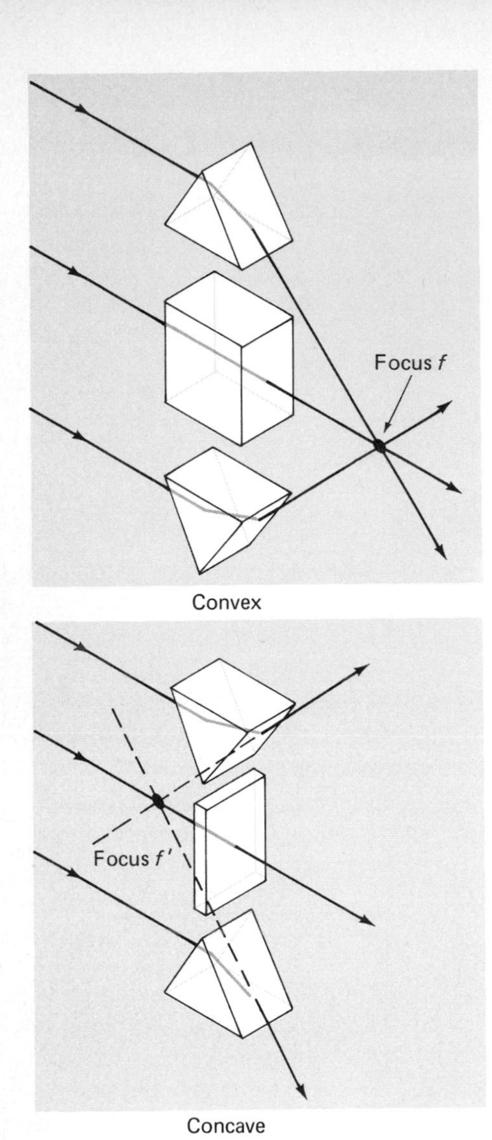

Fig. 8-22
The action of lenses can be imitated by a suitable arrangement of prisms and flat blocks of glass.

bigger. Hence, the image is magnified. The magnification will increase as the magnifying glass is moved farther from the object. When the object is placed beyond the focus, there is no further increase in magnification. Instead, the image becomes blurred if the magnifying glass stays near your eye. Every magnifying glass shows this behavior, and so each has only a certain range of usefulness.

The most important "optical instrument," although not the simplest, is the human eye, shown schematically in Fig. 8-25. The eye is basically a double-lens system. Both of the lenses are converging lenses. The first lens is the *cornea,* that is, the front surface of the eye exposed to the air or the eyelid.

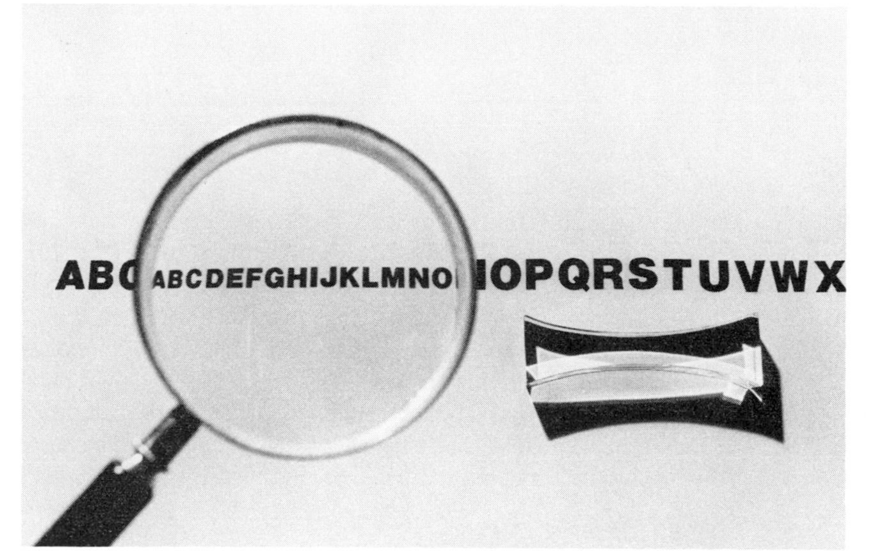

Fig. 8-23
At the top, the image seen through a convex lens is magnified. At the bottom, the image seen through a concave lens is diminished in size. The cross section of each lens is shown beneath the lettering. (*Photographs by Ted Rozumalski. From Black Star.*)

Most of the refraction of the light that enters the eye occurs at the cornea. The second lens is an elastic transparent organ. Medical doctors call it simply the *lens,* although "auxiliary lens" would be a much more appropriate name. Its function is to make certain adjustments as the eye attempts to view nearby objects. Another part of the eye is the *iris.* The iris is an opaque contractile diaphragm which forms the colored portion of the eye and is located between the cornea and the lens. The hole in the middle of the iris is called the *pupil* and allows light to enter the interior of the eyeball.

When we look at something, the image is formed by the cornea and the lens on the *retina.* The retina is covered by a multitude of light-sensitive

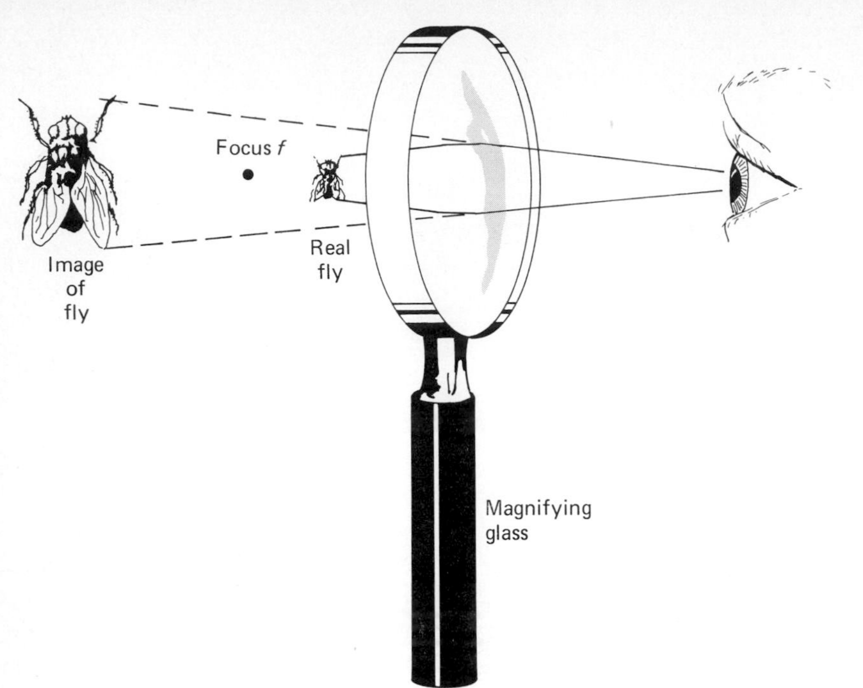

Fig. 8-24
Rays coming from the real fly are refracted by the convex lens, and so they appear to be coming from a larger fly.

nerve endings. The retina transmits electrical signals to the brain by means of optic nerves. The image formed on the retina is upside down (see Fig. 8-26). But do not misunderstand the meaning of the inverted image. Since all images are formed this way in our eyes, we are quite accustomed to it, and of course everything appears to be normal. The brain correctly interprets the orientation of the scene.

A complicated system of muscles and ligaments can change the curvature of the lens. Thus, the eye can focus on objects located at different distances. When the muscles contract, the tension lessens on the lens. The elastic lens

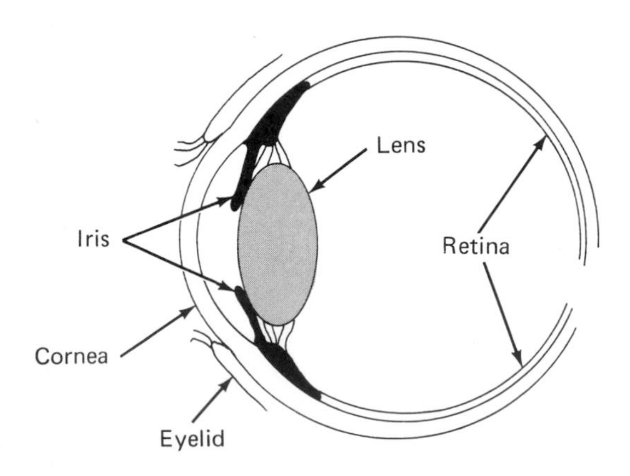

Fig. 8-25
The principal parts of a human eye shown in a cross-sectional view.

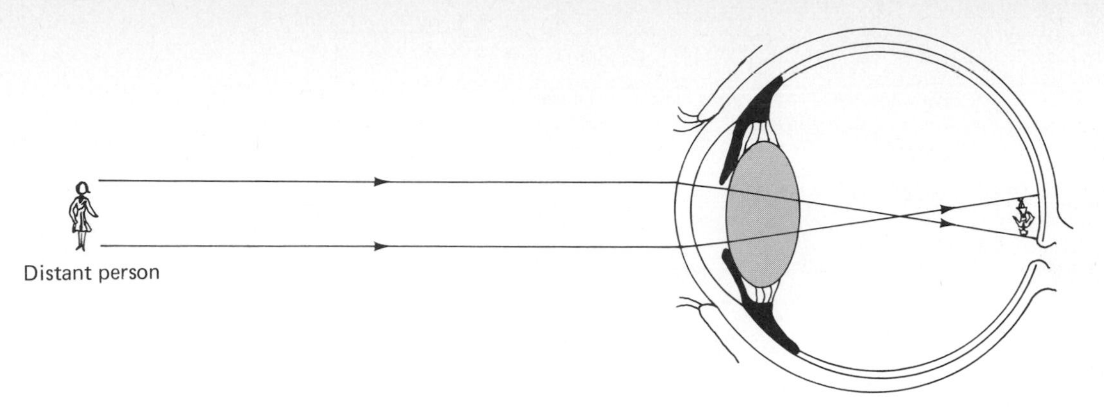

Fig. 8-26
The eye forms an inverted image on the retina of everything you see.

then bulges, or becomes more convex. When the muscles relax, the tension is restored on the lens. This causes the lens to flatten, or become less convex. For near vision the lens is bulging, and for far vision the lens is comparatively flat. Continual use of the eyes for near work produces eyestrain because of the prolonged effort of the eye muscles.

Eyes that cannot see normally can often be corrected with glasses. All prescription glasses are simply lenses that correct whatever visual defect might be present. *Nearsighted* eyes bring the image to a focus *in front* of the retina (see Fig. 8-27). The image is blurred. (A sharp image is seen when the image is focused directly on the retina.) Nearsightedness may occur when either the eyeball is too long or the lens is too convex. Glasses made with concave lenses effectively increase the focal length of the eye. As a result, the image is brought to a focus on the retina.

Fig. 8-27
Both nearsightedness and farsightedness are caused by the failure of the eyes to focus the image *on* the retina.

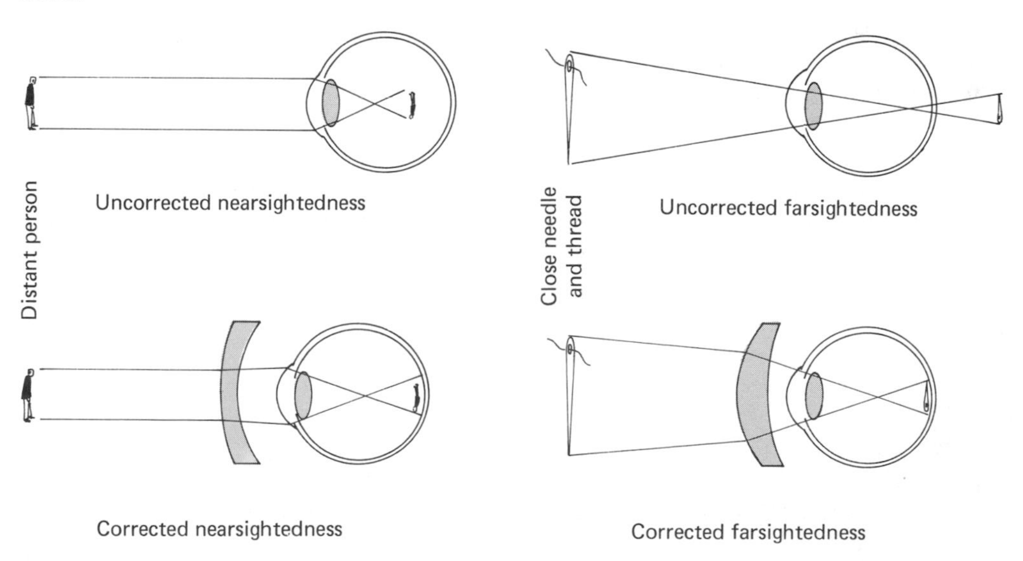

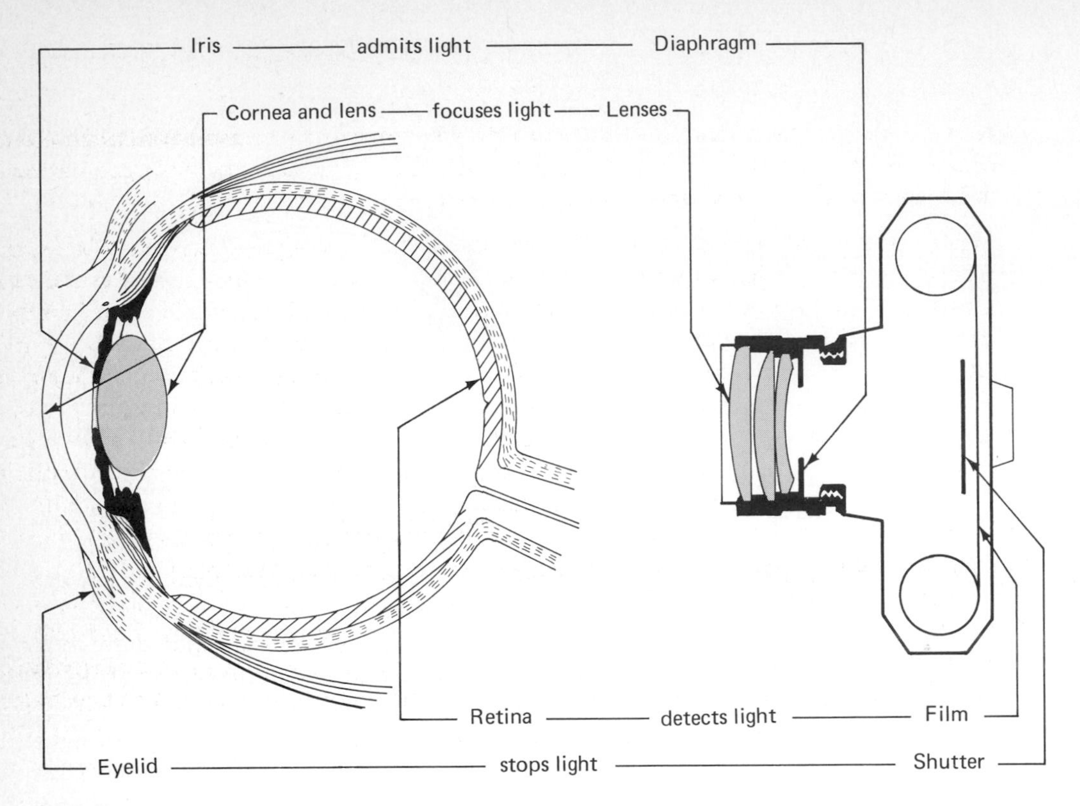

Fig. 8-28
A comparison of the human eye and a camera.

The opposite kind of visual problem is the *farsighted* eye. In farsightedness the eye brings the image to a focus *behind* the retina (see Fig. 8-27). This again results in a blurred image. Farsightedness may occur when either the eyeball is too short or the lens is too flat. Convex lenses in glasses effectively decrease the focal length of the eye. Consequently, the image is brought to a focus on the retina.

An understanding of the physics of the eye makes it easy to understand the camera. For each of the principal parts of the eye there is a similar part to a camera, as shown in Fig. 8-28. Light is admitted through a diaphragm in a camera, similar to the function of the iris. The common 35-millimeter camera has a system of three to seven different lenses. These focus the image just as the cornea and lens focus the image in the eye. The light is focused on the film in a camera. The film is analogous to the retina in the eye.

Light is admitted briefly into a camera by opening a shutter. This differs from the eye in that the eye is normally open. However, if you want to use your eyes like a camera, repeatedly open and close your eyelids.

8.5 DISPERSION

When sunlight passes through crystals and diamonds, brilliant colors of the spectrum of light are produced. Experiments show that the colors already present in white sunlight can be separated by the process of *dispersion.*

Table 8-2
Speed (in km/sec) of the colors of the spectrum in a vacuum and various transparent materials

Material	Red	Orange	Yellow	Green	Blue	Violet
Vacuum	299,793	299,793	299,793	299,793	299,793	299,793
Ice	229,550	229,199	229,076	228,675	228,222	227,633
Water	225,238	225,069	224,732	224,395	223,893	223,226
Glass	197,232	197,024	196,908	196,456	195,815	194,923
Diamond	124,395	124,137	124,024	123,575	122,669	121,966

Dispersion *Dispersion is the refraction of white light into the spectrum of colors.*

A crystal or diamond refracts the various colors in white light into the visible spectrum of colors. The effect of a crystal or diamond on white light can be understood by thinking about a prism.

In Sec. 8-4 the refraction of light by a prism was discussed. A prism was shown in Fig. 8-20 refracting a ray of light at two surfaces. White light is composed of the spectrum of colors, and each color in the spectrum has a different speed in a transparent material. Among the spectrum of colors, red light travels the fastest and violet light travels the slowest. The speeds of all other spectral colors are between these two extremes. In air there is very little difference in the speed of the various colors, and so all colors travel together. Precise values of speeds for spectral colors in different materials are given in Table 8-2.

The different speeds of the colors which are the components of white light result in each color being refracted a bit more or a bit less than its

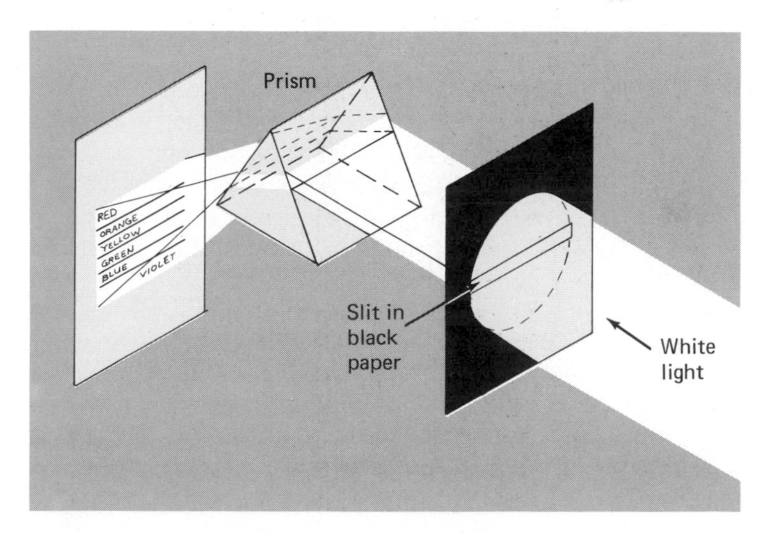

Fig. 8-29
A narrow beam of white light is dispersed by a prism into the spectrum of colors.

neighbor in the spectrum. This difference in behavior is the basis of the *law of dispersion.*

Law of dispersion **Whenever there is dispersion of white light into the spectrum of colors, red light is refracted the least, violet light is refracted the most, and the other spectral colors are refracted between these two extremes.**

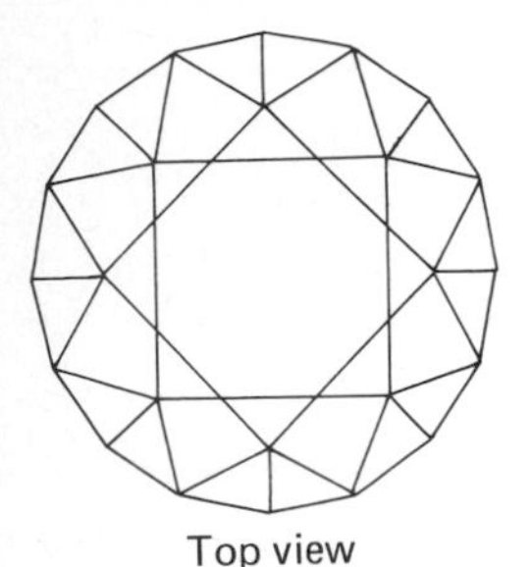
Top view

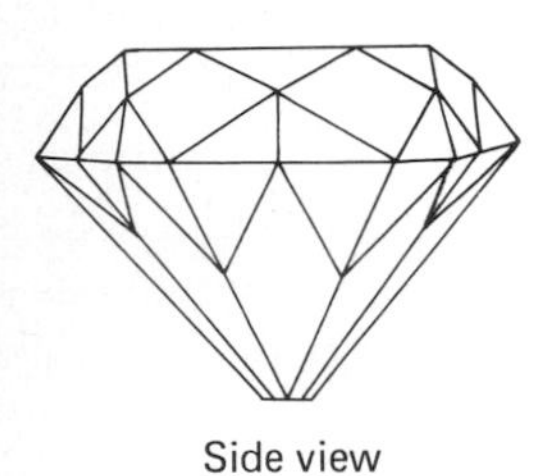
Side view

Fig. 8-30
Two views of an unmounted cut diamond.

The law of dispersion describes how a prism produces the spectrum of colors. If a wide beam of sunlight strikes a prism, many separate spectra are produced and their overlapping colors recombine into white light. But a narrow beam of light will produce the spectrum of colors (see Fig. 8-29). A sheet of white paper or a movie screen provides a surface on which the spectrum of colors can then be viewed.

The beauty of a diamond ring or crystal chandelier is the ability of the diamond or crystal to sparkle. Fig. 8-30 shows a diamond cut in such a way as to expose many small flat surfaces. These surfaces are called *facets.* Light striking a diamond from any direction can be refracted by many different facets. Each facet causes the dispersion of light, or in other words, a sparkling glitter of colors. One of the popular designs for cutting a diamond yields 56 tiny sparkling facets. Visit a jewelry shop and ask to see an unmounted diamond. You will then see dispersion at its finest!

There are other times when the spectrum of colors can be seen around your house or apartment. The edges of ornate cut-glass windows or glass panels in a door cause a beam of sunlight to be dispersed into the spectrum of colors. Sometimes, if sunlight strikes the edge of a mirror just right, a tiny "rainbow" can be seen on the wall of a bathroom. It is, of course, not water that causes this rainbow, but rather the dispersion of light through the edge of a piece of glass.

8.6 ATMOSPHERIC OPTICS

The everyday term for the spectrum of colors is the *rainbow.* The most familiar rainbows are the ones we usually see in the sky after it has been raining. Every spectrum of color is caused by refraction of white light according to the law of dispersion. What then makes a rainbow in the sky?

The refraction of light occurs at the surface of water droplets. And, in addition, the rays undergo reflection at the inner surface of the droplets. In Fig. 8-31 a beam of white light is shown being refracted at the upper point marked U. (The light reflected off the droplet at this point is not shown.) Dispersion of the white light occurs inside the droplet. The rays of the extremes of the spectrum—red light and violet light—are shown. All the rays of color are reflected off the opposite inner surface of the droplet. At this point some light is refracted out of the droplet. But this is also not shown. Instead, we continue to follow the path of the reflected light.

Arriving at the lower surface marked L, each ray of color is again

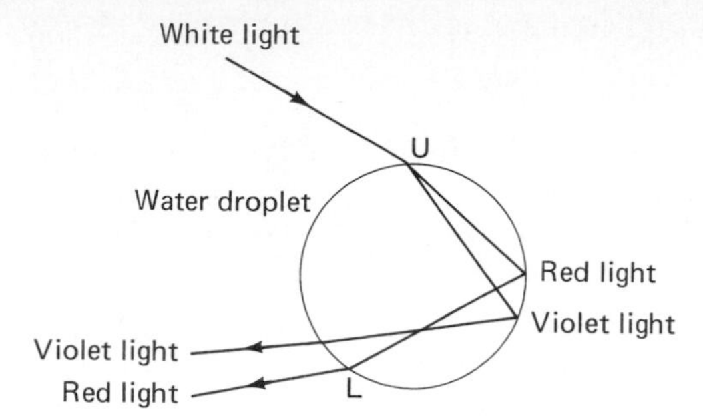

Fig. 8-31
A single droplet of water causes dispersion of white light.

refracted as it passes out of the droplet. (The light reflected again within the droplet is not shown.) Coming out of the water droplet is the spectrum of colors with red at the bottom and violet at the top. This analysis is correct for a single drop. We now consider what happens when the sky contains many drops.

From each drop, the rays of the different colors are headed in different

Fig. 8-32
When a rainbow is seen in the sky, the rays of different colors come from many different droplets.

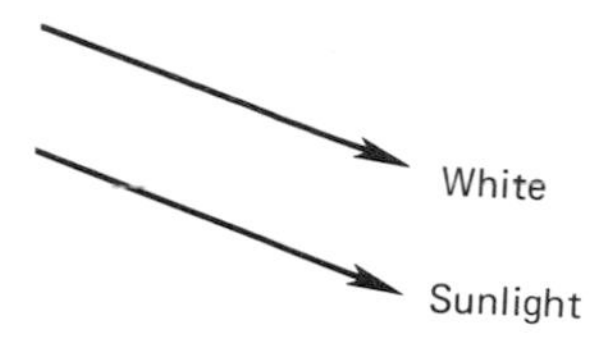

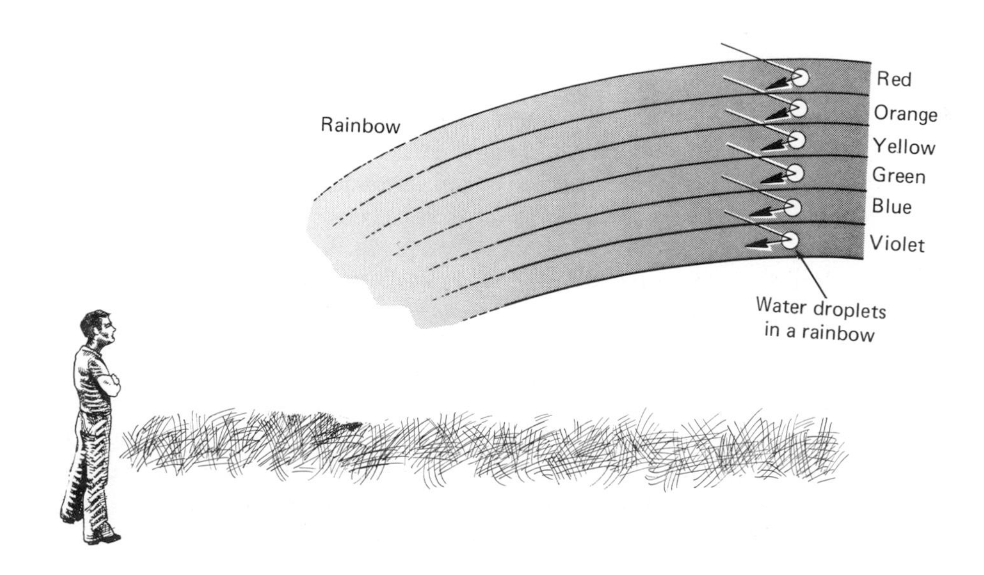

Fig. 8-33
Which color is seen on the upper edge of a rainbow?
(Photograph by John Hendry, Jr. From Photo Researchers, Inc.)

directions. For an observer, the ray of red light appears highest in the sky. All the other rays *from that particular raindrop* go over your head and are not seen. But a second raindrop just below the first one sends its red ray below your eyes, the orange ray is seen, and all the other rays again go over your head. In similar fashion a third raindrop below the other two sends its yellow ray to your eyes, and so on through the spectrum, as shown in Fig. 8-32. So although a *single raindrop* produces a spectrum running from violet on the top to red on the bottom, the rainbow that you see is caused by a *multitude of raindrops* producing red on the top of the rainbow and violet on the bottom of the rainbow.

Another optical phenomenon in the atmosphere, less common than the rainbow, is the sun pillar, shown in Fig. 8-34. The sun pillar appears as a vertical plume of light above a setting (or rising) sun. Sun pillars are caused by the reflection of light from microscopic ice crystals, and so they are common in cold weather. But even in summer there are ice crystals high in the atmosphere. So sun pillars can be seen throughout the year.

When a flat ice crystal falls through still air, it falls lying down just as a leaf or playing card falls. If the atmosphere were filled with *perfectly* horizontal ice crystals, the crystals would act like a multitude of horizontal mirrors and a faint image of the sun would be reflected from the underside of the ice crystals (see Fig. 8-35). The resulting reflected image of the sun would be the same as that formed by a flat surface of water. But the horizontal orientation of ice crystals is *not perfect.* So the reflection is spread out. In the same way, sunlight reflected from rippling water appears as a spread-out patch of light. You can also see "sun" pillars rising above the lights of shopping centers on winter evenings when the air is still.

Fig. 8-34
A sun pillar can be seen in the air above the sun as well as in the reflection below the sun. (*Photograph by Jerome Wyckoff.*)

So common is the blue of the sky and the red of the sunset that most people take their beauty for granted. Perhaps your enjoyment will be increased by learning how these colors are produced from the white light coming from the sun.

Sunlight travels without interruption through the empty void between

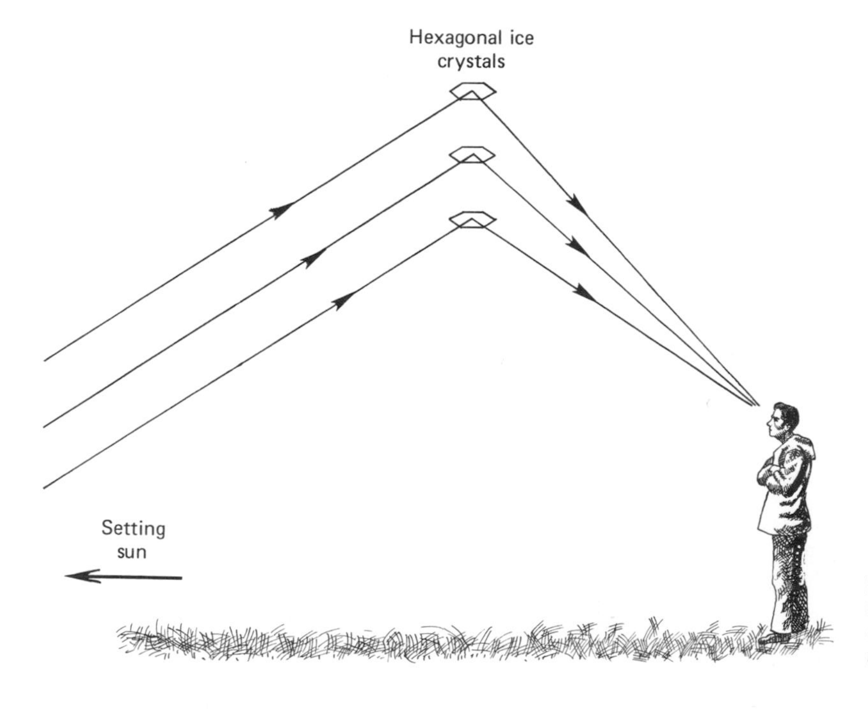

Fig. 8-35
Flat horizontal ice crystals act like tiny mirrors to form a sun pillar.

the sun and the earth. But as the sunlight enters the earth's atmosphere, much of the light is absorbed by air molecules and fine dust particles. After having absorbed the sunlight, the molecules of air and dust re-emit the electromagnetic waves of the visible spectrum. The light emitted by the molecules sometimes proceeds in the original direction of the light from the sun, but more often the light is *scattered,* or sent in every direction.

The scattering, however, is not the same for all colors of the visible spectrum. Shorter-wavelength colors, such as blue and violet, are scattered by molecules much more than the longer-wavelength colors of red and orange. The greater the number of molecules in the pathway of white sunlight, the greater the scattering of the blue light from the beam. When the sun is overhead, the sunlight passes through relatively few molecules in the earth's atmosphere. So a quick glance at the midday sun on a clear day shows the sun to be brilliantly white.

When the sun is just above the horizon, the sun has a characteristic reddish-orange color. At sunrise or sunset the light from the sun travels a long distance through the atmosphere. Many interactions of the sunlight with molecules scatter more and more blue and violet light out of the white sunlight. Removing the blue end of the spectrum of colors results in the remaining light being predominately reddish-orange in color. Hence, the

Fig. 8-36
Blue light is scattered more by molecules and dust particles than red light.

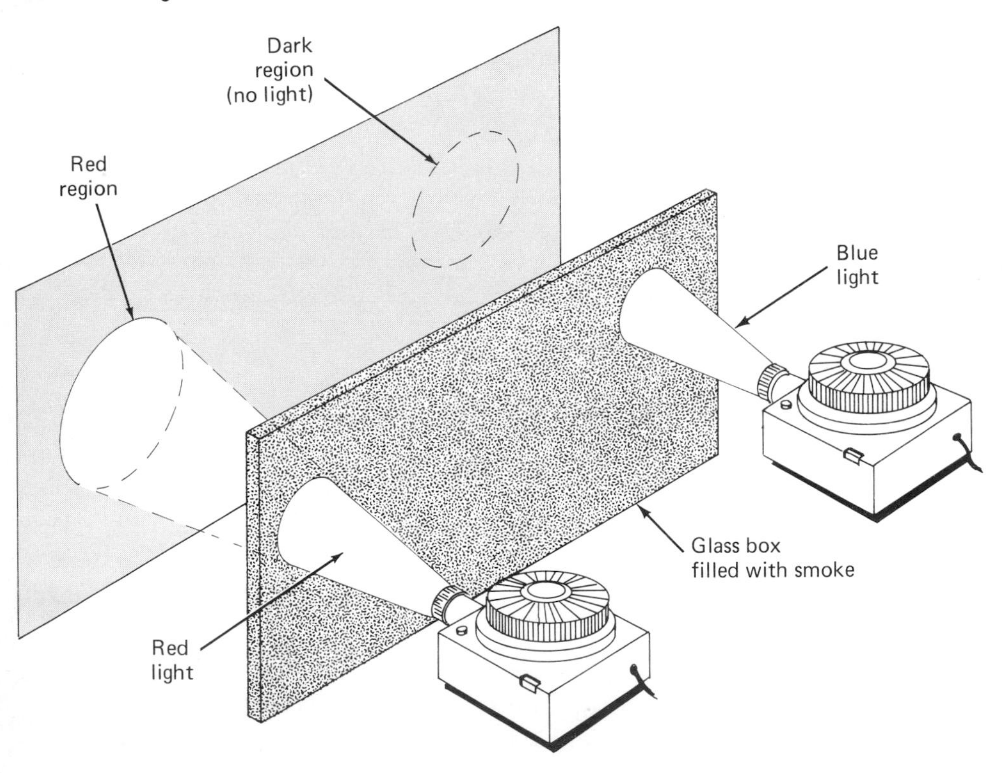

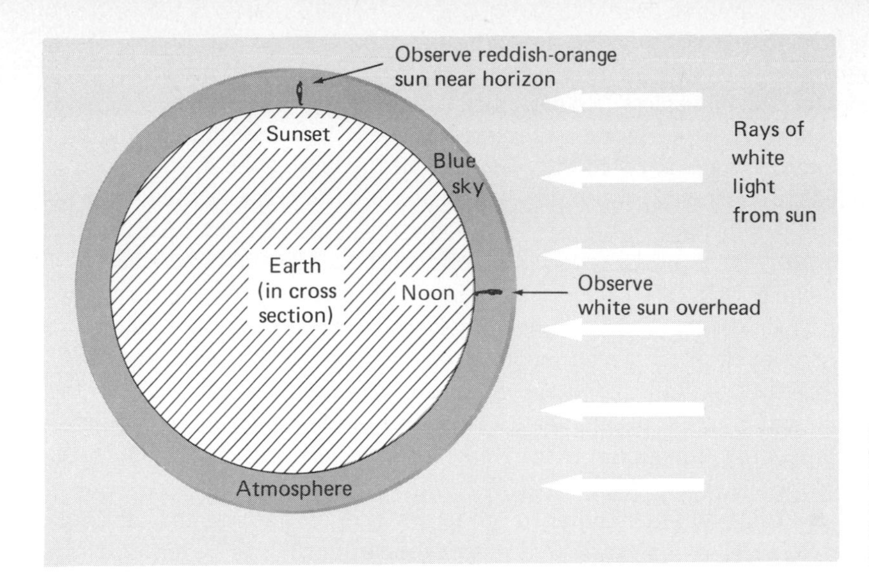

Fig. 8-37
The sun appears reddish-orange near the horizon because much blue light has been scattered from the sunlight.

light of the rising or setting sun not only gives the sun itself a reddish-orange color, but also gives a reddish-orange tint to the clouds and the landscape.

The bluish light scattered out of the sunlight spreads throughout the atmosphere. Over short distances this bluish color of the atmosphere is not noticeable. But increasing the distance to objects being viewed increases the bluish color. Mountains in the distance take on a bluish or purplish tint, even when covered completely with green trees.

To look up and see a blue sky is a special treat for people on the earth. Astronauts in outer space (or on the moon) look up and see a pitch-black sky. Except for pinpoints of light for stars and discs of light for planets in our solar system, the sky is everywhere black. There are no red sunsets or blue skies in outer space, because there is little or no atmosphere to scatter the light rays from the sun.

8.7 TELESCOPES AND BINOCULARS

Do you remember the last time you were sitting in a large football or baseball stadium? You could see players moving, but the details were not clear. If you looked through binoculars, everything seemed so much closer. The details of the scene were now visible to you. The same problem with visibility happens when we sit 200 yards from the stage at a rock concert. With unaided eyes we see only blurred images. But with the aid of some optical instrument we may be able to see a wrinkled brow or the strained neck muscles of a powerful singer.

Binoculars do not of course bring real objects any closer to us. They simply magnify the *image* of the object *on the retina of our eye.* When a larger fraction of the retina is covered by an image, we can see more details. The incident light reacts with a larger number of cells on the retina. More information is sent on to the brain. A magnified image produced by a telescope or binoculars lets more detail be seen.

One type of telescope, a *terrestrial telescope,* has three lenses. The lens closest to the eye is called the *eyepiece,* and the lens at the opposite end is called the *objective.* Between these two lenses is a third lens, as shown in Fig. 8-38. By following incoming rays in the diagram, you can visualize the essential features of the telescope. Notice that each of the three lenses is a converging lens. The incident light rays are spreading out, or diverging, before they strike the three lenses. The lenses refract these rays so that there is not so much spreading. In combination the lenses cause the rays to change direction several times. After passing through the eyepiece, the rays enter the eye in such a way as to illuminate a large area of the retina. The illuminated portion of the retina is larger than it would have been without the telescope, and so we see more detail.

By using different numbers of lenses and by varying the distances between them, a wide variety of practical instruments can be made. *Astronomical telescopes* are made with two lenses. When you look through an astronomical telescope, the image appears upside down. Most astronomers do not mind an inverted image of the stars, but you would not want to borrow one of their telescopes to watch a sailboat race! For objects on the earth, we use *terrestrial telescopes* with upright images. When someone speaks of a *scope* on a hunting rifle, they always mean a terrestrial telescope. The spotting scopes used in target practice and by surveyors are also terrestrial telescopes.

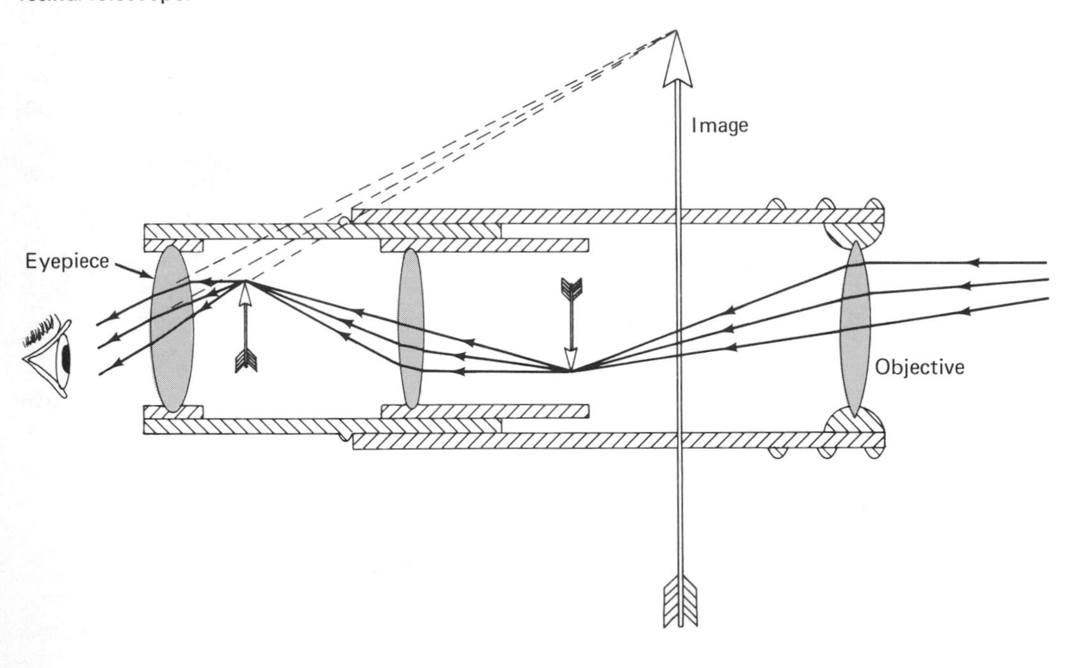

Fig. 8-38
A diagram showing the formation of an image in a terrestrial telescope.

Opera glasses are designed to be used with both eyes. A magnified image is sent to each eye by a separate set of lenses. These glasses produce a small magnification. The magnification is about 3 diameters. A circle 1 centimeter in diameter appears as a circle 3 centimeters in diameter. Stated another way, each linear dimension is made to look 3 times larger. So a square which is 1 centimeter on a side is made to look 3 centimeters high *and* 3 centimeters wide. Opera glasses are used where high magnification is not needed.

Prism binoculars are preferred to opera glasses by most people who enjoy the outdoors. Higher magnifications are easily attained, but there is an increased complexity and hence an increased cost. Most binoculars contain six or more lenses and two prisms on each side. The interior construction is shown in Fig. 8-39. By using such an arrangement, prism binoculars squeeze a long optical path into a small package. Incoming light is reflected twice by each prism. So the total distance traveled is longer than the outside dimension of the binoculars. This feature allows binoculars to achieve magnifications in the range of 6 to 20 without becoming too large to handle.

Common binocular sizes are called 7X35, 7X50, 8X20, 8X40, and so on. The first number in each set is the magnification. The designation 8X20 means that the magnification is *8 diameters*. When you buy binoculars, the first number is always your guide to magnification. Choose higher numbers if you want more magnification. An 8X20 binocular will make all objects appear *8* times closer.

Fig. 8-39
A cutaway view of one side of prism binoculars.
(*Courtesy of Bausch and Lomb.*)

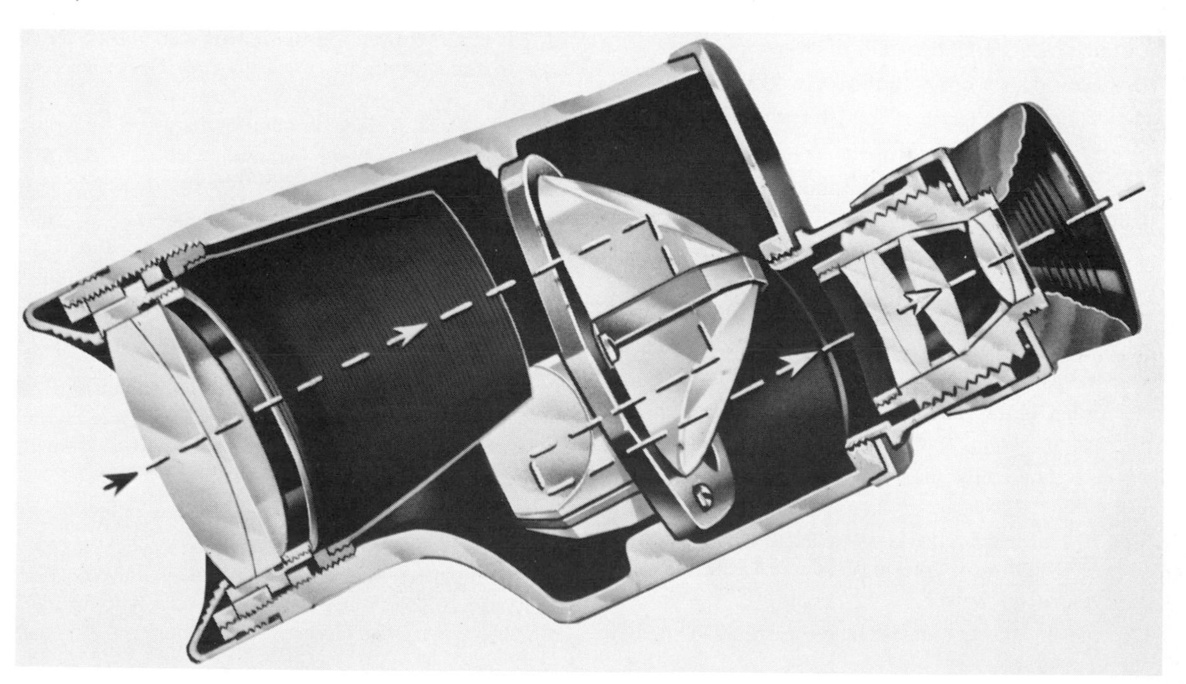

The second number in each set refers to the diameter of the objective lens. An 8X20 binocular has an objective lens which is 20 millimeters in diameter. The size of the objective lens is related to how much light can enter the binoculars. The bigger the lens, the more light that enters. If you wanted binoculars for viewing in full daylight, you might choose objective lenses 20 millimeters in diameter. But if you want to view objects at night or in dim light, you should choose bigger objective lenses. Objective lenses 35 millimeters or more in diameter are usually recommended for nighttime use.

REVIEW You should be able to define or describe the following terms: ray, mirror surface (including convex and concave), refraction, lens (including convex and concave), and dispersion.

You should be able to state the following laws and give examples of their application: law of reflection, law of refraction, and law of dispersion.

DISCUSSION QUESTIONS

1. Describe two processes by which an object can emit light.
2. What is the purpose of the small mirror mounted beneath the slide holder on a microscope?
3. How does a one-way mirror work?
4. What is the cause of light refraction?
5. What is meant by dispersion of light?
6. Draw a diagram showing the dispersion of white light through a prism.
7. How is the apparent time of the setting sun affected by refraction in the atmosphere?
8. Why do objects appear to shimmer when they are viewed through air just above a hot toaster? Would the objects shimmer if the toaster were very cold?
9. What is the cause of a mirage?
10. When a lens is used to magnify an object, the object is not really made larger. What does the lens *really* do?
11. If a clear drinking glass is half full of pure water (no bubbles, dirt, or coloring), how does the eye detect the presence of water?
12. Have you ever seen a concave mirror? Do you have one in your house?
13. Make a comparison between white light and a single spectral color—say, green light—passing through a prism.
14. It is sometimes said that a lens made of ice can be used to focus sunlight to start a fire. Does this seem correct?
15. When the surface of water is disturbed by ripples and waves, light can be focused when it is refracted through the curved surfaces. Where can these effects be seen?
16. Where does refraction take place in the human eye?
17. How does the eye adjust for seeing at different distances?
18. What is the purpose of bifocal glasses?
19. How does the scattering of light by particles in the atmosphere produce the reddish-orange color of sunsets?
20. Why does smoke look blue when it rises from the burning end of a cigarette?
21. Describe how both refraction and reflection are important in the formation of a rainbow.
22. Compare two pairs of binoculars that are rated 7X50 and 10X35. Which of the two admits more light? Which provides the larger magnification?

ANNOTATED BIBLIOGRAPHY

The Life Science Library published by Time-Life Books, New York, has a deservedly good reputation. One of the series, *Light and Vision,* is the best overall reference for this chapter. An unusually complete index makes it an almost universal reference. Color photographs and diagrams illustrate each idea. Two smaller paperbacks with clear discussions are I. M. Freeman, *Light and Radiation,* Random House, Inc., New York, 1968, and C. Rainwater, *Light and Color,* Golden Press, New York, 1971.

A wealth of good material is available on the eye and brain and how we make visual interpretations. If you have an inclination toward biology, don't miss the chapter "Believing Is Seeing" in J. S. Wilentz, *The Senses of Man,* Thomas Y. Crowell Company, New York, 1968. For a psychological leaning, try *Image, Object and Illusion* (Readings from *Scientific American*), W. H. Freeman and Company, San Francisco, 1971. Also, R. L. Gregory has produced two small books available in paperback which read more easily than some novels! His style is delightful in *The Intelligent Eye,* 1970, and *Eye and Brain,* 1966, both by McGraw-Hill Book Company, New York. Two short articles on human vision are "The Perception of Surface Color," by J. Beck, *Scientific American,* August 1975, p. 62, and "Negative Aftereffects in Visual Perception," by O. E. Favreau, *Scientific American,* December 1976, p. 42.

For atmospheric optics, an older book, *The Nature of Light and Colour in the Open Air,* by M. Minnaert, is a classic. It has been republished in paperback by Dover Publications, Inc., New York, 1954. Its 362 pages are often full of detail, even enough for the expert. Just for fun you might read the 9-page chapter "Luminous Plants, Animals and Stones." Several excellent and rare photographs of mirages are in "Mirages" by A. Fraser and W. Mach in *Scientific American,* January 1976, p. 102. One of the best-written articles on sun pillars and haloes around the moon is R. Greenler's "The Origin of Sun Pillars," *American Scientist,* vol. 60, 1972, pp. 292–302.

Another place of departure from this chapter would be into the area of optical illusions. Particularly good—not the standard figures that every schoolchild has seen—are the three chapters, "Painting and Decoration," "Architecture," and "Camouflage," in M. Luckiesh, *Visual Illusions,* Dover Publications, Inc., New York, 1965.

Anyone interested in photography will benefit from the article "The Photographic Lens" by W. Price, *Scientific American,* August 1976, p. 72. An artificial lens to be inserted into the human eye is described in *Time,* Oct. 17, 1977, p. 68.

9 ATOMIC PHYSICS

This chapter will discuss the electronic structure of the atom, the energy of electrons, and the interactions between atoms and electromagnetic radiation. Certain discrete physical quantities, which are the basis for quantum physics, will be introduced. In addition, certain aspects of special relativity will be discussed.

9.1 THE ENERGY LEVELS OF ELECTRONS

In Chap. 6 it was suggested that the atom has a structure similar to that of the sun and the planets of our solar system. Like the sun at the center of the solar system, the nucleus is at the center of the atom. Like the planets revolving around the sun, the electrons are in motion around the nucleus. But *unlike* the energy of the planets in orbit around the sun, the energy of the electrons shows a remarkable characteristic. Planets or satellites can have any value of total energy. This total energy is made up of kinetic energy and *gravitational* potential energy. But an electron can possess only certain values of total energy. The total energy for an electron in an atom is the sum of its kinetic energy and its *electrical* potential energy. In an atom the total energy of an electron is said to be *quantized.*

To say that the energy of an electron is quantized means that the total energy may have only *discrete* values. An electron in an energy level cannot move continuously from one energy level to another. Instead, any energy added to or subtracted from the atom must come in "bundles" or "chunks." Comparing a ramp to a stairway, we could say that the stairway is quantized. A ramp has a continuously changing height. A stairway has individual steps. On a ramp, you can put your foot at any level you choose. But a stairway allows you to put your foot only at one level or at the next.

So, too, an electron in an atom has only certain allowed energies, and these energies differ from one another by finite discrete amounts. An electron in an atom may possess only one of these energies. It is not possible for the electron's energy to be partway between two quantized values.

The idea of quantization is not new, although the words may not be familiar. Charge is quantized, even though these words were not used in Chap. 6. The proton has a charge of $+1.6 \times 10^{-19}$ coulombs—no more, no less. The electron has a charge of -1.6×10^{-19} coulombs—again, no more, no less. And all charges consist of integral multiples of this amount. There is some evidence that these charges are made up of still smaller charged particles called *quarks.* The charges of the quarks are thought also to be quantized.

Electrons cannot be in any energy level, but only in certain definite separate levels. The simplest atomic structure which exhibits quantization is found in the hydrogen atom. A hydrogen atom consists of one proton as the nucleus and one electron in the energy levels. Normally the electron occupies the energy level nearest the nucleus. This level has the lowest energy. Surrounding the lowest energy level are other energy levels which are not occupied by electrons. Figure 9-1 shows an electron occupying the lowest energy level of a hydrogen atom.

The atoms of the chemical elements can be arranged according to their *atomic number.*

Atomic number *The atomic number Z of an element is defined as the number of protons in the nucleus of one of its atoms.*

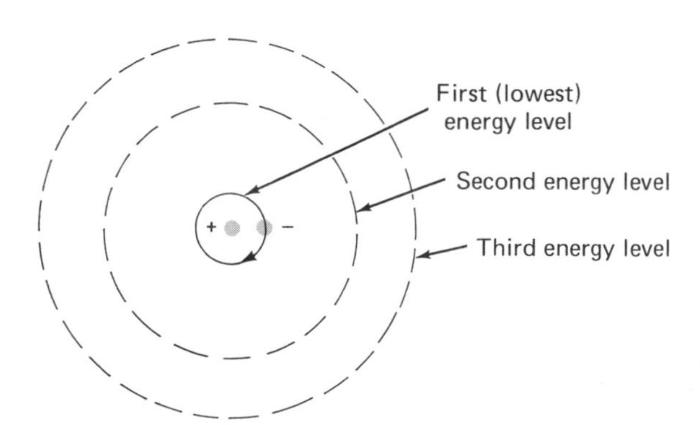

Fig. 9-1
Quantized energy levels surround the nucleus of a hydrogen atom. An electron occupies the lowest energy level in this schematic diagram.

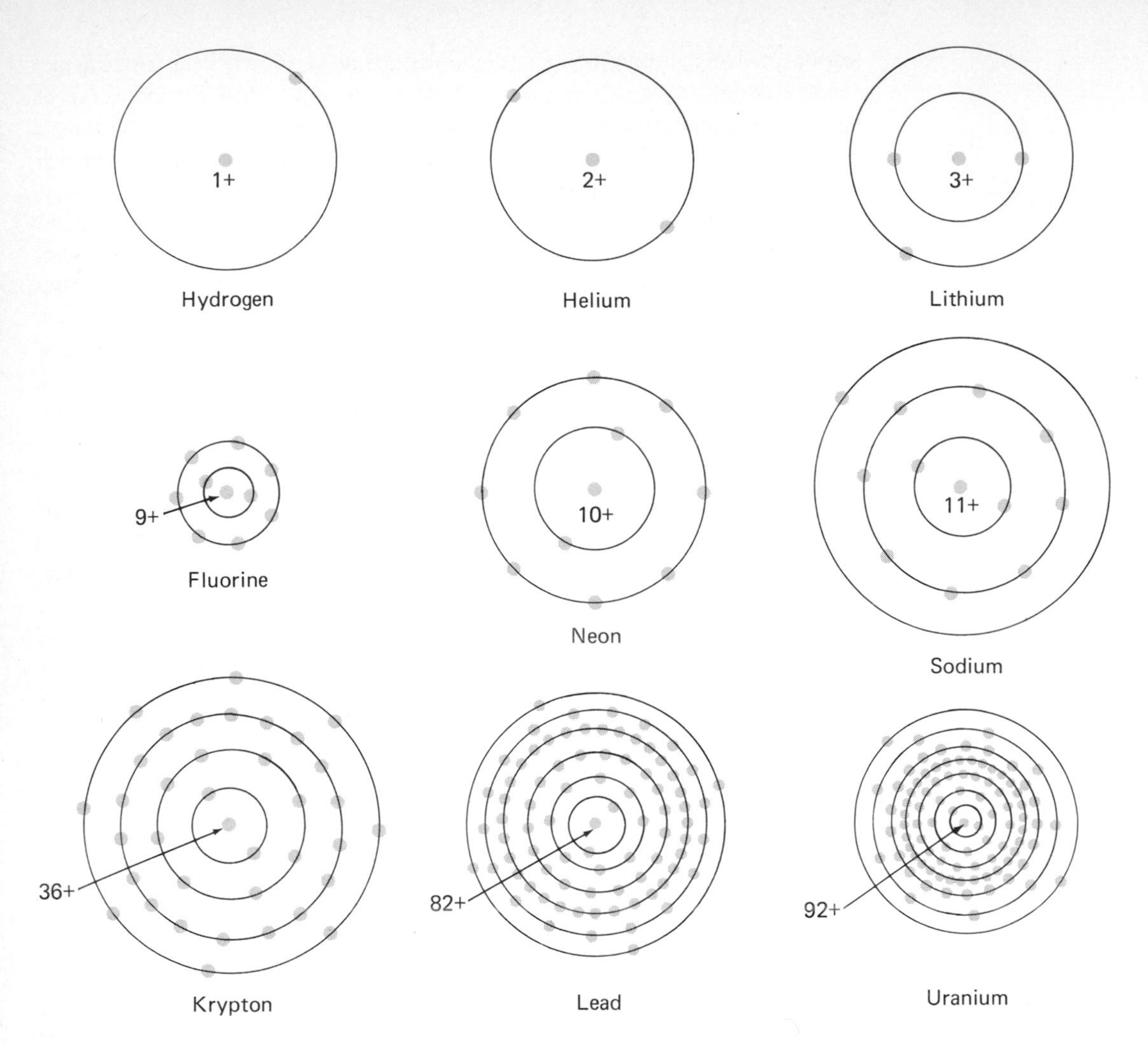

Fig. 9-2
Every chemical element has its own unique and characteristic energy levels, as shown for some elements in this schematic drawing.

A hydrogen atom with one proton has an atomic number equal to 1. A helium atom with two protons has Z equal to 2. Uranium atoms with atomic number 92 contain 92 protons.

The chemical identity of an atom is established by the number of protons in the nucleus. In an *atom* there is charge neutrality in the sense that the number of electrons equals the number of protons. But an *ion* possesses a net electric charge. A net charge occurs whenever the number of electrons differs from the number of protons. Nevertheless, whether in an atom or in an ion, the electrons must occupy certain energy levels according to definite rules.

One such rule establishes an upper limit on the number of electrons which may occupy a given energy level. The level of lowest energy is numbered 1, and the higher energy levels are numbered 2, 3, 4, 5, 6, and so forth. These energy levels are sometimes called *shells.* For the first energy level, or shell, the maximum number of electrons is 2. For the second level the maximum number of electrons is 8; for the third level, 18; for the fourth level, 32; for the fifth level, 50; and for the sixth level, 72. In fact, in the nth level we may place up to a maximum of $2 \cdot n \cdot n = 2n^2$ electrons.

We can think of building up atoms step by step. Starting with hydrogen and proceeding through the elements, the periodic chart arranges elements according to their atomic number. The first energy level is filled with two electrons for the helium atom ($Z = 2$). The second shell starts filling with lithium ($Z = 3$) and is filled with eight electrons for neon ($Z = 10$). The third shell starts being occupied with sodium ($Z = 11$) and is completely occupied with argon ($Z = 18$). The filling of shells continues as the atomic number of each succeeding element increases. The shells of several different atoms are represented in Fig. 9-2. For simplicity, the energy levels are drawn lying in a *plane* with circular shells, but actually they are three-dimensional in character.

Notice in Fig. 9-2 that all atoms are approximately the same size, regardless of the atomic number. For the elements with higher atomic numbers there are obviously more electrons in the energy levels about the nucleus. However, the energy levels are not equally spaced as we move away from the nucleus. Even in an element with a large number of electrons, the overall size of the atom is not much larger than that of an element with only several electrons.

9.2 BOHR'S ENERGY LAW

There are many possible energy levels which surround a nucleus. It is easy to picture these levels in our mind. Some of the levels are far from the nucleus, and others are close to the nucleus. In an atom, electrons occupy only some of these levels. The electrons usually remain close to the nucleus. The levels closer to the nucleus have the lower energies.

This tendency to seek the lower energy levels is much like the tendency of water to seek the lowest level by running downhill.

Ground state *When all the electrons are in their lowest energy levels, the atom is described as being in the ground state.*

Excited state *An atom is said to be in an excited state when one or more electrons are in higher energy levels.*

An electron can, of course, be completely removed from an atom. Then the electron is not merely in a higher energy level. The electron is, in fact, no longer bound to the atom. When an electron is removed from an atom, the atom is called an *ion.*

An atom can be placed in an excited state in several different ways. In each case energy must be added to the atom. One way is to send an electric spark through the atoms of a gas. Many atoms and molecules are in the gaseous state at room temperature. Common examples are hydrogen, oxygen, nitrogen, and neon. A spark passing through a gas excites the atoms or molecules. A spark is simply a short-duration electric current, usually through air. The electrons of a spark move at high speed and thus possess substantial kinetic energy. An electron of the spark collides with an electron in one of the atoms of the gas. Kinetic energy is transferred to the atom's electron, which is then raised to a higher energy level. An electron may, in fact, receive so much kinetic energy that it is completely separated from the atom. So a spark can raise a neon atom to an excited state or to an ionized state as shown in Fig. 9-3.

An atom in an excited state quickly returns to the ground state. The time for the transition is very fast, typically about a billionth of a second. In returning to the ground state the atom emits radiation. The visible light from a spark in air, or a neon light above a cafe, or a fluorescent light bulb are all caused by atoms of a gas returning to the ground state. How does the gas emit light? The answer lies in an investigation of the behavior of the electrons in their energy levels.

As long as an electron stays in a given energy level, there is no change in the energy of the atom. Electrons in higher energy levels, however, will make transitions to vacancies in lower energy levels. As the electron moves to a lower energy level, an atom must release some energy. Otherwise the electron could not occupy the lower energy level. The energy released by the atom cannot simply disappear. Such an event would violate the law of conservation of energy. An electron moving to a lower energy level releases a quantum of electromagnetic radiation known as a *photon.*

Photon *A photon is defined as a quantum of electromagnetic radiation.*

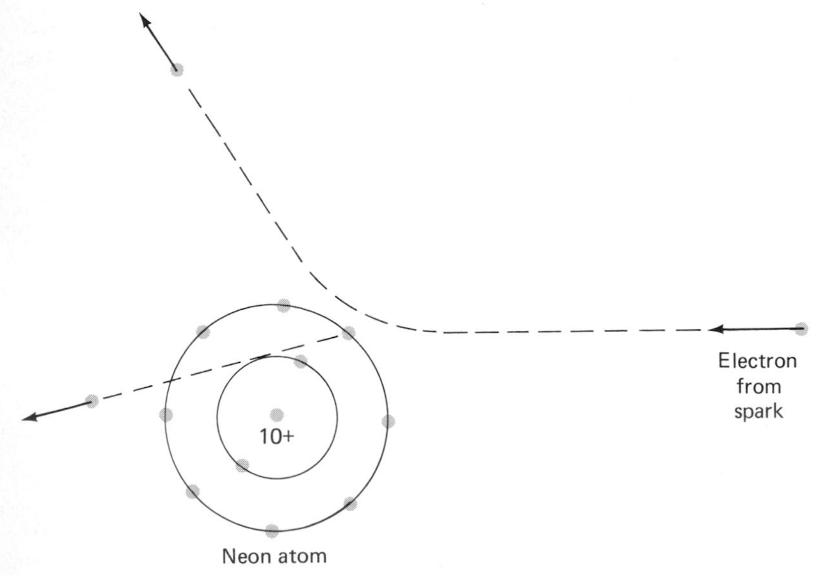

Fig. 9-3
The electron from a spark is shown ionizing a neon atom.

Fig. 9-4
The neon light is caused by excited atoms of the gas returning to the ground state. (*Photograph by Korrine Nusbaum.*)

In general, a photon is the quantum (or "bundle") of energy for *all* forms of electromagnetic radiation. In the case of the outer electron of an atom returning to the ground state, the electromagnetic radiation can take the form of infrared radiation, visible light, or ultraviolet light. Figure 9-5 shows a hydrogen atom emitting a photon because of the transition of an electron from a higher energy level to a lower energy level.

All electromagnetic radiation has a particular frequency. Thus, every photon has a particular frequency. The frequency of the photon emitted or absorbed by an atom is determined by the difference in energy between the two levels involved in the transition. This relationship is called *Bohr's energy law.*

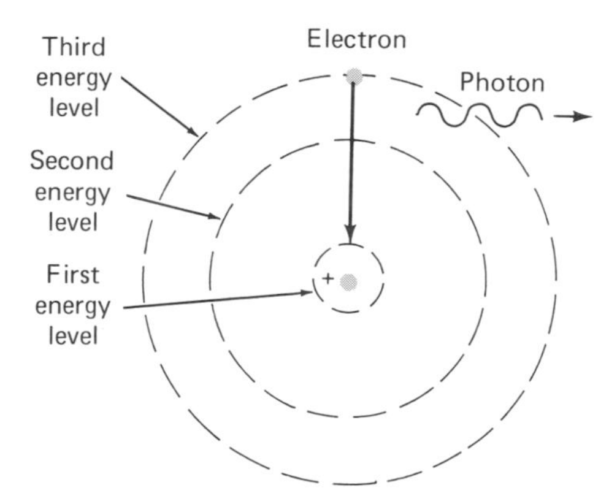

Fig. 9-5
When an electron transfers, or "jumps," from a higher energy level to a lower energy level, a photon is emitted.

Bohr's energy law **Electrons may "jump" from a higher energy level to a lower energy level. The frequency of the emitted photon is determined by the difference in energy between the initial and final energy levels.**

Bohr's energy law can be expressed in a formula,

$$(\text{Difference in energy levels}) = (\text{a constant})(\text{frequency})$$
$$\Delta E = hf$$

where ΔE = difference in energy between the initial and final levels (measured in joules)
h = Planck's constant, which is equal to 6.62×10^{-34} J·sec
f = frequency (measured in hertz)

When an electron moves between energy levels, it need not just make one jump and emit a single photon. Electrons will often jump from level to level, finally landing in the level of lowest energy. In such a process of multiple jumps, several photons are emitted. Each photon has a frequency consistent with Bohr's energy law. The result is that an excited atom may emit not just a single photon, but a series of photons. Some of the photons may have a frequency in the visible region of the electromagnetic spectrum. But it is also possible that some of the photons have frequencies in the infrared or ultraviolet regions of the electromagnetic spectrum.

Although all atoms have a series of higher and higher energy levels, the arrangement of these levels is different for the atoms of each chemical element. One element, such as hydrogen, emits a series of photons with certain frequencies or, equivalently, certain wavelengths (see Fig. 9-6). A different element, such as neon, emits a series of photons with different wavelengths. In this way, each chemical element has its own unique group of emission wavelengths.

The specific wavelengths of photons emitted by all the elements have been cataloged. Catalogs listing the wavelengths of electromagnetic waves

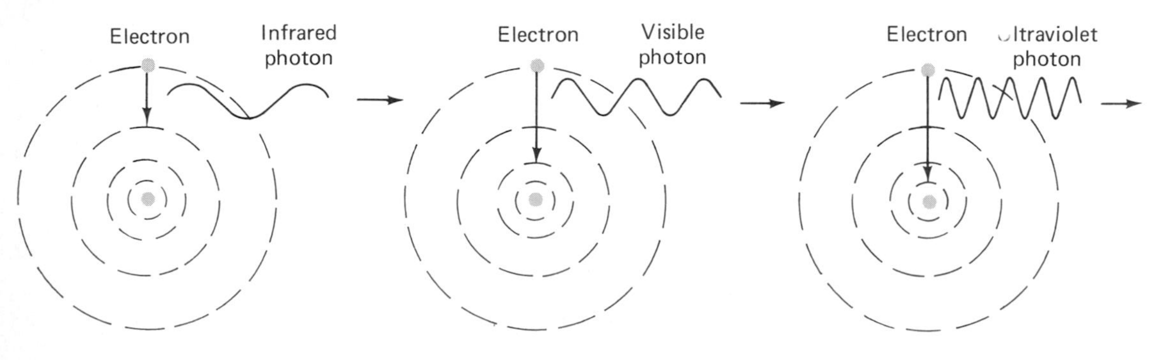

Fig. 9-6
Each emitted photon from a hydrogen atom has a characteristic frequency. The change in energy level and the frequency of the emitted photon are related by Bohr's energy law.

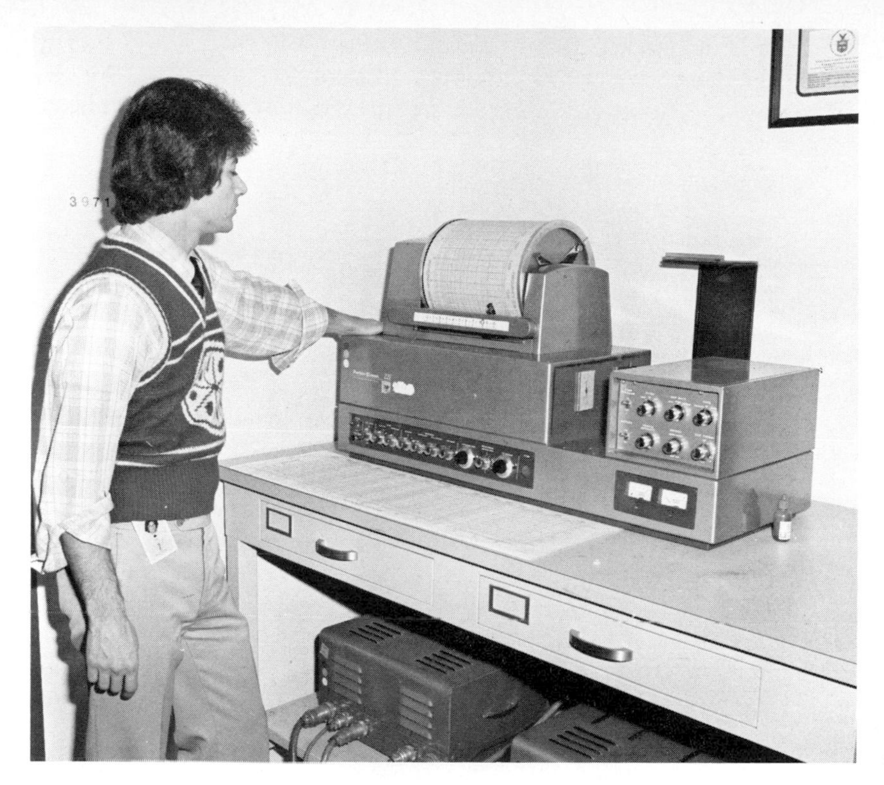

Fig. 9-7
The excited atoms of every element emit a different series of photons of light. An unknown chemical can be identified in a police laboratory by comparing the emission frequencies with a catalog of the known frequencies of various chemicals.
(*Photograph courtesy of FBI.*)

emitted by excited atoms can be used to identify unknown substances. To identify an unknown substance, one uses the following procedure. The unknown substance is supplied energy to excite some of the atoms. Then, as the atoms return to their ground state, the emission wavelengths are measured. When these wavelengths are compared to the catalog listing and a match is found, the unknown has been identified. It is quite common for a police laboratory to use this method of analysis in identifying poisons, as shown in Fig. 9-7. The same method can be used to identify explosives or unknown chemicals.

9.3 THE LASER

One of the most exciting scientific developments for science fiction fans has been the laser. The laser has taken the idea of a ray gun out of the realm of the imagination and made it into a reality. Although lasers are not made in the shape of a gun, some are capable of burning holes in solid objects. This and a few other properties that we will discuss make lasers seem like the ray guns seen in science fiction movies. Lasers also have several other practical uses. In the process of holography, visible laser light can be used to make a truly three-dimensional photograph.

The electromagnetic waves emitted by a laser are similar in some ways to

Fig. 9-8
If light sabers were made from lasers, could there be light beams a meter long? Could the light beams make noise as they move? Could two "light sabers" crash against each other? (*Photograph from Springer/Bettmann Film Archive.*)

ordinary light. The light from a laser has the same speed as all other electromagnetic radiation. Laser light can be reflected and refracted by mirrors and lenses. But unlike ordinary light, laser light is produced by a somewhat rare kind of excited atom.

A common demonstration laser uses a mixture of helium and neon gases to produce red laser light. Excited atoms normally emit a photon in returning to a less excited state. Helium, however, can release the excess energy of one of its higher energy levels in a different way. When an excited helium atom collides with a neon atom in its ground state, the helium atom transfers its energy to the neon atom. The helium atom returns to its ground state without emitting a photon. At the same time the neon atom is excited. In a laser the helium atoms supply energy to excite neon atoms, and the excited neon atoms then produce laser light.

In a laser an excited neon atom will spontaneously emit a photon with a wavelength of 0.6328 millionths of a meter. This wavelength corresponds to a photon of red light. This single photon causes another excited neon atom to return to a lower energy state and emit a photon. Now two identical photons of red light are moving through the laser. As the two photons pass other excited neon atoms, more atoms are stimulated to emit photons of red light (see Fig. 9-9). The amount of laser light is increased. Or phrased differently, there is an amplification of the electromagnetic radiation. The word laser comes from the first letters of the description: *l*ight *a*mplification by *s*timulated *e*mission of *r*adiation.

A laser is constructed by filling a glass tube with a mixture of 10 times as

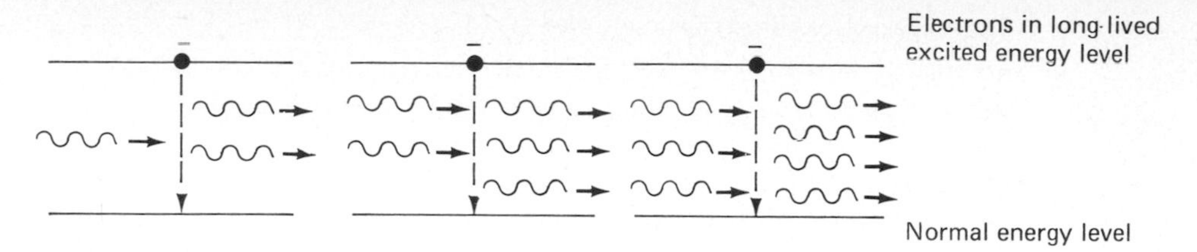

Fig. 9-9
Stimulated emission of photons occurs in a laser. One photon triggers the release of a second photon, two photons trigger the release of other photons, and so on.

much helium as neon. Attached to one end of the tube is a regular mirror. Attached to the other end is a half-silvered mirror similar to the one-way mirror described in Sec. 8-2. Laser light is reflected back and forth between the mirrors and stimulates the emission of photons. At first the laser light is weak and only a small amount penetrates the half-silvered mirror. Very quickly, however, as more photons join the beam, the laser light becomes bright. Then many photons pass through the half-silvered mirror. This beam which comes out of the laser housing is the beam which is seen during a demonstration.

The light emitted by a laser is extremely bright. You should never look directly into a laser. If you did, the light would appear as bright as the sun. Laser light is so bright that it can seriously injure your eyes. To view a laser safely, always look at it from the side. Figure 9-10 shows a laser in operation.

The brightness of a light wave is very similar to the loudness of a sound wave discussed in Sec. 5-7. Both brightness and loudness are *subjective* characteristics of waves. The *objective* word related to brightness and loud-

Fig. 9-10
A laser beam should always be viewed from the side to protect the eyes.

ness is intensity. Most lasers used for demonstrations do not have sufficient intensity to burn holes in objects. For cutting purposes it is common to use a carbon dioxide laser. This kind of laser emits an extremely high intensity beam of infrared radiation. When laser light is absorbed by a material, the internal energy of the material increases. As a result, the temperature of an object can be raised at the spot where the laser light strikes. If the laser is of sufficiently high intensity, the material vaporizes at this spot. High-intensity lasers can drill holes or cut objects by burning through them, as shown in Fig. 9-11.

A medical use of a laser is to "weld" a detached retina back in place (see Fig. 9-12). A retina that has become separated from the wall of the eye receives a short, intense exposure to laser light. The absorbed energy raises the temperature some 15 or 20 C deg at a small spot on the retina and adjacent tissue. The burned tissue forms a scar that holds the retina and adjacent wall of the eye together. The time of exposure to the laser light is typically a thousandth of a second, the shortest surgery on record.

Photography with laser light is certainly a different photographic technique than has been used for the past hundred years. With ordinary light a camera produces a two-dimensional image of a three-dimensional object.

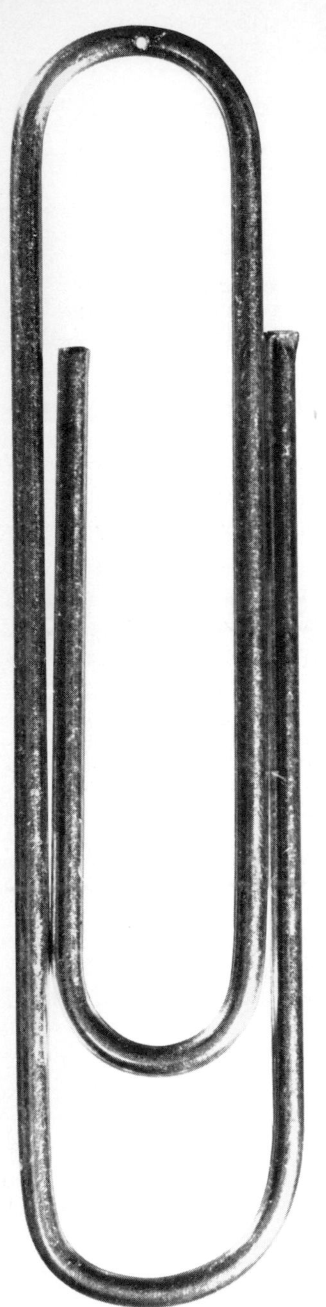

Fig. 9-11
A high-intensity infrared laser beam burned a hole through the top of the paper clip. (*Photograph courtesy of DOE.*)

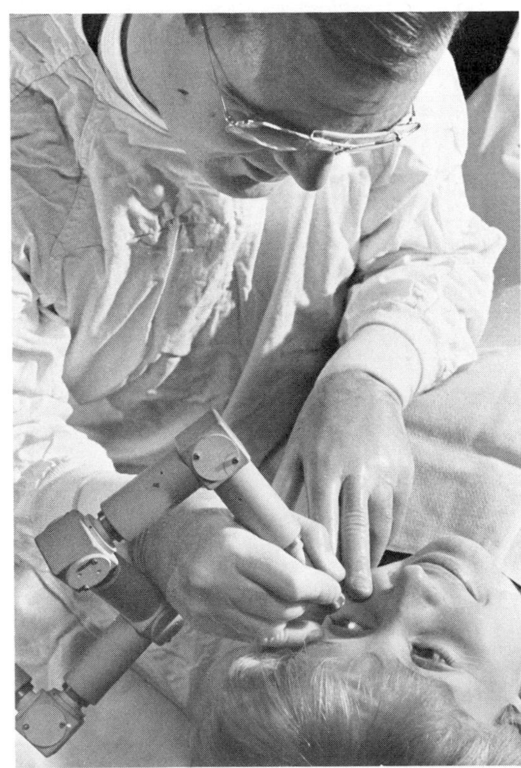

Fig. 9-12
A laser ophthalmoscope is used to examine a patient's eye before the detached retina is welded back in place. (*Photograph courtesy of AT & T.*)

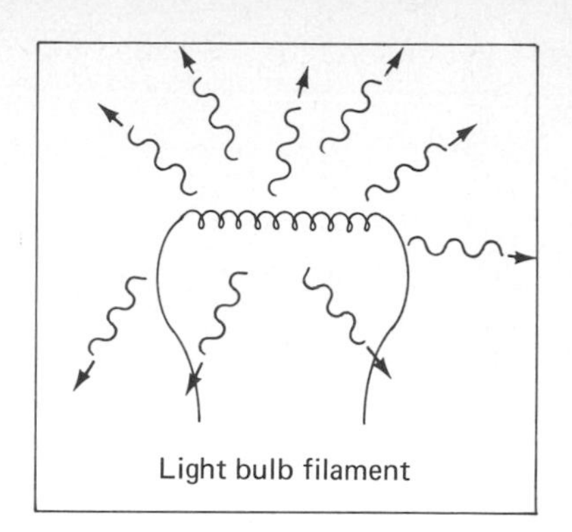

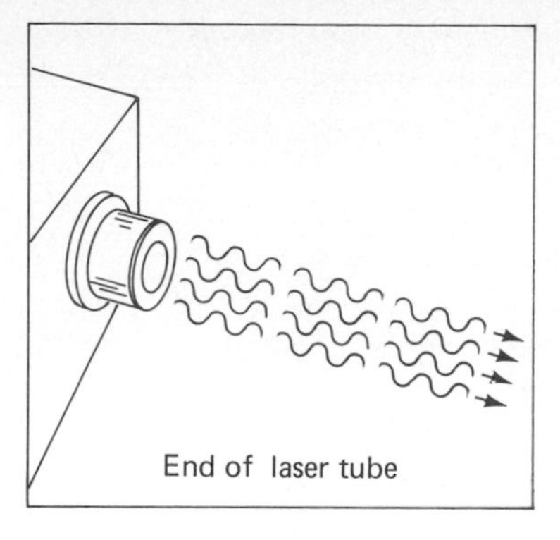

Fig. 9-13
Schematic representations of incoherent radiation from a light bulb and coherent radiation from a laser are shown.

With laser light, however, it is possible to reproduce three-dimensional images of an object. This improvement in photography is due to the use of *coherent* laser light. In a coherent light beam, the crests of every photon are lined up exactly with the crests of every other photon. Light from any source other than a laser is not coherent. For example, there is no lining up of the individual photons in sunlight. Thus, sunlight is not coherent radiation. The difference between coherent and incoherent light is shown in Fig. 9-13.

An object to be photographed in three dimensions is illuminated by laser light. Some light is reflected from the surface of the object and strikes photographic film. This light contains the three-dimensional image of the object and is no longer coherent. In addition, some light from the laser strikes a mirror, which does not change the coherent nature of laser light. The light reflected from the mirror is also recorded in the negative. The two light beams produce a complicated exposure in the negative, which is called a

Fig. 9-14
A hologram viewed with ordinary incoherent light appears as a complicated pattern. (*Photograph by Ted Rozumalski. From Black Star.*)

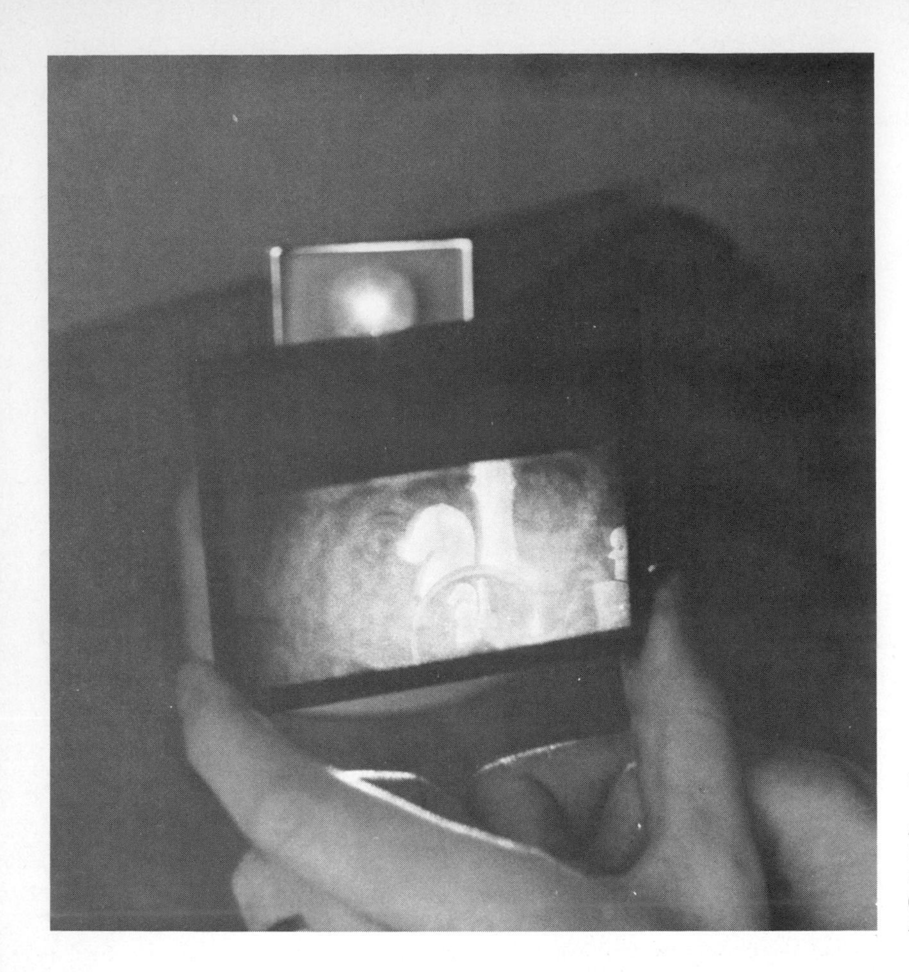

Fig. 9-15

A hologram viewed with coherent laser light appears as a three-dimensional image. (*Photograph by Ted Rozumalski. From Black Star.*)

hologram. A hologram does not look like a typical negative when viewed in ordinary light, as shown in Fig. 9-14.

The hologram records a complicated mixture of wave patterns. The actual three-dimensional image is mixed together with the coherent laser beam. By mixing these two waves, the negative contains different exposures not only in the height and width of the light-sensitive chemicals, but also through the thickness of these chemicals. To unmix the two sets of wave patterns, laser light is shone on a hologram. The negative absorbs some of the coherent laser light while reproducing the original wave. The three-dimensional image appears to leave the hologram and can be viewed from many different angles. An image from a hologram allows you to look around edges or sides of the object, just as you could do if the object were really in front of you. The object can no longer be touched, but the complete wave pattern of a three-dimensional object is there to be seen. Figure 9-15 is a photograph of a hologram illuminated with laser light.

9.4 PHOTOELECTRIC EFFECT

The first part of this chapter discussed the process of an electron in an excited atom returning to a lower energy level. When the energy of the atom is decreased, a photon is created and escapes from the atom. The opposite process may also occur. A photon may be completely absorbed by an atom, and when it is, the atom is placed in an excited state. If the photon has enough energy to completely separate the electron from the atom, the process is known as the *photoelectric effect.*

Photoelectric effect *The photoelectric effect is defined as the liberation of electrons by electromagnetic radiation incident on a substance.*

In the photoelectric effect, a photon disappears and an electron is liberated from an atom, as shown schematically in Fig. 9-16.

The photoelectric effect is the basis of the "electric eye." In the operation of an electric eye a beam of light strikes a specially coated metal surface inside a glass tube. The light causes electrons to be emitted from the metal surface. The stream of liberated electrons then strikes another metal surface, where it moves through the metal as an electric current. The current can be led by a wire to other devices or to circuits outside the electric eye. As long as light strikes the electric eye, electricity flows in the circuit. If the light beam is stopped, then electricity no longer flows. So interrupting the light beam is equivalent to switching off the electricity.

An everyday example of the use of the electric eye is found in many elevators. The doors of an elevator should not close on anyone entering or leaving the elevator. So, as a safeguard, an electric eye is on one door and a small lightbulb is on the other (see Fig. 9-17). As long as no one steps into the light beam, the light strikes the electric eye. Charge flows in the control circuit, and the elevator door closes. But if someone steps into or out of the elevator, the light beam is interrupted. Within the electric eye electrons stop moving. The charge stops flowing, and the doors do not close. An electric eye is an important part of the circuit for automatic elevators.

Electric eyes can also be used in detection circuits to open doors that are normally closed. Many supermarket doors open when a light beam is broken

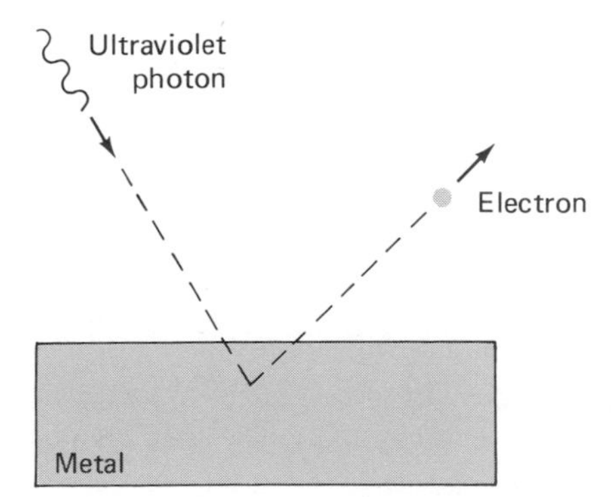

Fig. 9-16
In the photoelectric effect a photon ejects an electron from an atom of the metal.

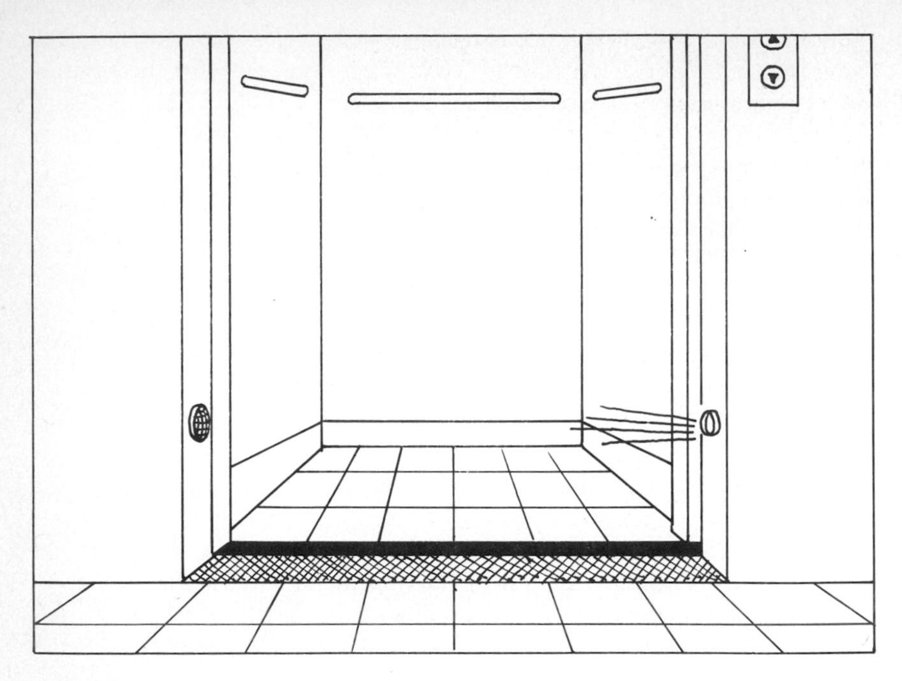

Fig. 9-17
Elevator doors can be controlled by a detection circuit. On one door is a bright light and on the other door is an electric eye.

by a customer. Once again, an electric eye stops generating electricity when a person passes through the beam of light. The detection circuit causes the supermarket doors to open. Instead of opening doors, a photoelectric detection circuit can also be used to set off burglar alarms. Many museums use electric eyes to detect the presence of unauthorized personnel after normal exhibition hours.

The problem of supplying energy for society in the future may be solved to some extent by using the photoelectric effect. The sun sends tremendous amounts of energy to the earth in the form of visible light and ultraviolet radiation. When specially prepared materials are exposed to sunlight, the energy of the photons is converted directly to electricity. Such devices are called *solar cells.* Solar energy is free, but the manufacture of solar cells is quite expensive. Electricity from solar cells costs much more than the electricity now being sold. Many solar cells, covering several square miles, would be needed to generate as much as a present facility. Nevertheless, solar cells are being used successfully to power satellites orbiting the earth (see Fig. 9-18). If the price of solar cells can be significantly reduced, they may eventually be used to generate electricity for homes.

9.5 X-RAYS

The atomic processes discussed in this chapter are based on Bohr's energy law. Photons are produced when atoms lose energy, and atoms gain energy

Fig. 9-18
Solar cells on an orbiting satellite convert light energy directly into electrical energy. (*Photograph courtesy of NASA.*)

when photons disappear. In the operation of the electric eye the photons come from the visible and ultraviolet portions of the electromagnetic spectrum. At a higher frequency in the electromagnetic spectrum are x-rays. Described in another way, x-rays have shorter wavelengths than ultraviolet radiation. Bohr's energy law states that the higher the *frequency* of a photon, the higher the *energy* of the photon. Since x-rays have higher frequencies than ultraviolet radiation, x-ray photons have more energy than ultraviolet photons (see Table 9-1).

To produce x-rays, electrons must lose much more energy than is necessary to produce ultraviolet radiation. One way to provide electrons with very large amounts of energy is to use a high-voltage source of electricity. Electric energy can be transferred to the electrons. When the electrons later give up this energy, x-rays are produced. This is done in a device known as an x-ray tube, shown in Fig. 9-19.

An x-ray tube is quite simple in construction. At one end of the x-ray tube there is a glowing hot wire. When the wire is red-hot, some conduction electrons escape from the surface of the metal. At the other end of the x-ray

ELECTRON MICROSCOPES

Optical instruments have been used for centuries. Powerful telescopes have scanned the sky searching for faint stars. And small magnifying glasses have been used to learn about the details of the wing of a fly or the engraving on a postage stamp. All these activities use visible light (electromagnetic waves) in order to form a magnified image. But electrons can be used similarly to form enlarged images. An electron microscope uses electrons to produce a visible image on a screen. The images in electron microscopes can have magnifications up to 1 million. A magnification of 1 million means that every linear dimension in the image is 1 million times longer than in the real object. This is a great advantage when compared with the magnification of a strong optical microscope—about 1,000.

If we trace the paths of electrons as they move through an electron microscope, we begin with the electron gun. The electron gun is the source of electrons. Usually it consists of a tungsten filament heated to incandescence. As the tungsten glows, it emits electrons. These electrons are then accelerated across a region of vacuum by a high voltage. A very low air pressure must be used. Electrons scatter from any molecules of gas which remain. The pressure must be maintained at about one ten-millionth of atmospheric pressure. The voltages in common use are in the range 100,000 to 1,000,000 volts.

Electrons must be accurately guided through the electron microscope. Just as visible light is focused by lenses, electrons too are focused by magnetic lenses. Current is sent through coils of wire to produce magnetic fields. These fields exert forces and bend the paths of electrons. In this way the electrons are focused, as shown in the figure. An electron microscope has at least three or four magnetic lenses. A diagram of a magnetic lens is shown below.

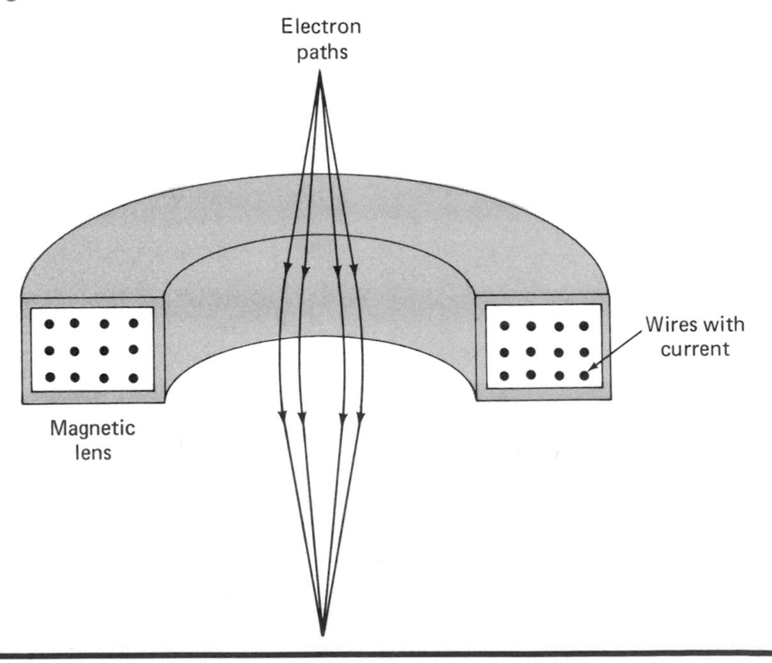

In one kind of microscope the electrons are sent through the specimen. Various parts of the specimen vary in density and absorb some of the electrons. Differing absorption of electrons provides the contrast in the final image. Only very thin slices of material can be used in such a transmission electron microscope. The electrons which get through the specimen are brought to a focus on a light-emitting fluorescent screen or a photographic plate.

In a scanning electron microscope the beam of electrons is sent to a small region of the sample (see photograph below). The beam is then directed to different regions of the sample until all regions have been scanned. This type of microscope is often used on thick materials. When the material is so thick that electrons cannot be sent through, an alternative arrangement is used. Electrons are directed at an angle to the sample. The electrons collide with the sample and cause other electrons to be emitted. These secondary electrons are then focused to get an image of the surface.

(Photograph from Wide World Photos.)

(Photograph: Scanning electron micrograph of an erythrocyte enmeshed in fibrin (about 20,500 X). Emil Bernstein and Eila Kairinen, Gillette Research Institute, Rockville, Maryland.)

The photograph to the left was taken with an electron microscope. It shows a red blood cell. The extreme enlargement can be judged by knowing that a red blood cell is about $\frac{7}{1,000,000}$ of a meter in diameter. A red blood cell looks like a round pillow which has been squashed in on each side. The red blood cell in this photograph is covered with fibrin strands, which is the substance that forms a blood clot. Fibrin strands form a web which stops the flow of red blood cells.

Electron microscopes have been used to study many viruses and bacteria. They are also exceptionally valuable for determining the nature of metal surfaces. Several experimental microscopes have been built employing voltages greater than 3 million volts. It will be interesting in future years to observe the new details revealed by these machines.

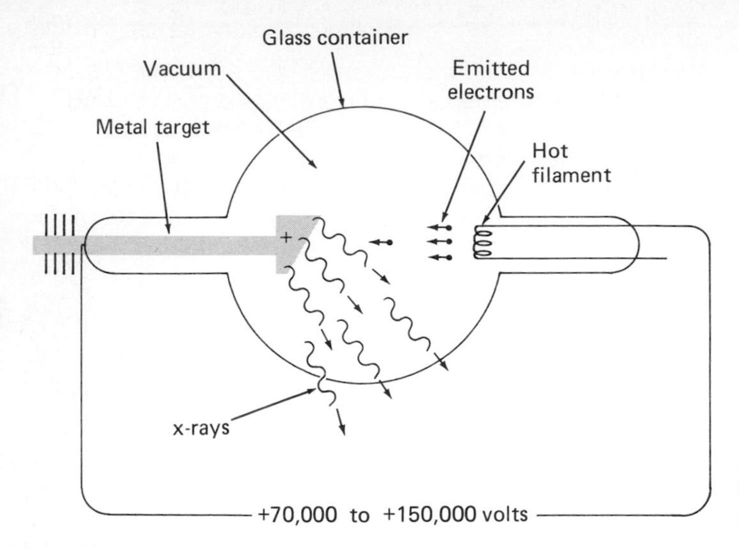

Fig. 9-19
In an x-ray tube electrons colliding with a metal target cause the emission of x-rays.

tube a large piece of metal serves as a target for the electrons. Putting a large positive voltage on the target causes the negative electrons to be strongly attracted. The voltage of the target is typically +70,000 to +150,000 volts in dental and medical x-ray tubes. The strong attraction of the target for the electrons causes the electrons to gain a correspondingly high kinetic energy. The electrons then strike the target and lose their kinetic energy. In losing kinetic energy the electrons produce x-rays.

In some ways x-rays are similar to visible light, and in other ways x-rays are different. Like visible light, x-rays will expose photographic film. The chemicals in the film are affected by x-rays, and so pictures can be taken using x-rays as easily as using visible light. But unlike visible light, x-rays can pass through many objects which are opaque to visible light.

Table 9-1
Comparison of wavelengths, frequencies, and energies within the electromagnetic spectrum

	Common name of radiant energy	
Decreasing wavelength ↓	Radio waves	Increasing frequency and energy ($\Delta E = hf$) ↓
	Infrared radiation	
	Visible light	
	Ultraviolet radiation	
	X-rays	
	Gamma rays	

Table 9-2a
Periodic chart of elements

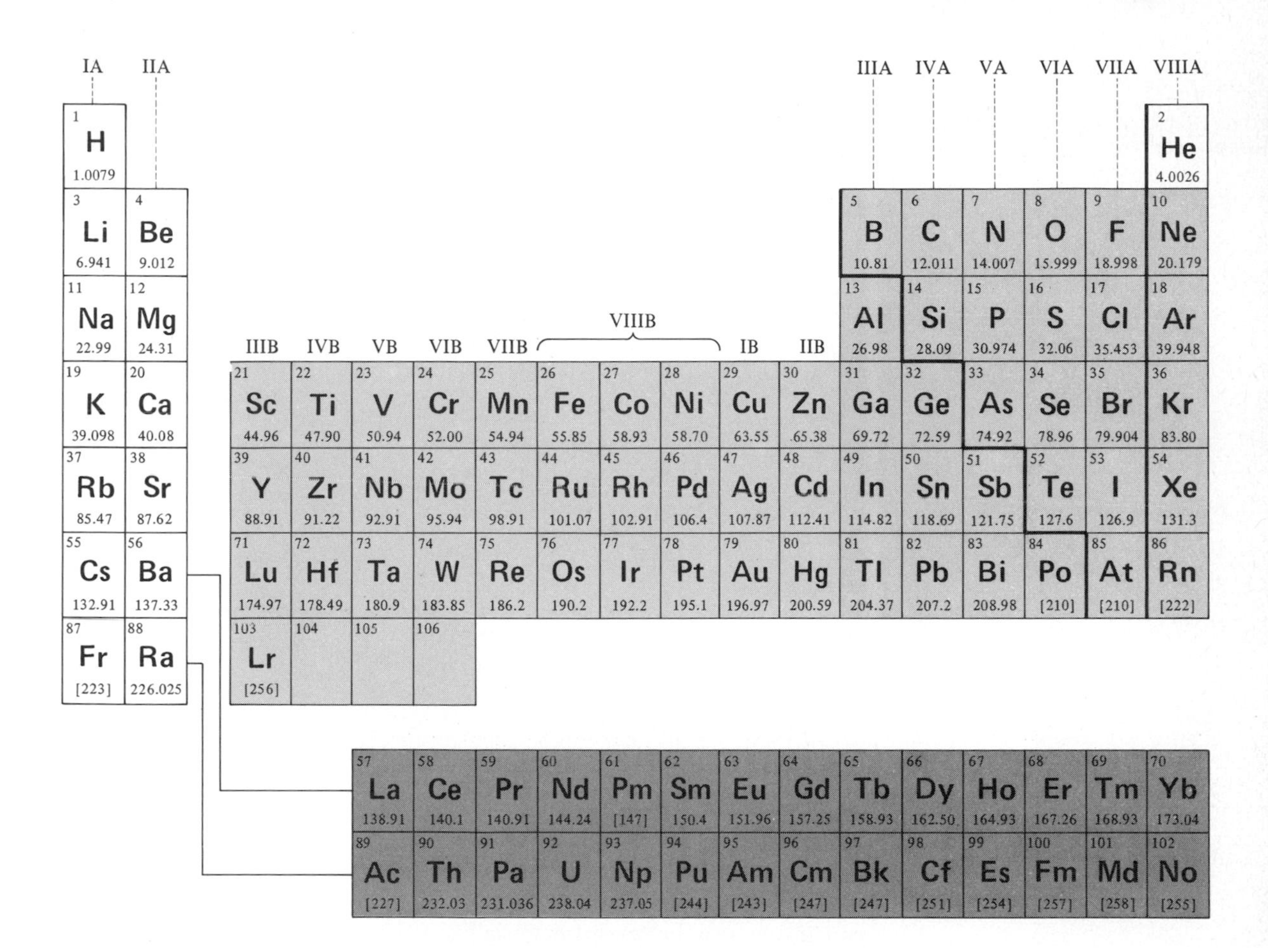

Representative elements: H, B
Transition elements: Fe
Inner transition elements: Ce

IA	IIA	IIIB	IVB	VB	VIB	VIIB	VIIIB	VIIIB	VIIIB	IB	IIB	IIIA	IVA	VA	VIA	VIIA	VIIIA
1 H 1.0079																	2 He 4.0026
3 Li 6.941	4 Be 9.012											5 B 10.81	6 C 12.011	7 N 14.007	8 O 15.999	9 F 18.998	10 Ne 20.179
11 Na 22.99	12 Mg 24.31											13 Al 26.98	14 Si 28.09	15 P 30.974	16 S 32.06	17 Cl 35.453	18 Ar 39.948
19 K 39.098	20 Ca 40.08	21 Sc 44.96	22 Ti 47.90	23 V 50.94	24 Cr 52.00	25 Mn 54.94	26 Fe 55.85	27 Co 58.93	28 Ni 58.70	29 Cu 63.55	30 Zn 65.38	31 Ga 69.72	32 Ge 72.59	33 As 74.92	34 Se 78.96	35 Br 79.904	36 Kr 83.80
37 Rb 85.47	38 Sr 87.62	39 Y 88.91	40 Zr 91.22	41 Nb 92.91	42 Mo 95.94	43 Tc 98.91	44 Ru 101.07	45 Rh 102.91	46 Pd 106.4	47 Ag 107.87	48 Cd 112.41	49 In 114.82	50 Sn 118.69	51 Sb 121.75	52 Te 127.6	53 I 126.9	54 Xe 131.3
55 Cs 132.91	56 Ba 137.33	71 Lu 174.97	72 Hf 178.49	73 Ta 180.9	74 W 183.85	75 Re 186.2	76 Os 190.2	77 Ir 192.2	78 Pt 195.1	79 Au 196.97	80 Hg 200.59	81 Tl 204.37	82 Pb 207.2	83 Bi 208.98	84 Po [210]	85 At [210]	86 Rn [222]
87 Fr [223]	88 Ra 226.025	103 Lr [256]	104	105	106												

57 La 138.91	58 Ce 140.1	59 Pr 140.91	60 Nd 144.24	61 Pm [147]	62 Sm 150.4	63 Eu 151.96	64 Gd 157.25	65 Tb 158.93	66 Dy 162.50	67 Ho 164.93	68 Er 167.26	69 Tm 168.93	70 Yb 173.04
89 Ac [227]	90 Th 232.03	91 Pa 231.036	92 U 238.04	93 Np 237.05	94 Pu [244]	95 Am [243]	96 Cm [247]	97 Bk [247]	98 Cf [251]	99 Es [254]	100 Fm [257]	101 Md [258]	102 No [255]

The ability of x-rays to penetrate solid objects depends upon the atomic nature of the object. Chemical elements with low atomic number Z are easily penetrated by x-rays. Some of these low atomic number elements such as hydrogen, carbon, and oxygen make up human flesh. X-rays easily pass through flesh. On the other hand, x-rays that pass through flesh, for example, do not penetrate deeply into bone or steel. It is the calcium atoms in bones and the iron atoms in steel that absorb x-rays. Each of these chemical elements has a higher atomic number than carbon, hydrogen, or oxygen.

One of the best shields against x-rays is lead. Lead has a very high atomic number. And lead is relatively inexpensive. Look at the periodic chart of the elements shown in Table 9-2*a*. There are many chemicals with a high atomic

Table 9-2b
Names and symbols of elements

Actinium	Ac	Erbium	Er	Mercury	Hg	Samarium	Sm
Aluminum	Al	Europium	Eu	Molybdenum	Mo	Scandium	Sc
Americium	Am	Fermium	Fm	Neodymium	Nd	Selenium	Se
Antimony	Sb	Fluorine	F	Neon	Ne	Silicon	Si
Argon	Ar	Francium	Fr	Neptunium	Np	Silver	Ag
Arsenic	As	Gadolinium	Gd	Nickel	Ni	Sodium	Na
Astatine	At	Gallium	Ga	Niobium	Nb	Strontium	Sr
Barium	Ba	Germanium	Ge	Nitrogen	N	Sulfur	S
Berkelium	Bk	Gold	Au	Nobelium	No	Tantalum	Ta
Beryllium	Be	Hafnium	Hf	Osmium	Os	Technetium	Tc
Bismuth	Bi	Helium	He	Oxygen	O	Tellurium	Te
Boron	B	Holmium	Ho	Palladium	Pd	Terbium	Tb
Bromine	Br	Hydrogen	H	Phosphorous	P	Thallium	Tl
Cadmium	Cd	Indium	In	Platinum	Pt	Thorium	Th
Calcium	Ca	Iodine	I	Plutonium	Pu	Thulium	Tm
Californium	Cf	Iridium	Ir	Polonium	Po	Tin	Sn
Carbon	C	Iron	Fe	Potassium	K	Titanium	Ti
Cerium	Ce	Krypton	Kr	Praseodymium	Pr	Tungsten	W
Cesium	Cs	Lanthanum	La	Promethium	Pm	Uranium	U
Chlorine	Cl	Lawrencium	Lr	Protactinium	Pa	Vanadium	V
Chromium	Cr	Lead	Pb	Radium	Ra	Xenon	Xe
Cobalt	Co	Lithium	Li	Radon	Rn	Ytterbium	Yb
Copper	Cu	Lutetium	Lu	Rhenium	Re	Yttrium	Y
Curium	Cm	Magnesium	Mg	Rhodium	Rh	Zinc	Zn
Dysprosium	Dy	Manganese	Mn	Rubidium	Rb	Zirconium	Zr
Einsteinium	Es	Mendelevium	Md	Ruthenium	Ru		

number near lead. All these elements are very good at stopping x-rays, but most of them are also very expensive to purchase.

It is not possible to state a definite thickness of material which will stop all x-rays. Lead and other chemicals merely stop a portion of the x-rays. For example, a piece of lead 4.5 mm thick will absorb about 99.99 percent of the x-rays from an x-ray tube which operates at 200,000 volts. If the thickness of lead were less, a lower percentage of x-rays would be stopped. Also, relative comparisons between several materials can be made. For example, 2 mm of lead will absorb about the same fraction of x-rays as 16 mm of concrete or 22 mm of dirt.

The x-ray photographs used in medicine, dentistry, or industry all depend on the difference in penetrating ability of x-rays in different materials. Sometimes this creates a problem. For example, a physician might want to check for an ulcer in the stomach. The difficulty is that the stomach and the ulcer easily let x-rays pass through. So the patient drinks a liquid containing a high atomic number element. One compound that is frequently used contains barium, since barium has atomic number $Z = 56$. Barium compounds are good at stopping x-rays. After a person drinks such liquids, clear x-ray photographs can be made. These pictures will reveal the shape of the stomach and any disorder which might be present (see Fig. 9-20).

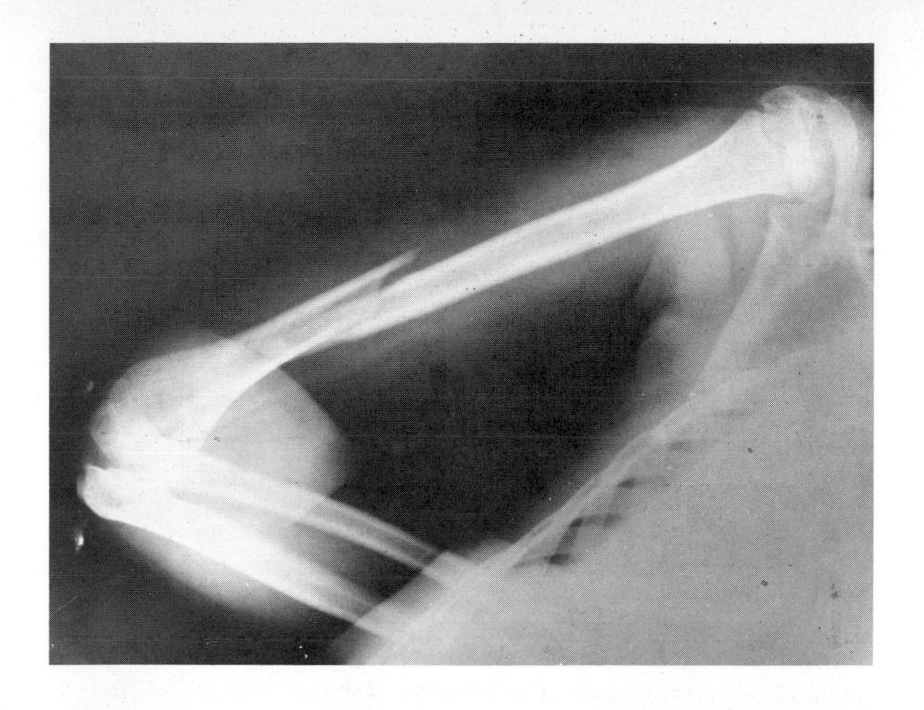

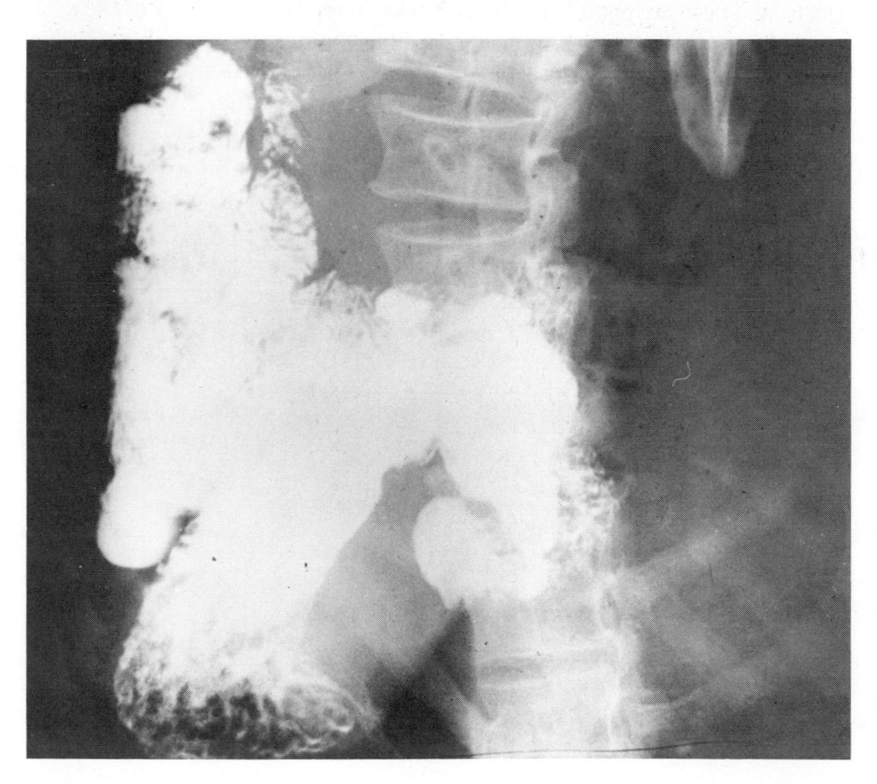

Fig. 9-20
Common uses of x-rays are medical diagnosis of broken bones and improper functioning of internal organs. (*Top photograph from L. V. Bergman and Associates, Inc. Bottom photograph courtesy of Armed Forces Institute of Pathology.*)

In dentistry and industry x-rays are used to find holes or flaws. A cavity in your tooth will allow x-rays to pass through more easily than the tooth itself. So a dentist looks for changes in contrast on the x-ray photograph (see Fig. 9-21). An unexpected difference in contrast at the surface of a tooth locates the position of a cavity. There is always a change in contrast at the center of the tooth, but that is natural. The fleshy portion in the center of each tooth is easily penetrated by x-rays.

In industrial applications, cavities can be located in metal castings as easily as cavities can be located in teeth. Any crack or opening within a solid casting will show up on an x-ray photograph. An example of this is shown in Fig. 7-16.

Examination of luggage and handbags at airports also depends upon x-rays. The security checks are designed to search for metallic weapons concealed inside suitcases. X-rays do not penetrate deeply into the metal parts of a gun but easily pass through leather or plastic. In this application, instead of producing a photograph, the x-rays are used to form an image electronically on a television screen (see Fig. 9-22).

The use of x-rays to make photographs of the interior of objects is possible only because not all the x-rays get through. The process by which an x-ray photon is absorbed and disappears is the photoelectric effect. A photon interacts with an atom such as calcium in your bones or teeth and ejects an electron. The x-ray photon completely disappears, but the energy of the x-ray photon frees an electron from the atom. The result of the photoelectric effect for x-rays is not only to ionize a calcium atom, but also to give considerable

Fig. 9-21
In a dental x-ray photograph the dense fillings absorb a higher percentage of x-rays than the surrounding healthy teeth. Cavities absorb fewer x-rays than healthy teeth.
(Photograph courtesy of Armed Forces Institute of Pathology.)

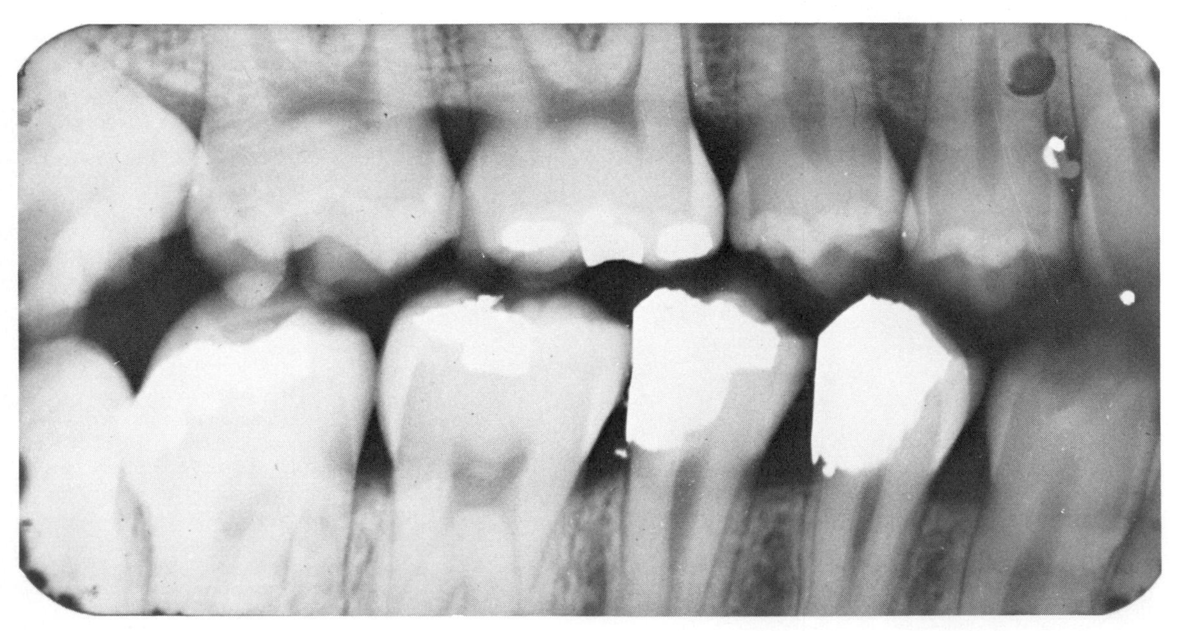

Fig. 9-22
All carry-on luggage is x-rayed at airport security checkpoints. (*Photograph from United Press International.*)

kinetic energy to an ejected electron, as shown in Fig. 9-23. The production of high-energy electrons is an unavoidable consequence of being exposed to x-rays in dental or medical examinations.

A single x-ray photon ionizes one atom by ejecting one highly energetic electron. This electron loses energy by ionizing *hundreds* and sometimes thousands of other atoms (see Fig. 9-24). In a human cell these ionized atoms

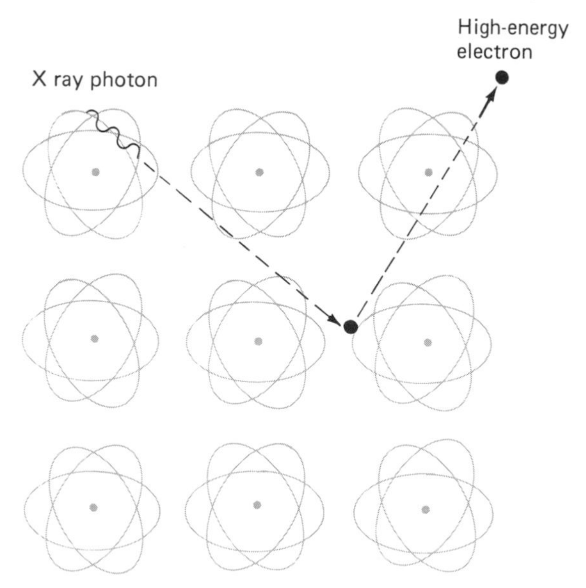

Fig. 9-23
When an x-ray photon produces ionization of an atom, the ejected electron possesses considerable kinetic energy.

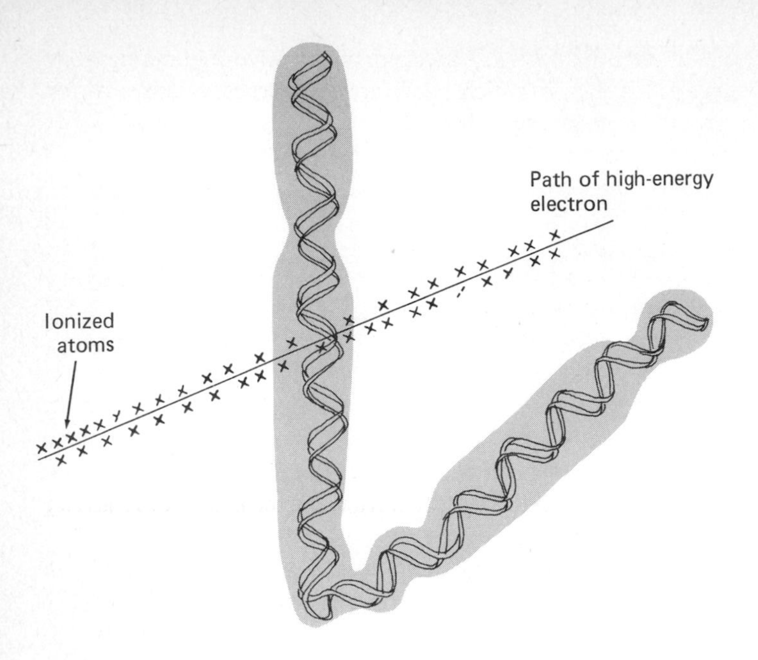

Fig. 9-24
A high-energy electron produces ionized atoms when passing through matter. The damage can be severe if the ionization occurs in delicate human chromosomes.

can be members of a simple molecule such as water or an exceedingly complex molecule such as a chromosome. The biological effect may be inconsequential or disastrous depending on the nature of the damage. The specific effect of x-ray exposure depends upon the amount of exposure and the region exposed. There is agreement, however, that x-rays can cause the growth of malignant tumors. For this reason, x-rays are only used on people when absolutely needed. Even then the exposure should be as low as possible in order to minimize risks.

In everyday conversation the word radiation conveys the suggestion of danger. But in the strict sense of the word radiation applies to all electromagnetic waves. All electromagnetic waves radiate outward from their source. And much of this radiation is not considered dangerous. The radiation known as AM, FM, and TV is all around us. And apparently it causes no ill effects.

The more energetic rays of ultraviolet radiation and x-rays are capable of ionizing atoms by means of the photoelectric effect.

Ionizing radiation *Ionizing radiation is defined as photons (or particles) that have sufficient energy to ionize atoms in their passage through a substance.*

It is *ionizing* radiation that is dangerous. The danger in ionizing radiation is the production of highly energetic electrons. These electrons can damage the chromosomes of cells. The less energetic rays, which have frequencies less than visible light, do not cause ionization.

Ultraviolet radiation does not penetrate deeply into the skin. Overexposure to ultraviolet radiation, usually because of excessive suntanning, may cause skin cancer. But x-rays, which penetrate the deepest recesses of the body, are capable of causing cancer throughout the body. Some cancer is not easily detected until the malignant tumors are well developed.

9.6 SPECIAL THEORY OF RELATIVITY

Probably no other name is more closely associated with physics in the popular mind than that of Albert Einstein. Einstein was awarded the Nobel prize in physics "for his services to theoretical physics and in particular for his discovery of the law of the photoelectric effect." (Relativity was not specifically mentioned in the official citation because it was still regarded as too controversial.) We have seen that the photoelectric effect is based on photons being quanta of radiant energy. In other words, electromagnetic waves are a part of quantum physics as well as the physics of continuous media. Although quantum physics was revolutionary, the theory of relativity probably had an even greater impact in the world of ideas.

Newton's laws of motion are based on the idea that space and time are absolute. Newtonian physics states that all clocks measure the same time between events. And all rulers measure the same distance between two locations. When these statements are true, space and time are said to be absolute.

For much of what we see and do the laws of Newton are sufficiently accurate. But under certain circumstances more accurate laws are needed. Einstein wondered how physical laws might be changed when objects move at extremely high speeds—speeds comparable to the speed of light. Of course, all electromagnetic waves travel at the speed of light. But what would be the behavior of mass traveling at or near the speed of light?

The truly remarkable outcome of Einstein's thoughts was the theory of relativity. One part of this theory is known as the *special* theory of relativity, which deals with space, time, and matter in the absence of a gravitational field. The *general* theory of relativity is a refinement of the special theory of relativity and includes the effect of gravitational fields. In this section, we will limit our discussion to the special theory of relativity.

Fig. 9-25
Albert Einstein was the creator of much of relativity theory. (*Photograph from United Press International.*)

The special theory of relativity denies the existence of absolute space and time. Any distance or time that might be measured depends on *relative* motion. Hence, the theory of *relativity.*

Time can be measured by different clocks in relative motion. If you are taking an automobile trip, you would normally measure elapsed time on your wristwatch. But you could also measure the interval of time by checking clocks at each end of the trip. One clock could be checked at the beginning of your trip (see Fig. 9-26), and another clock could be checked at your destination (see Fig. 9-27). Then a simple subtraction of the two clock readings would give you the elapsed time of travel. If this time of travel were compared to the time as determined by a wristwatch, they would agree.

In newtonian physics it is assumed that all such clock experiments yield

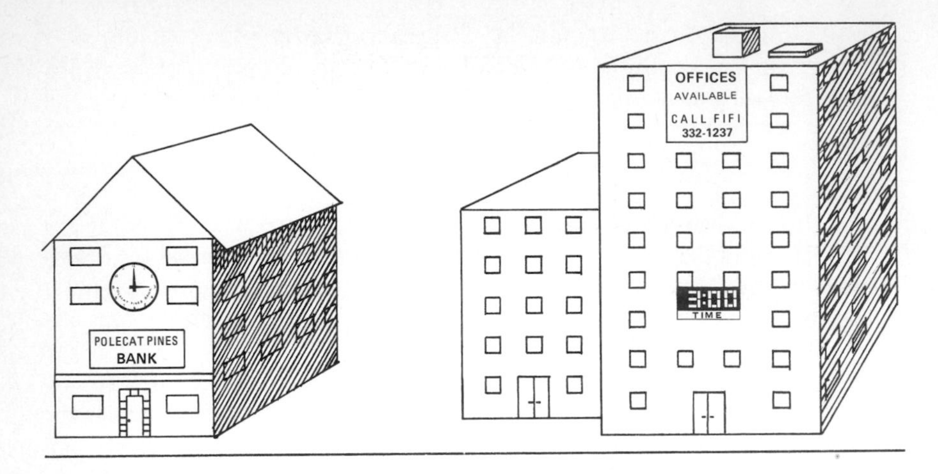

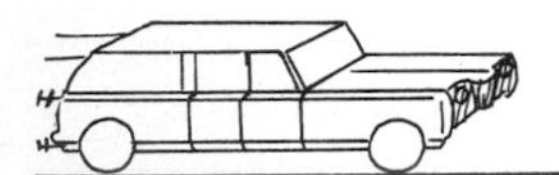

Fig. 9-26
A trip begins at 3:00 PM as noted by the clock face on the bank building.

the same results. This is summarized by saying that newtonian time is absolute. Any set of accurate clocks will measure equal time intervals.

In the special theory of relativity time is not absolute. Different clocks can measure *different* intervals of time. And these different time intervals are measured for the same physical events. In our example above, the time measured by the two clocks at rest on the earth would yield one value for the duration of the trip. And the single clock, the wristwatch, would yield a *different* value (see Figs. 9-28 and 9-29).

When we first meet such statements, they are difficult to accept. The idea of an absolute time seems to be a basic part of our thinking. And yet the theory of relativity denies absolute time. How can this be reconciled with our everyday experiences?

In the theory of relativity there are only *small* differences from newtonian theory when ordinary speeds are involved. Since our experiences in cars and even jet planes are at small or moderate speeds, we do not notice anything except an absolute time. However, if we could travel at much larger speeds (comparable to the speed of light), relativistic effects would be noticeable.

We can describe relativistic time measurements by returning once again to our car making a trip. But let us imagine the car to be traveling at high speed relative to the earth. In this case, we need the theory of relativity to predict the results. We must use the theory of relativity whenever speeds comparable to the speed of light are involved. At these speeds the results of newtonian physics are not correct. They do not agree with experiments.

A wristwatch at rest in the car (but moving at speeds approaching that of

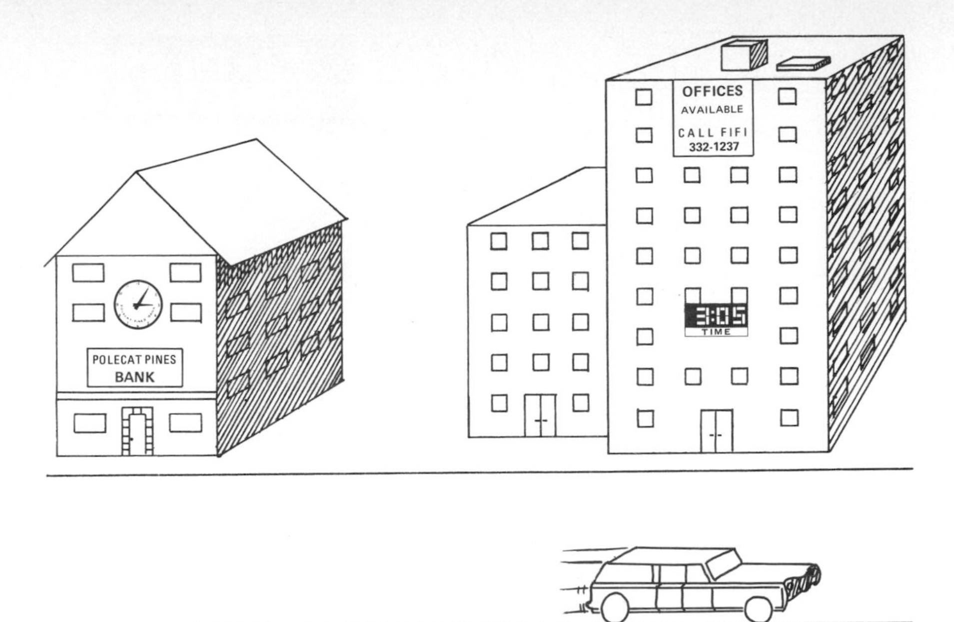

Fig. 9-27
A trip ends at 3:05 PM as noted by the clock face on the office building.

light relative to the earth) measures *less* time than that of the clocks on the earth. This is sometimes summarized by saying that when traveling at speeds approaching that of light, *time slows down.* A person in the car would not notice the slowing of time, but an observer on the earth would notice that time is running slower for the traveler in the car. So Newton's idea of absolute time is overthrown by Einstein's idea of relative time.

There is a relativistic formula for comparing time intervals as measured by different clocks. It is written

$$t_0 = \sqrt{1 - \frac{v^2}{c^2}}\; t$$

where t_0 stands for the interval of time as measured by a *single* clock, and t stands for the time interval measured by *two* clocks which are some distance apart. Referring to Figs. 9-28 and 9-29, t_0 would represent the time interval read by the wristwatch. And t would represent the time interval read by the two clocks on the buildings. In the formula c stands for the speed of light, and v stands for the speed of the single clock relative to the other clocks.

Not only is time not absolute, but also space is not absolute. Again, the effect is noticeable only for speeds near the speed of light. A ruler moving at high speed with respect to other rulers will be measured to be shorter, as shown in Fig. 9-30. In other words, *rulers shrink* when traveling at speeds near that of light. This is not only true for rulers but also for every other physical object. Everything becomes shorter by moving at high speeds. So

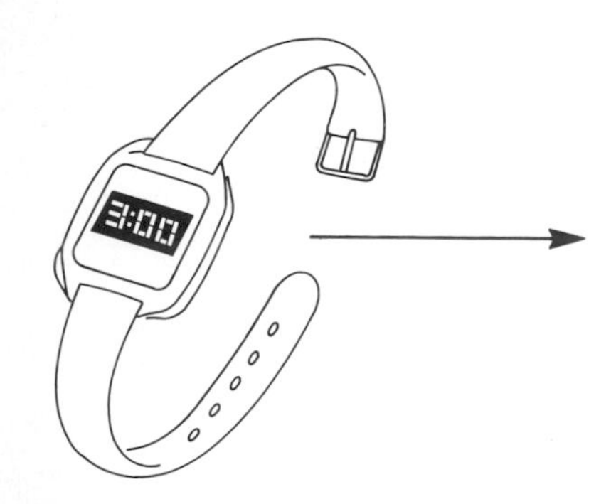

Fig. 9-28
A moving wristwatch passes the bank clock. The office clock is some distance away.

Newton's idea of absolute space is also overthrown by Einstein's idea of relative space.

The speed of light is extremely important in the special theory of relativity. It acts as an upper speed limit for all physical velocities. In newtonian physics one speed can be added to another. And these speeds are added directly in the obvious way. A bus (shown in Fig. 9-31) might be moving at 100 kilometers per hour when a passenger leans out and throws a baseball forward at 50 kilometers per hour. Someone standing at the roadside would observe the baseball moving forward at 150 kilometers per hour.

Trying the same experiment with light would give a different result. Suppose a spaceship could move at half the speed of light with respect to earth. Imagine a flashlight shining from the spaceship in the forward direction, as shown in Fig. 9-32. It might be expected that an observer on the earth would see the light moving at $1\frac{1}{2}$ times the speed of light. But this is not the case. Light always travels (in vacuum) at 299,790 kilometers per second. The speed of light *is* an absolute quantity.

The speed of light is also an upper speed limit for matter. There has been speculation that particles of mass might travel *faster* than the speed of light, but none have been found. All particles of mass (including spaceships) travel at *less* than the speed of light. Theoretically, it would take an infinite amount of energy to accelerate a spaceship to travel *at* the speed of light.

Single particles, such as electrons in an x-ray tube, achieve tremendous

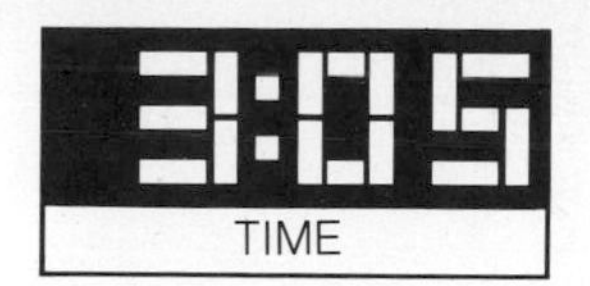

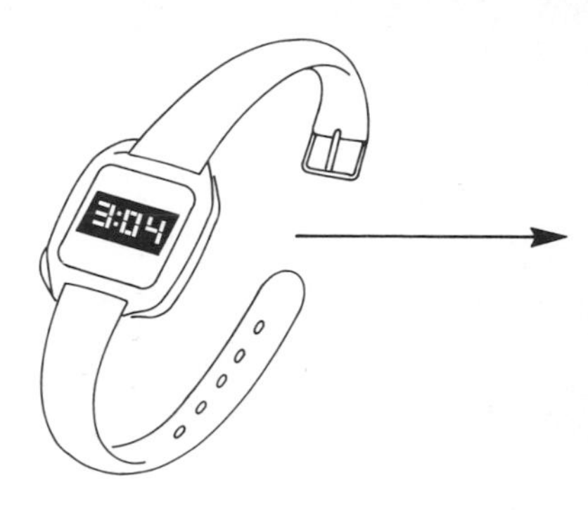

Fig. 9-29
The moving wristwatch passes the office clock, and they no longer read the same time.

speeds, but always less than the speed of light. For a target voltage of +100,000 volts the speed of the electrons in a vacuum is 164,400 kilometers per second. For any experiment involving moving masses, the speeds involved are less than the speed of light.

There is yet another surprise in the special theory of relativity. The law of conservation of energy, which is fundamental to all of physics, is found to have a different, but equivalent, form. The law of conservation of energy states that in any process all the energy must be accounted for. A transfer of energy can occur between potential energy, kinetic energy, work, heat, internal energy, sound energy, electrical energy, radiant energy, or any combination of these energies. But the total energy is conserved.

The special theory of relativity asserts that all kinds of energy have mass. And every mass has energy. There is an equivalence between mass and energy. Einstein's formula for this equivalence is

$$\text{Energy} = (\text{mass})(\text{speed of light})^2$$

$$E = mc^2$$

Fig. 9-30
A ruler shrinks in length when moving at speeds comparable to the speed of light.

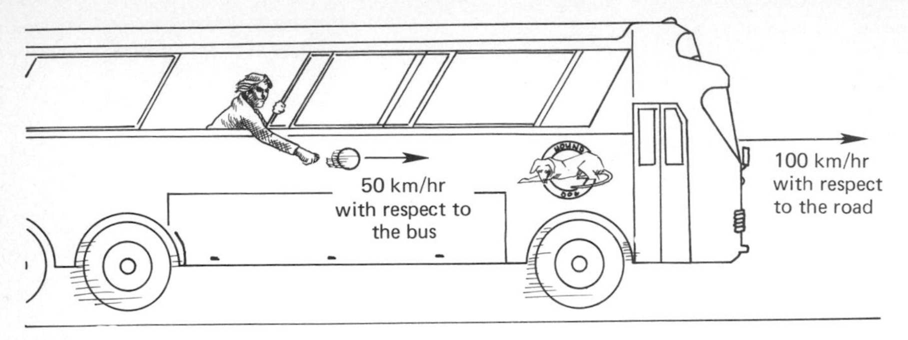

Fig. 9-31
What is the speed of the baseball with respect to the road?

where E = energy
m = mass
c = numerical value of the speed of light

For a calculation, the energy would be in joules, the mass in kilograms, and the speed of light in meters per second.

Because of this equivalence between mass and energy we can state the law of conservation of energy in another way. We can say that there is a law of conservation of mass. Either conservation law is equivalent to the other.

Prior to the development of relativity theory, chemists had often spoken of a law of conservation of mass. They meant that the total mass of all materials before some chemical reaction was equal to the total mass of all

Fig. 9-32
What is the speed of the beam of light with respect to the earth?

Fig. 9-33
Radiant energy carries mass away from a candle. (*Photograph by Roger Appleton. From Photo Researchers, Inc.*)

materials after the reaction. The law had been experimentally checked, and it agreed with measurements of mass.

For example, one could check a match before and after burning. Burning is a chemical reaction involving oxidation. In making the check, chemists knew that they must take into account *all the materials* involved in the reaction. One could not simply compare the mass of the match with the mass of the ashes. Instead, one must include the mass of the oxygen used in the reaction. And proper account must be taken of all the gases produced by the burning match. By sealing a match totally inside a glass jar, all materials could be handled properly. The results seemed clear. The law of conservation of mass was experimentally verified.

And yet one item was missing. In the chemical reaction of burning a match, light is emitted. Radiant energy is given off. This radiant energy has mass which was "lost" through the walls of the glass jar. According to relativity theory the mass of the radiation must be taken into account or we will not have an *exact* law of conservation of mass (see Fig. 9-33). The older chemical measurements were not accurate enough to detect the small differences.

If a body gives off an amount of energy E in the form of radiation, its mass decreases by E/c^2. For chemical reactions, such as burning a match, this mass decrease is exceptionally small. It could not be detected by a modern chemical balance. The older law of conservation of mass (ignoring radiation) is *approximately* true for chemical reactions. The law of conservation of mass is *exactly* true in every situation if *all* mass is taken into account.

9.7 THE WAVE MODEL OF ELECTRONS

Chapter 9 would not be complete without a discussion of the modern description of the atom. The statements in Sec. 9-1 and 9-2 about energy levels of electrons in atoms cautiously avoid the idea that the energy levels are the actual paths or orbits of the electrons. The illustrations in those sections, however, may leave the idea that the electrons move in well-defined paths. Such a description, or *model* as it is called in physics, was proposed by N. Bohr in the early part of this century and accepted until the model could be improved.

Electrons in the Bohr model of the atom are considered to be particles. Electrons have mass just as marbles, balls, or planets. So the *corpuscular* (that is, particlelike) model or description is easily understood. The Bohr atom is an extension of our knowledge of known particles to the behavior of electrons in the atom. An improvement in the corpuscular model of the electron comes with a clearer understanding of the nature of electromagnetic radiation.

In Chap. 7 electromagnetic radiation was described in terms of wavelike properties: frequency and wavelength. The wave model of electromagnetic radiation is excellently suited to explain many everyday phenomena. But in the discussion of atomic events, such as the transition of an electron between energy levels or the photoelectric effect, the wave model of electromagnetic radiation does not explain the observations. So a new model is introduced. Electromagnetic radiation has a corpuscular behavior that leads to the model of the photon. Depending on the physical events, the description of electromagnetic radiation uses either the wave model or the corpuscular model.

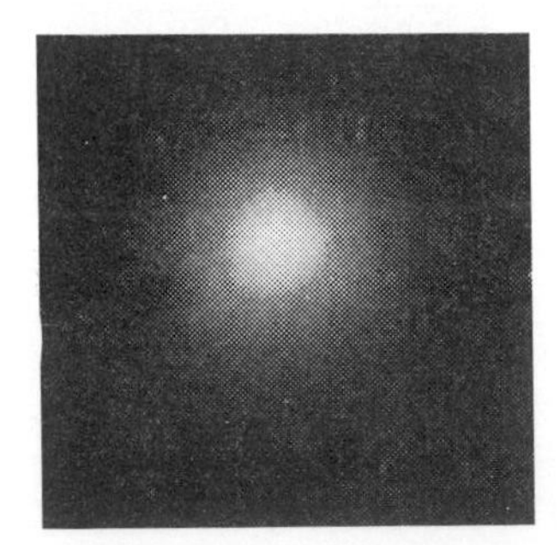

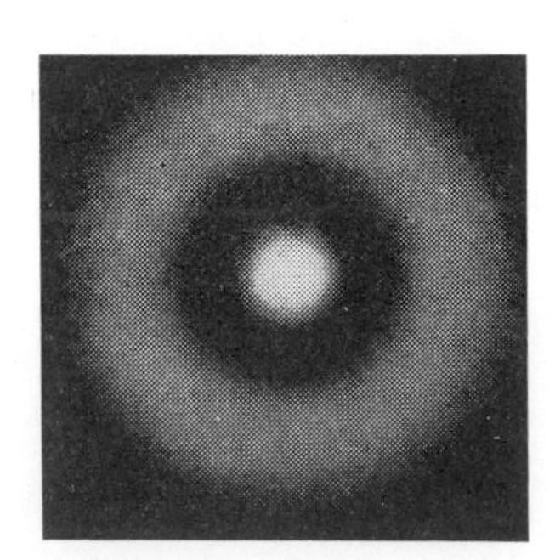

Fig. 9-34
Photographic representation of the electron cloud of the hydrogen atom in the ground state and next higher energy level. (*Photographs from* Principles of Modern Physics *by Robert B. Leighton. Copyright 1959, McGraw-Hill Book Company. Used with permission of McGraw-Hill Book Company.*)

The fact that electromagnetic radiation can be described using two different models led physicists to speculate about electrons and masses in general. In 1925 L. de Broglie suggested that the same dualism (two different models) of wave and corpuscle used for electromagnetic radiation may also apply to mass. A material particle will have a wave corresponding to it, just as a photon has a wave corresponding to it. So the modern description of an electron also uses either the corpuscular model or the wave model, depending upon the circumstances.

The simpler model of an electron in an atom is the *corpuscular model.* The atom is described in terms similar to the description of the solar system. The orbits or paths of both electrons and planets are well defined. But in the less familiar *wave model* of electrons in an atom, the orbits are *not* well defined. In fact, the electrons themselves are smeared out into what is called an *electron cloud,* as shown in Fig. 9-34. Although the electron itself is not localized, in its shell, the energy of the electron remains well defined. Bohr's energy law requires that the change of energy, and so each energy level, be well defined.

The questions naturally arise, what *is* an electron and what *is* electromagnetic radiation? Physics is unable to answer the question in philosophical terms. The answer given in physics depends upon the observation or experiment. In large-scale or macroscopic events the corpuscular model of mass and the wave model of electromagnetic radiation are the preferred descriptions. In atomic or microscopic events the wave model of mass and the corpuscular model of electromagnetic radiation are the preferred descriptions.

REVIEW

You should be able to define or describe the following terms: atomic number, ground state, excited state, photon, photoelectric effect, and ionizing radiation.

You should be able to state and give an application of Bohr's energy law.

PROBLEMS

1. An electron in an atom makes a transition from one energy level to another. In the process the electron loses 3.96×10^{-19} joules of energy. What is the frequency of the emitted electromagnetic radiation?

2. An electron in an excited atom returns to the ground state, releasing a photon of frequency 9×10^{14} hertz. How much energy did the electron lose in the transition between the excited state and the ground state?

3. An atom of mercury is excited by adding 7.8×10^{-19} joules of energy. Calculate the wavelength of the radiation which is emitted as the atoms return to the ground state.

4. In the identification of an unknown element it is observed to emit radiation of wavelength 4×10^{-7} meters. What change in energy levels within an atom is associated with this wavelength?

5. A classic experiment was performed in 1941 to test the idea of time change in relativity theory. In the experiment two different time intervals were measured. The time intervals were times of travel for muons in the atmosphere. (A muon is a particle produced by cosmic rays entering the atmosphere of the earth from outer space.) If we assume that the two times, 6.5×10^{-6} seconds and 0.7×10^{-6} seconds, correspond to the time intervals t and t_0, calculate the speed of the muons.

6. A rocket ship of the future makes a straight pass over New York City and Detroit. A passenger notes that clocks on the ground are recording more time than her wristwatch. Each 5 minutes of "wristwatch time" corresponds to 6 minutes as measured by the clocks on the ground. What is the speed of the rocket ship?

7. The planet-destroyer from the movie *Star Wars* is used to obliterate an entire planet of mass 6×10^{24} kilograms. (This is about the mass of the earth.) What is the total energy in all forms sent streaming off into space?

8. A light bulb sends out 300 joules of visible radiation during a 2-minute period. During these 2 minutes, how much mass is radiated away?

9. The sun gives off 2.4×10^{28} joules of radiant energy in 1 minute. How much mass is lost by the sun during this time?

DISCUSSION QUESTIONS

1. What is meant by saying that a certain quantity is *quantized?*

2. In a hydrogen atom, what quantity is quantized?

3. Which levels in an atom have the largest energies?

4. When an atom emits a photon, what determines the frequency of the photon?

5. List some practical uses for lasers.

6. What is a hologram?

7. Describe the operation of a laser.

8. How does monochromatic blue light from a laser differ from the blue light from a lake?

9. What is meant by coherent light?

10. In a television set the viewing screen is coated with certain chemical phosphors. In the operation of the set, these phosphors are struck by electrons. Describe the result.

11. In what processes are photons created? In what processes are photons destroyed?

12. Does a mirror create photons when we see a reflection?

13. What is ionizing electromagnetic radiaton?

14. How is the photoelectric effect used in an electric eye to control the doors to an elevator?

15. How are x-rays related to the electrons in the inner energy levels of an atom?

16. List various physical characteristics of x-rays.

17. What is the difference between "hard" x-rays and "soft" x-rays?

18. How is it possible to identify one organ from another in an x-ray photograph of your body?

19. How is the word *relative* used in the theory of relativity?

20. How is the word *absolute* used in the theory of relativity?

21. Is time an absolute quantity in newtonian mechanics? In the theory of special relativity?

ANNOTATED BIBLIOGRAPHY

A general reference for this chapter is *The Atom and its Nucleus* by G. Gamow. Published by Prentice-Hall, Inc., Englewood Cliffs, N.J., in 1961, it contains clear diagrams and excellent prose on quanta, Bohr's energy law, and x-rays. For an additional reading on atoms, see J. Heilbron, "J. J. Thomson and the Bohr Atom," *Physics Today,* April 1977, p. 23.

The best reference for Sec. 9-3 is R. Brown, *Lasers: Tools of Modern Technology,* Doubleday & Company, Inc., Garden City, N.Y., 1968. A large number of colored photographs and diagrams make for pleasant reading. Numerous applications are given. An alternative book is N. Sobolev, *Lasers and Their Prospects,* Mir Publishers, Moscow, 1974. If you enjoy science fiction and thoughts about the future, don't miss Sobolev's section on communication with extraterrestrial civilizations (pages 228–241). Holograms that can be viewed in ordinary light are described in E. Leith's article "White Light Holograms," *Scientific American,* October 1976, p. 80.

If you enjoy tinkering with gadgets, why not try a solar cell to charge a battery for you? Almost any local electronics store sells them. Or write for a catalog from Edmund Scientific Co., 430 Edscorp Bldg., Barrington, N.J. 08007. The Edmund catalog lists solar cells as well as booklets on solar energy.

Most books on relativity contain a great deal of mathematics along with their description. This is perhaps almost natural and unavoidable. To make calculations within the theory of special relativity, formulas are necessary. One introduction without mathematics was written by Einstein. It is contained in the book *Relativity, the Special and General Theory,* Crown Publishers, Inc., New York, 1961. The first 30 pages are excellent, showing Einstein's considerable talent for writing. Another book filled with excellent descriptive material on relativity is a biography by B. Hoffman, *Albert Einstein Creator and Rebel,* The Viking Press, Inc., New York, 1972. Two other especially good books on relativity are P. Davies' *Space and Time in the Modern Universe,* Cambridge University Press, New York, 1977, and M. Gardner's *Relativity Explosion,* Vintage Books, Random House, Inc., New York, 1976.

A brief discussion of x-ray dangers is given in F. Warshofsky's article "Warning: X-Rays May Be Dangerous to Your Health," *Reader's Digest,* August 1972, p. 173. A readable technical article on x-rays is R. Gordon, G. Herman, and S. Johnson, "Image Reconstruction from Projections," *Scientific American,* October 1975, p. 56.

10 NUCLEAR PHYSICS

The nucleus of an atom is the main source of radioactivity. Radioactivity is used in medicine for diagnostic examinations and for the cure of some diseases. Also, the nuclei of certain atoms can be a source of energy through the processes of fission and fusion.

10.1 NUCLEAR STRUCTURE

In our discussion of the atomic theory of matter we have seen that electrons are kept in their energy levels by the electrical attraction to the nucleus. One component of the nucleus is a positively charged particle known as the *proton.* These protons are equal in number to the electrons in an atom. Since the charge of the proton is equal and opposite to the charge of the electron, an atom is electrically neutral. The nucleus is quite small, typically one ten-thousandth the diameter of an atom. It is difficult to realize that the small nucleus of the atom is surrounded mostly by empty space. Imagine the nucleus to be enlarged to the size of an eraser on a pencil. Place this "nucleus" at the center of the 50-yard line of a football field. Then the "electron cloud" would be enlarged to the size of a fog swirling no nearer the "nucleus" than the 15-yard lines and generally in the end zones and grandstands.

Within the nucleus are found the positively charged protons. In addition to protons, an uncharged particle known as the *neutron* (the neutral one) is usually found in the nucleus. The neutron is very similar to the proton, except that the neutron carries no charge.

One remarkable fact about the nucleus is that the protons and neutrons stay together. Protons would not be expected to remain together because of the electric force of repulsion between the like charges. But there is another force, a *nuclear* force, that holds the nucleus together.

The nuclear force is fundamentally different from electric or gravitational forces. An extremely strong force of *attraction* occurs between particles in the nucleus when they are close enough to each other. The nuclear force of attraction exists between pairs of particles, whether two protons, two neutrons, or a proton and a neutron. The nuclear force is different from the coulomb force, which tends to push protons apart rather than hold them together. (The gravitational force tends to hold protons together in the nucleus but is much too weak to be effective against the coulomb force of repulsion. See Sec. 6-2.) Figure 10-1 is a schematic illustration of an atom being held together by nuclear forces in the nucleus and electric forces between the proton and the electron.

Another property of the nuclear force that is different from electric or gravitational forces is the short range of the nuclear force. For example, if the distance between two charged particles is doubled, both the electric and gravitational forces decrease to one-fourth their former values. However, if the distance between two particles in the nucleus is doubled, the nuclear force between the particles effectively vanishes. As a result, the nucleus is stable only when the particles experience a delicate balance between the nuclear force of attraction and the electric force of repulsion.

Precise measurements have shown that the neutron is slightly more massive than the proton. Each of these two particles is about 1,840 times more massive than an electron. Due to the large masses of the neutron and

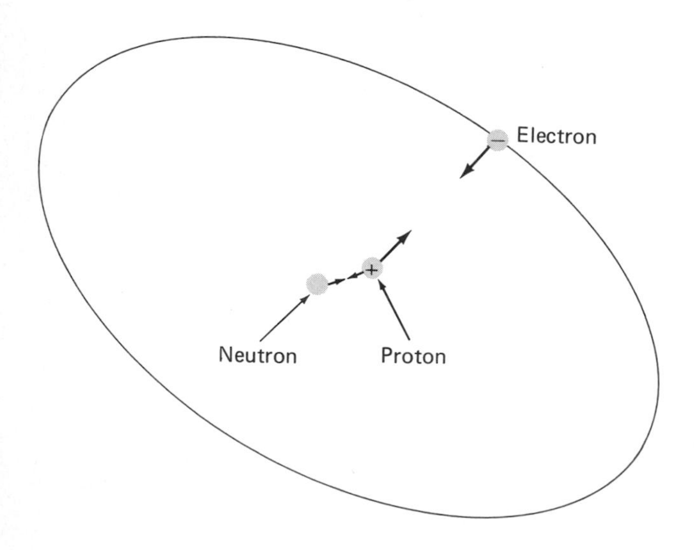

Fig. 10-1
An electron, represented as a point charge to simplify the drawing, is shown being attracted to the nucleus. The nucleus is held together by the nuclear force.

proton in comparison to the electron, the mass of an atom is concentrated almost entirely in the nucleus.

Every nucleus can be identified by a number known as the mass number A.

Mass number *The mass number* A *is defined as the total number of protons and neutrons in the nucleus.*

Notice that the mass number A is different than the atomic number Z. The atomic number Z is the number of protons in the nucleus. It is quite simple to find the number of neutrons in the nucleus. The mass number minus the atomic number gives the number of neutrons in the nucleus. In equation form

$$\text{(Number of neutrons)} = \text{(mass number)} - \text{(atomic number)}$$

$$N = A - Z$$

where N = number of neutrons
A = mass number
Z = atomic number

The number of neutrons in the nucleus plays no role in the chemical properties of an atom. For example, most of the helium found in nature has two protons and *two* neutrons in the nucleus. But some rare helium is found with two protons and only *one* neutron in the nucleus (see Fig. 10-2). This difference does not change the chemical properties of helium. Helium atoms with either of these two nuclei have exactly the same chemical properties because both atoms have two electrons. Chemical behavior is determined by the electronic structure of an atom, not by the nucleus.

Differences in nuclei due to different numbers of neutrons lead to the concept of an *isotope.*

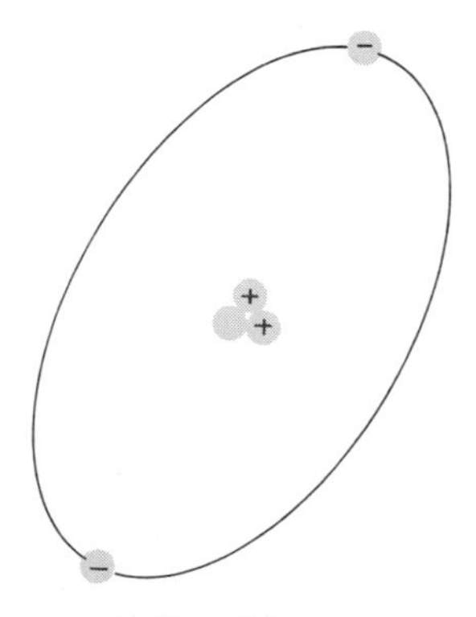
Helium-3 isotope

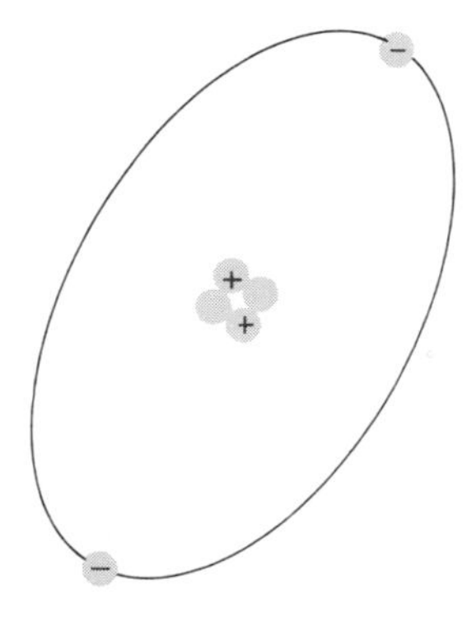
Helium-4 isotope

Fig. 10-2
Each helium nucleus contains two protons, but the number of neutrons varies. A helium-3 nucleus contains only one neutron, while a helium-4 nucleus contains two neutrons.

Isotopes *Isotopes of an element are defined as atoms having the same number of protons in their nuclei but different numbers of neutrons.*

All the isotopes of any given element have the same *chemical* properties. But not all isotopes of an element have the same *nuclear* properties. In the example of the helium nuclei just mentioned, we speak of the two kinds as isotopes of helium. Each of the two isotopes behaves identically in chemical experiments. But in experiments which involve the nucleus, the two isotopes can be identified and separated from each other.

Some of the isotopes of an element are stable, and some are *radioactive.* The nucleus of a radioactive isotope is capable of spontaneously releasing energy. A radioactive isotope is known as a *radioisotope.* All isotopes are labeled with chemical symbols and appropriate subscripts and superscripts. The subscript to the left of the chemical symbol is the atomic number Z. The superscript to the left is the mass number A. So the two main isotopes of helium are labeled ${}^{3}_{2}He$ and ${}^{4}_{2}He$. The subscript 2 stands for two protons. The superscript 3 stands for a total of three protons and neutrons. Thus, there is one neutron in ${}^{3}_{2}He$. How many neutrons are in ${}^{4}_{2}He$?

10.2 RADIOACTIVITY

A radioactive nucleus, whether naturally occurring or artificially produced, will release energy. Energy is released from a radioactive nucleus because the nucleus is unstable. The reason for this lack of stability lies in the number of neutrons. All *stable* nuclei have the number of neutrons equal to or only slightly different from the number of protons. If the number of neutrons is substantially different from the number of protons, then the nucleus is radioactive. It makes no difference whether there are substantially more or fewer neutrons than protons. In either case, the nucleus becomes unstable and radioactivity occurs. Table 10-1 compares the isotopes of several elements and identifies which isotopes are radioactive.

There are three major ways in which radioactive nuclei can spontaneously release their energy: by alpha particle emission, by beta particle emission, and by gamma ray emission. A given radioisotope emits only one type of particle or ray and is not free to go from one type of emission to another. All three types of emission can be detected by photographic film. Each of the emissions causes film to be exposed. We recall that visible light, ultraviolet radiation, and x-rays also expose film. All three types of radioactivity are called radiation because they expose film, but only the gamma ray is an electromagnetic wave.

ALPHA EMISSION

The composition of alpha particles was unknown when these particles were first discovered and named. Later, alpha particles were found to consist of two protons and two neutrons tightly bound together. Alpha particles are therefore exactly the same as a helium nucleus. So the symbol for an alpha particle is simply ${}^{4}_{2}He$. If this information had been known at an earlier date, it is probable that the name alpha particle would never have been used.

Table 10-1
Comparison of nonradioactive and radioactive isotopes

Name	Symbol	Number of protons	Number of neutrons	Radioactive?	Comments
Hydrogen	$^{1}_{1}H$	1	0	No	Natural
	$^{2}_{1}H$	1	1	No	Natural
	$^{3}_{1}H$	1	2	Yes	Artificial
Helium	$^{3}_{2}He$	2	1	No	Natural
	$^{4}_{2}He$	2	2	No	Natural
	$^{5}_{2}He$	2	3	Yes	Artificial
	$^{6}_{2}He$	2	4	Yes	Artificial
Lithium	$^{5}_{3}Li$	3	2	Yes	Artificial
	$^{6}_{3}Li$	3	3	No	Natural
	$^{7}_{3}Li$	3	4	No	Natural
	$^{8}_{3}Li$	3	5	Yes	Artificial
	$^{9}_{3}Li$	3	6	Yes	Artificial
Uranium	$^{227}_{92}U$	92	135	Yes	Artificial
	$^{228}_{92}U$	92	136	Yes	Artificial
	$^{229}_{92}U$	92	137	Yes	Artificial
	$^{230}_{92}U$	92	138	Yes	Artificial
	$^{231}_{92}U$	92	139	Yes	Artificial
	$^{232}_{92}U$	92	140	Yes	Artificial
	$^{233}_{92}U$	92	141	Yes	Artificial
	$^{234}_{92}U$	92	142	Yes	Natural
	$^{235}_{92}U$	92	143	Yes	Natural
	$^{236}_{92}U$	92	144	Yes	Artificial
	$^{237}_{92}U$	92	145	Yes	Artificial
	$^{238}_{92}U$	92	146	Yes	Natural
	$^{239}_{92}U$	92	147	Yes	Artificial
	$^{240}_{92}U$	92	148	Yes	Artificial

A typical example of radioactivity in which an alpha particle is emitted occurs in $^{238}_{92}U$. Uranium-238 is a radioisotope. Written in the form of a nuclear reaction, the emission of an alpha particle from uranium-238 is

$$^{238}_{92}U \longrightarrow ^{234}_{90}Th + ^{4}_{2}He$$

This reaction, illustrated in Fig. 10-3, shows that the uranium-238 radioisotope breaks down into a thorium-234 radioisotope plus an alpha particle. Notice that the subscripts and superscripts in the nuclear reaction balance. (The atomic numbers balance because of the conservation of *electric charge.* Mass numbers balance because of the conservation of *mass.*) The 92 protons in the uranium nucleus become 90 protons in the thorium nucleus and 2 protons in the alpha particle. The 238 protons and neutrons in uranium become 234 protons and neutrons in thorium and 4 protons and neutrons in the alpha particle. Thus, the number of protons on each side of the reaction is the same, and the number of neutrons on each side is the same. No protons or neutrons are created or destroyed in the process of alpha particle emission.

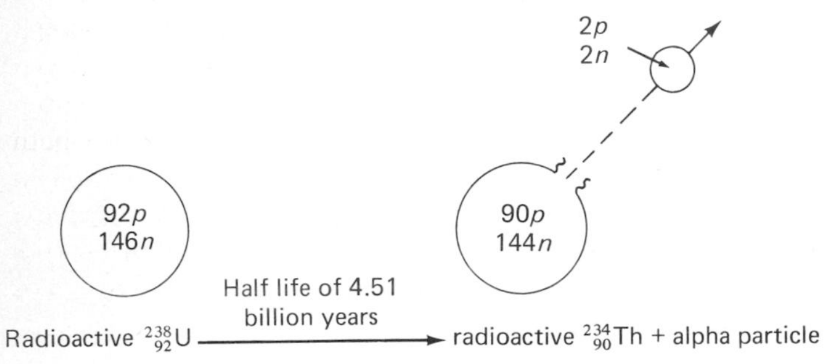

Fig. 10-3
Radioactive uranium-238 emits an alpha particle to become thorium-234.

The breakdown of a radioactive nucleus is often called *nuclear decay.* The chemical element uranium decays into the elements thorium and helium. A nuclear reaction is basically different from a chemical reaction. In a chemical reaction the atoms rearrange themselves into new molecules, while the nuclei of the atoms remain unchanged. In a nuclear reaction the nuclei of one atom are changed into nuclei of a different atom. The elements produced in nuclear reactions may be radioactive or stable. The thorium-234 produced in the nuclear decay of uranium-238 is itself radioactive and decays by the emission of a beta particle.

BETA EMISSION

The emission of a beta particle from a radioisotope is perhaps the most surprising of the three processes of radioactivity. The beta particle is simply an *electron,* no different than the electrons in a wire which carries current or the electrons surrounding the nucleus of the atom. A beta particle comes from a radioactive nucleus, but a beta particle does not exist in the nucleus! Surprising as it may seem, the beta particle is *formed* in the act of releasing energy from the nucleus. The neutron, which is usually stable in the nucleus, is capable of disintegrating into a proton and an electron. The electron is immediately ejected and carries kinetic energy away from the nucleus. The proton remains in the nucleus.

Do not make the mistake of considering the neutron to be a tightly bound proton and electron pair waiting to be split apart. The proton and beta particles do not exist until the neutron disintegrates.

The symbol for a beta particle in a nuclear reaction is ${}_{-1}^{0}e$. The letter e stands for the electron. The subscript of minus one indicates a negative charge. That is, the charge of the electron is equal and opposite to the positive charge of the proton. The superscript of zero indicates that the electron is not made up of any protons or neutrons.

As an example of beta particle emission we choose thorium-234. The nuclear reaction for the decay of thorium-234 is written as

$${}_{90}^{234}\mathrm{Th} \longrightarrow {}_{91}^{234}\mathrm{Pa} + {}_{-1}^{0}e$$

This reaction, illustrated in Fig. 10-4, shows that thorium-234 becomes protactinium-234 by the emission of a beta particle. Again the subscripts and superscripts in the nuclear reaction balance. The 90 positively charged protons in thorium become 91 positively charged protons in protactinium plus one negatively charged beta particle. Each of these units of charge, both positive and negative, is of the same magnitude. Thus, during the reaction no electric charge is either created or destroyed. Charge is conserved for a nuclear decay involving beta emission. (Charge is conserved in all three types of emission, but the number of protons changes only in the nuclear reaction involving beta emission.)

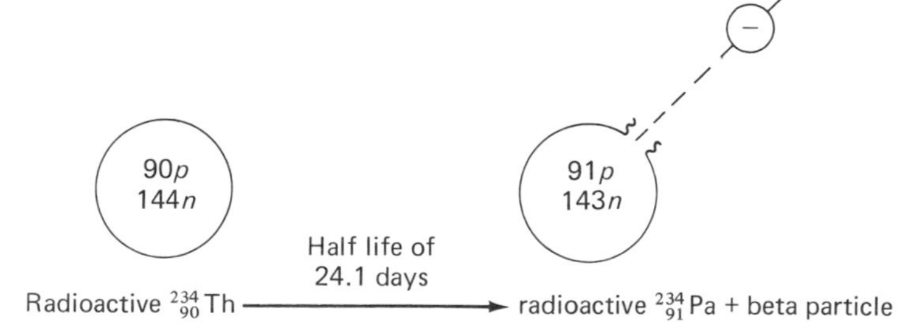

Fig. 10-4
Radioactive thorium-234 emits a beta particle to become protactinium-234.

The superscripts show that the number of neutrons and protons together is also conserved. There are 234 on the left side of the reaction and 234 on the right side. But the number of protons or neutrons taken individually is not conserved. There are 90 protons on the left and 91 on the right. Thus one proton has been produced during beta emission. (How many neutrons are there on each side?)

GAMMA EMISSION

A third way in which some radioactive nuclei release energy is through the emission of a gamma ray. When some radioisotopes emit a beta particle, the resulting decay products often have an excess of energy. The excess energy can be released as gamma rays, which are very similar to x-rays. The difference between x-rays and gamma rays lies in the *origin* of the rays and not in the *energy* of the rays. A gamma ray is electromagnetic radiation that accompanies the release of energy from the nucleus. An x-ray is electromagnetic radiation that accompanies the release of energy from electrons, as in an x-ray tube.

Usually gamma rays have more energy (hence, higher frequency or shorter wavelength) than x-rays, but this is not always the case. The gamma rays emitted by a radioactive nucleus are no different in their physical characteristics than the x-rays emitted from an x-ray machine. Most gamma rays, however, are comparable to x-rays generated using electrons with hundreds of thousands or millions of volts.

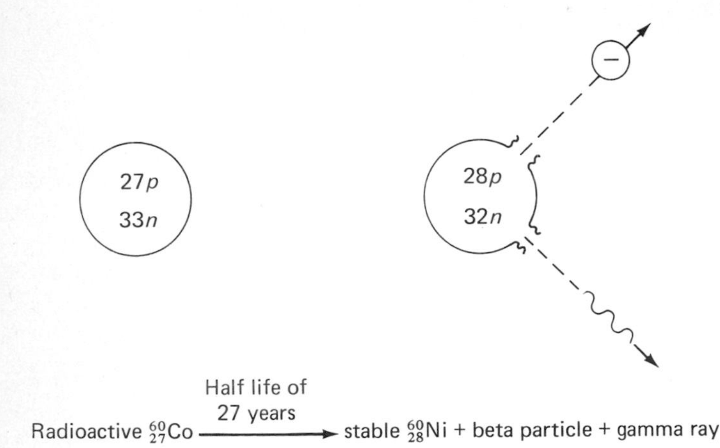

Fig. 10-5
Radioactive cobalt-60 emits a gamma ray after the emission of a beta particle to become nickel-60.

Not all radioisotopes which emit beta particles also emit gamma rays. The naturally occurring radioisotope thorium-234 emits a beta particle in decaying to protactinium-234, but does not emit a gamma ray. The artificial radioisotope cobalt-60 is an example of a radioactive nucleus which emits a gamma ray in the process of emitting a beta particle (see Fig. 10-5).

$$^{60}_{27}\text{Co} \longrightarrow {}^{60}_{28}\text{Ni} + {}^{0}_{-1}\text{e} + \text{gamma ray}$$

The gamma ray emitted from the cobalt-60 radioisotope is *not* stored in the nucleus. Instead, the gamma ray is produced during the reaction. When the masses of the nickel-60 nucleus and the beta particle are added together, their sum is found to be less than the mass of the radioactive cobalt-60 nucleus. Since mass is conserved in this reaction (and all nuclear reactions), there must be other mass associated with the products of the reaction. This mass is linked with the kinetic energies of the decay products (nickel-60 nucleus and beta particle) and with the radiant energy of the gamma ray. Every kind of energy has mass, which can be calculated using Einstein's famous formula: $E = mc^2$.

10.3 HALF-LIFE

The fact that a radioisotope of one element can decay into another element raises the question of lifetime. How long do radioisotopes exist before decaying into another element? For a specific radioactive nucleus no law of physics can explain or predict when the nuclear decay might occur. But for a sample containing a very large number of radioactive nuclei, certain experimental regularities are observed. These experimental results are obtained using an instrument known as a *Geiger counter,* shown in Fig. 10-6.

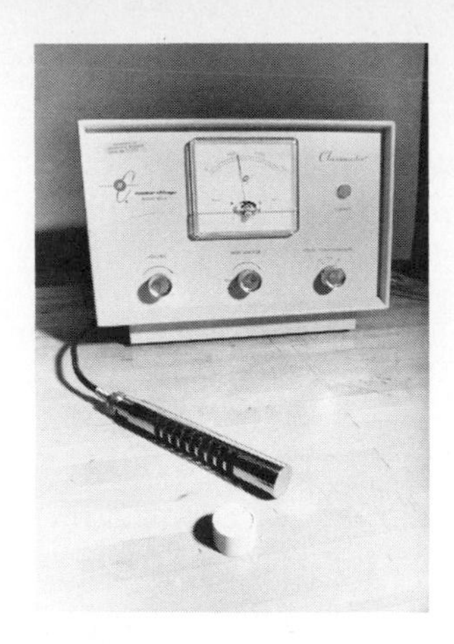

Fig. 10-6
A Geiger counter measures emissions from radioactive materials. It detects alpha particles, beta particles, and gamma rays. (*Photograph by Ted Rozumalski. From Black Star.*)

A Geiger counter measures the ionizing radiation released during nuclear decay. Whenever an alpha particle, beta particle, or gamma ray passes through the detector of a Geiger counter, a pulse of electricity is generated. Sometimes the pulse is used to make an audible "click" for each passage of some form of ionizing radiation. At other times, the number of pulses in a time interval are simply recorded by a meter. The meter reading is then a measure of the amount of radioactivity.

A typical experiment to measure nuclear decay begins by measuring the amount of radioactivity from a pure sample. At frequent intervals of time the radioactivity is again measured. As time passes, the radioactivity decreases if the radioisotope being investigated does not decay into an isotope that is also radioactive. A graph can be made of the data. The *relative* amount of radioactivity is plotted against the elapsed time from the beginning of the measurements. The relative amount of radioactivity is given in fractions of the original amount of radioactivity. For thorium-234, the graph would look like Fig. 10-7.

Several remarkable features of radioactivity are shown in Fig. 10-7. Unlike chemical reactions, radioactivity decreases as time passes but does not stop after a well-defined time. So in nuclear physics, the question of the "total" lifetime of a radioisotope is avoided. Of most significance is the measured time for *one-half* of the radioisotope to decay. This amount of time is known as the *half-life.*

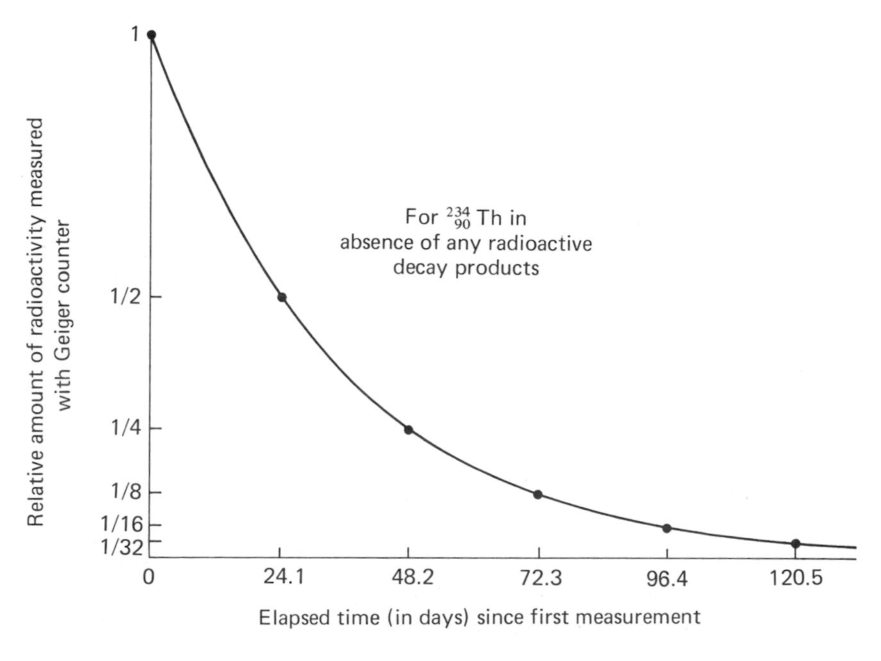

Fig. 10-7
When the relative amount of radioactivity from a sample of a radioisotope is graphed against time, a curve is obtained. The curve for thorium-234 shows a half-life of 24.1 days.

Half-life *The half-life of a radioisotope of an element is defined as the time required for one-half of a given quantity of the radioisotope to decay into a different element.*

The half-life is a measurable and meaningful way to compare the times that radioisotopes are radioactive. Table 10-2 compares the half-lives of the same isotopes listed in Table 10-1.

Table 10-2
Comparison of half-lives of isotopes

Name	Symbol	Half-life
Hydrogen	$^{1}_{1}H$	Stable (infinite half-life)
	$^{2}_{1}H$	Stable (infinite half-life)
	$^{3}_{1}H$	12.26 years
Helium	$^{3}_{2}He$	Stable (infinite half-life)
	$^{4}_{2}He$	Stable (infinite half-life)
	$^{5}_{2}He$	0.000000000000000000002 seconds
	$^{6}_{2}He$	0.82 second
Lithium	$^{5}_{3}Li$	About 0.000000000000000000001 seconds
	$^{6}_{3}Li$	Stable (infinite half-life)
	$^{7}_{3}Li$	Stable (infinite half-life)
	$^{8}_{3}Li$	0.84 second
	$^{9}_{3}Li$	0.17 second
Uranium	$^{227}_{92}U$	1.3 minutes
	$^{228}_{92}U$	9.3 minutes
	$^{229}_{92}U$	58 minutes
	$^{230}_{92}U$	21 days
	$^{231}_{92}U$	4.3 days
	$^{232}_{92}U$	74 years
	$^{233}_{92}U$	162,000 years
	$^{234}_{92}U$	250,000 years
	$^{235}_{92}U$	710,000,000 years
	$^{236}_{92}U$	23,900,000 years
	$^{237}_{92}U$	6.75 days
	$^{238}_{92}U$	4,510,000,000 years
	$^{239}_{92}U$	23.5 minutes
	$^{240}_{92}U$	14 hours

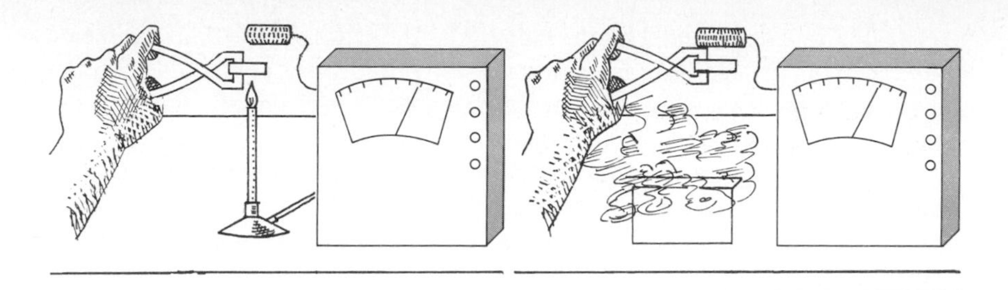

Fig. 10-8
The half-life of a radioisotope cannot be changed. A sample of uranium will show the same radioactivity in a bunsen burner flame or at the temperature of liquid nitrogen.

This same decrease in radioactivity with time is observed for all radioisotopes. The only difference between the half-life of two different radioisotopes is the actual numerical value for the two half-lives. Some radioisotopes have half-lives measured in times ranging up to millions of years or more. A sample of these isotopes emits radiation for millions of years. The radioisotopes uranium-235 and uranium-238 both have very long half-lives. These radioisotopes were present when the earth was formed and are still present in sizable quantities. When an isotope has an infinite half-life, it lasts indefinitely—that is, the isotope is not radioactive but stable.

Half-lives of other radioisotopes are measured in times down to millionths of a second or less. Of course, a half-life of $\frac{1}{1,000,000}$ of a second means that after only 1 second, the amount of radioactivity has been cut in half 1,000,000 times. In other words, a short half-life radioisotope decays very quickly. Many of the radioisotopes of uranium have relatively short half-lives (in the range of minutes or days). Any of these radioisotopes present when the earth was formed have decreased by one-half so often that these radioisotopes are effectively not found in nature. These short-lived radioisotopes, however, can be artificially produced.

The half-life of a given radioisotope cannot be changed. If a sample is heated or cooled, it makes no difference in the half-life (see Fig. 10-8). Whether a pure chemical element or a part of a large molecule, the radioisotope always has the same half-life. The length of the half-life depends upon the nature of the nucleus. So any chemical tampering does not affect the speed of nuclear decay. Even bombarding a nucleus in a so-called atom smasher does not change its half-life. Other changes, such as forcing a proton into the nucleus, are possible. But then a different nucleus has been formed. Radioactivity is a spontaneous phenomenon that can be neither sped up nor slowed down.

10.4 NUCLEAR CLOCKS

One use of radioisotopes is in establishing the age of archaeological remains and the age of the earth itself. The nuclear decay of a radioisotope is not affected by any natural process. Some radioisotopes were present when the earth was first formed. These isotopes are often part of a piece of rock or other specimen. The radioisotopes locked in a specimen decay at an unchanging rate determined by the half-life of the radioisotope. Various measurements of the radioisotope and its decay product are compared with the known decay curve to decide when the radioactivity began. In this way the age of the specimen can be calculated.

The fact that radioactive minerals exist at all in nature means that the earth has not existed for an infinite length of time. Several billion years ago the earth's molten sphere contained a certain quantity of uranium. Cooling formed a crust that trapped most of the original uranium on the earth's surface. Some minerals in the earth's crust contain uranium radioisotopes that were present when the earth was formed. A part of this uranium has decayed through a succession of radioisotopes. The end product of this decay process is a stable isotope of lead, which is locked in the minerals with the remaining uranium.

The nuclear decay of ${}^{238}_{92}U$ eventually produces nonradioactive ${}^{206}_{82}Pb$, and the decay of ${}^{235}_{92}U$ eventually produces nonradioactive ${}^{207}_{82}Pb$. If no lead were near the uranium-bearing mineral at the time of formation of the earth's crust, then all the lead found must be the result of nuclear decay. Measuring the amount of ${}^{238}_{92}U$ relative to ${}^{206}_{82}Pb$ and the amount of ${}^{235}_{92}U$ relative to ${}^{207}_{82}Pb$ permits the calculation of the time when no lead was present. In other words, the age of the earth can be determined. The result is that the age of the earth is about 4,550 million (4.550 billion) years old.

The use of the decay curves of other radioisotopes permits measurement of times much less than billions of years. For example, it is important in archaeology to determine the age of artifacts when no written record has been kept. Or sometimes museums can detect fraudulent artwork that is claimed to be older or younger than it actually is. In these cases the decay curve of the radioisotope ${}^{14}_{6}C$ is used to determine the age. The half-life of carbon-14 is about 5,600 years. It is useful for dating specimens back about 20,000 years. But it would be useless in dating objects which are a million years old.

Isotopes of carbon are found in all living matter—plants, trees, animals, and humans—as part of the organic molecules of which it is composed. The carbon isotopes of ${}^{12}_{6}C$ and ${}^{13}_{6}C$ are not radioactive and have been present since the formation of the earth. Radioactive ${}^{14}_{6}C$, however, is being continually produced, but *not* by the decay of another radioisotope. Instead, ${}^{14}_{6}C$ is being produced by the bombardment of the upper atmosphere by cosmic rays.

Cosmic rays that enter the upper atmosphere are mostly protons, accompanied by some alpha particles and other heavier nuclei. These protons are ionized atoms of interstellar hydrogen gas. Between the center of our galaxy and the earth these protons are accelerated to speeds approaching that of light. The high-energy protons strike air molecules in the upper atmosphere and break their nuclei into pieces. Some of the many fragments from the shattered nuclei are neutrons that have been blasted free.

These liberated neutrons in turn strike the nuclei of nitrogen gas molecules. The nuclear reaction is

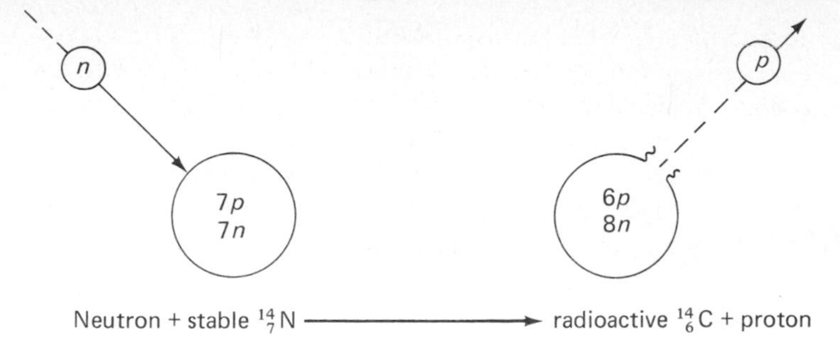

Fig. 10-9
Cosmic rays produce free neutrons which in turn produce carbon-14. The carbon-14 is useful for dating specimens up to an age of 20,000 years.

$$^{14}_{7}N + ^{1}_{0}n \longrightarrow ^{14}_{6}C + ^{1}_{1}p$$

where $^{1}_{0}n$ stands for the bombarding neutron and $^{1}_{1}p$ stands for a proton ejected from the nitrogen nucleus (see Fig. 10-9). The resulting nucleus is a radioisotope of carbon. The radioactive $^{14}_{6}C$ has a half-life of 5,568 years.

All living matter absorbs a very small amount of radioactive $^{14}_{6}C$ along with the stable carbon in its tissues. When the living thing dies, no new $^{14}_{6}C$ is taken in, and so the amount of radioactive carbon gradually decreases as time goes on. Since the half-life of $^{14}_{6}C$ is 5,568 years, the radioactivity will last thousands of years before it becomes too small to measure. By measuring the amount of $^{14}_{6}C$ relative to the unchanging amount of $^{12}_{6}C$ in a specimen, an archaeologist can calculate the age of the specimen. The ages of sandals and

Fig. 10-10
This sandal was found in a cave in Oregon. By the use of carbon-14 dating, the age was determined to be approximately 9,030 years. *(Photograph courtesy of Rainer Berger.)*

Fig. 10-11
Carbon-14 dating of this piece of rope from Peru places the age at about 2,630 years. (*Photograph courtesy of Rainer Berger.*)

Peruvian rope have been analyzed using this technique, as shown in Figs. 10-10 and 10-11.

10.5 FISSION AND FUSION

The discussion of radioactivity emphasized the fact that radioactivity is a process by which energy is released from the nucleus. The emission of an alpha particle, a beta particle, or a gamma ray carries energy away from the radioactive nucleus. In addition to these three processes of releasing nuclear energy, there are two more basic processes known as *nuclear fission* and *nuclear fusion.* Unlike the spontaneous process of radioactivity, fission and fusion require special conditions before nuclear energy will be released.

In order for energy to be released by nuclear fission, it is necessary to bombard the nuclei with neutrons. Usually a bombarding neutron simply enters the nucleus and produces a radioisotope. For example, radioactive carbon-14 is produced when nitrogen is bombarded with a neutron. But for other radioisotopes with large atomic number, bombardment with a neutron causes the nucleus to split in two. Examples of such nuclei are the radioisotopes uranium-235 and plutonium-239. When bombarded with neutrons, these radioisotopes split into two roughly equal pieces in the process of nuclear fission.

Nuclear fission *Nuclear fission is the splitting of a large nucleus into two comparably sized smaller nuclei.*

A typical nuclear fission reaction is

$$^{235}_{92}U + ^{1}_{0}n \longrightarrow ^{94}_{36}Kr + ^{139}_{56}Ba + ^{1}_{0}n + ^{1}_{0}n + ^{1}_{0}n$$

where the fission products are radioisotopes of krypton and barium plus three neutrons.

The fission reaction shown in Fig. 10-12 is typical, but it is not the only one that occurs for uranium-235. Just as in the cracking of a walnut, the splitting of a uranium-235 nucleus does not always produce exactly the same pieces. Not only are different pairs of nuclei produced in various nuclear fission reactions, but also the number of neutrons is different in each reaction. In some cases three neutrons are released, and in other cases only two neutrons are released. For uranium-235, *on the average,* there are $2\frac{1}{2}$ neutrons liberated from each fissioning nucleus. The production of neutrons can cause other nuclei to split and liberate still more neutrons. A rapidly growing avalanche, known as a *chain reaction,* may then occur.

For each separate reaction in the chain reaction, energy is liberated. The fissioning of $^{235}_{92}U$ into $^{94}_{36}Kr$, $^{139}_{56}Ba$, and three neutrons is balanced as far as atomic numbers and mass numbers are concerned. But mass numbers, such as 235, 94, and 139, are simply whole numbers counting the total number of protons and neutrons in each nucleus. The experimentally measured masses of single nuclei are *not* whole numbers. For example, the mass of the $^{235}_{92}U$ nucleus is 235.0439 atomic mass units, where one atomic mass unit equals 1.6600×10^{-27} kilogram. In the fission reaction shown in Fig. 10-12, some mass seems to disappear. Before the reaction the mass of one $^{235}_{92}U$ nucleus plus a neutron is 236.0526 atomic mass units. After a typical reaction the mass of the stable fission fragments plus three neutrons is only 235.8382 atomic mass units. As we have seen in Sec. 10-2, mass is conserved in all nuclear reactions. What appears to be "missing" mass needed to balance the

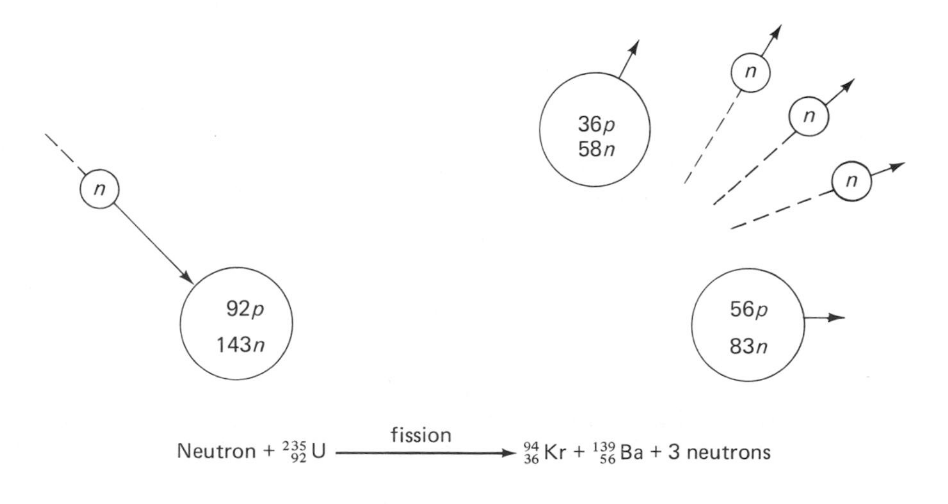

Fig. 10-12
Uranium-235 can undergo fission into many different fragments. The splitting nucleus is shown dividing into barium-139, krypton-94, and three neutrons. The barium and krypton nuclei are radioactive (not stable).

Table 10-3
Calculation of energy released in fission of uranium-235 nucleus

Mass before fission reaction (uranium-235 nucleus plus one neutron)	236.0526 atomic mass units
Mass of stable fragments after fission reaction (zirconium-94 and lanthanum-139 nuclei plus three neutrons)	235.8382 atomic mass units
Difference of masses before and after fission reaction	0.2144 atomic mass unit
Percentage of original mass released as kinetic energy and radiant energy	$\frac{0.2144}{236.0526} \times 100\% = 0.0908\%$

So for every kilogram (1,000 grams) taking part in the fission reaction, energy is released equivalent to 0.908 gram.

equation is mass associated with the kinetic energy of the fission products and radiant energy of gamma rays. For every kilogram of fissioning uranium-235, the resulting kinetic energy and radiant energy are equivalent to approximately $\frac{9}{10}$ of a gram of mass (see Table 10-3). This small mass represents a tremendous amount of energy. The energy equivalent to 1 gram of mass is the same as the heat produced from burning 7,000 *tons* of coal!

The energy released by the fissioning of uranium-235 appears mostly as kinetic energy of the fission fragments. The rapidly moving fission fragments collide with nearby atoms, which then vibrate more vigorously. In other words, the kinetic energy of the fission fragments raises the temperature of the surrounding material. A small percentage of the energy which is liberated in the fission reaction appears as gamma rays. These gamma rays are extremely energetic and, consequently, very dangerous. Typically, the gamma rays are comparable in effect to x-rays produced in an x-ray tube operating at 20 *million* volts.

The release of energy through nuclear fission is based on splitting a nucleus of a large atomic number element, such as uranium. Elements of small atomic number, such as hydrogen, release energy by "welding" nuclei together by nuclear fusion. Two hydrogen nuclei can be made to fuse together to form a new, more massive nucleus of helium.

Nuclear fusion *Nuclear fusion is the combination of two low-mass nuclei to form a more massive nucleus.*

A typical nuclear fusion reaction involves hydrogen, which is the nucleus with the lowest mass of all the elements. The isotope $^{2}_{1}H$ is the raw material for the nuclear fusion reaction. Instead of only one reaction, the nuclear fusion of hydrogen is made up of three reactions. The three reactions are

$$^{2}_{1}H + ^{2}_{1}H \longrightarrow ^{3}_{2}He + ^{1}_{0}n$$

$$^{2}_{1}H + ^{2}_{1}H \longrightarrow ^{3}_{1}H + ^{1}_{1}H$$

$$^{3}_{1}H + ^{2}_{1}H \longrightarrow ^{4}_{2}He + ^{1}_{0}n$$

These fusion reactions begin with the relatively plentiful nonradioactive isotope $^{2}_{1}H$. In order to make the nuclear fusion reactions take place, the hydrogen must have a temperature of several million degrees. For this reason, a nuclear fusion reaction is sometimes known as a *thermonuclear* reaction, which is shown in Fig. 10-13.

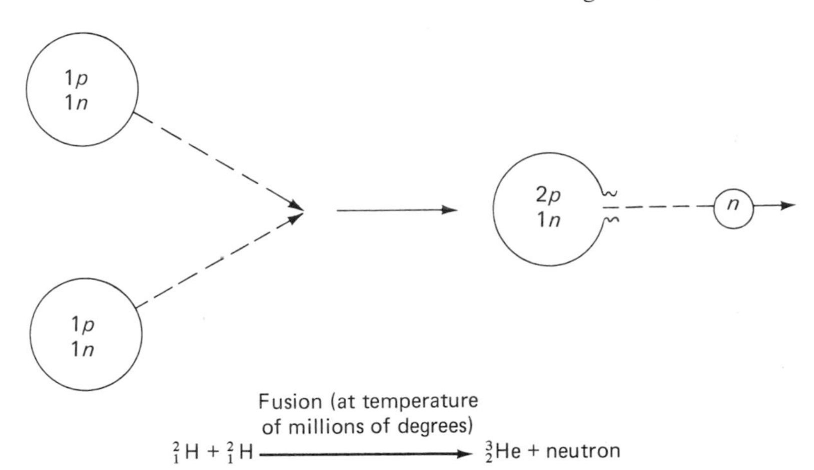

Fig. 10-13
At a very high temperature two hydrogen nuclei fuse together to become a nucleus of helium and a free neutron.

In nuclear fusion reactions, energy is released. The reactions are balanced in both their atomic numbers and mass numbers. Like the fission reaction, the experimentally measured masses of the hydrogen and helium isotopes are not whole numbers. The mass of the $^{2}_{1}H$ nucleus is 2.01410 atomic mass units, the mass of the $^{3}_{2}He$ nucleus is 3.01603 atomic mass units, and the mass of the $^{4}_{2}He$ nucleus is 4.00260 atomic mass units. The mass of the five hydrogen nuclei entering the fusion reaction is 10.07050 atomic mass units. After the completion of the thermonuclear reaction, the mass of the hydrogen nucleus, the two different helium nuclei, and the two neutrons is 10.04393 atomic mass units. The mass needed to balance the reaction (or in other words, needed to conserve mass) is associated with the kinetic energy of the fusion products and radiant energy of gamma rays. For every kilogram of hydrogen-2 involved in the fusion reaction, energy equivalent to about 2.7 grams of mass is released (see Table 10-4). Kilogram for kilogram, the nuclear fusion reaction of hydrogen produces 3 times as much energy as the fissioning of uranium. The fusion reaction is a source of more energy than the fission reaction.

An important thermonuclear reaction takes place in the sun. The interior of the sun has a temperature high enough to sustain thermonuclear reactions. A temperature of about 15 million degrees at the center of the sun fuses

Table 10-4
Calculation of energy released in fusion of hydrogen-2 nuclei

Mass before fusion reaction (five hydrogen-2 nuclei)	10.07050 atomic mass units
Mass after fusion reaction (hydrogen-1, helium-3, and helium-4 nuclei plus two neutrons)	10.04393 atomic mass units
Difference of masses before and after fusion reaction	0.02657 atomic mass unit
Percentage of original mass released as kinetic energy and radiant energy	$\frac{0.02657}{10.07050} \times 100\% = 0.273\%$

So for every kilogram (1,000 grams) taking part in the fusion reaction, energy is released equivalent to 2.73 grams.

hydrogen into helium in a complicated series of reactions. In the course of 1 million years the sun sends out radiant energy equivalent to about one ten-millionth of its total mass. This energy is necessary for the continued existence of life on the earth.

10.6 WEAPONS

When a nucleus of uranium-235 or plutonium-239 splits during nuclear fission, several neutrons are released. If the sample of fissionable material is large enough to prevent the escape of these neutrons, they will each trigger further reactions. The entire process is a chain reaction. When a chain reaction occurs, the sample of fissionable material is said to have a *critical mass.*

Critical mass *A critical mass is defined as the mass of fissionable material that is just able to sustain a nuclear chain reaction.*

A sample of fissionable material smaller than the critical mass is unable to sustain a nuclear chain reaction. When neutrons are liberated, some escape from the material and do not strike other fissionable nuclei. If the mass of the sample is large enough to be *exactly* critical, then the number of neutrons produced in each step of the process is the same as produced in the previous one. The result is a steady release of nuclear energy. If the mass of the sample is larger than the critical mass, then the chain reaction grows and becomes uncontrolled. The fissioning nuclei liberate so much energy that the process is explosive. The result is a fission bomb, commonly called the *atomic bomb.*

The construction of an atomic bomb is based on two important ideas. The first idea is to separate the uranium or plutonium into several portions so that each part is *subcritical.* The individual pieces of the fissionable material are each too small to be critical. As long as these subcritical pieces are separated, an uncontrolled chain reaction does not occur. However, when the subcritical parts are brought together, an uncontrolled chain reaction occurs in the now supercritical mass of fissionable material.

The second important idea in the construction of an atomic bomb is the *containment time.* As the supercritical mass of fissionable material liberates nuclear energy, the uranium or plutonium will begin to melt. As the temperature increases further, the material will begin to vaporize. The parts of the fissionable material might separate and become subcritical, and then the reaction would stop. Therefore, the rapidly heating mass of fissionable material must be held together long enough for the fissioning process to be completed.

Containment time *The containment time is defined as the time necessary to hold fissioning nuclei together to produce a nuclear explosion.*

In order to achieve a supercritical mass and contain it long enough to produce an atomic bomb, a highly explosive chemical drives together the pieces of subcritical fissionable material. One type of atomic bomb uses a gun-barrel device, shown in Fig. 10-14. An ordinary high explosive is used to blow one subcritical piece of fissionable material from one end of the gun barrel into another subcritical piece of fissionable material held firmly at the other end. The atomic bomb dropped on the city of Hiroshima, Japan, during World War II was a gun-barrel device using uranium-235.

Another type of atomic bomb uses a spherical arrangement of chemical high explosives (see Fig. 10-15). At the center of the bomb is placed a subcritical sphere of fissionable material. When the high explosive is set off, an inwardly directed "implosion" wave is produced. The fissionable material

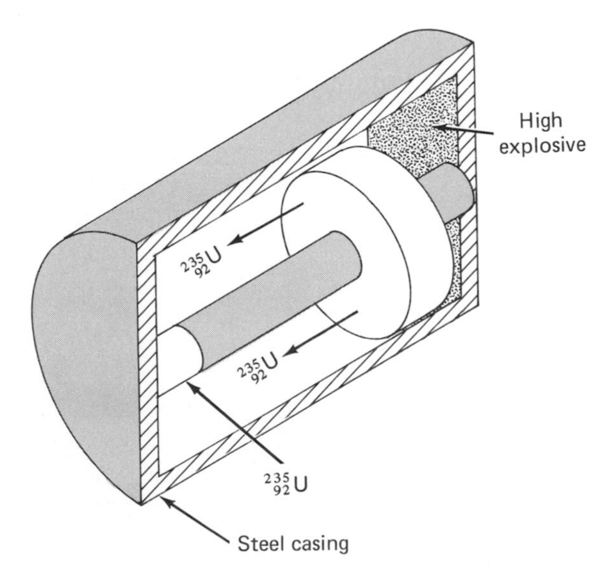

Fig. 10-14
Several subcritical pieces of uranium-235 are brought together to form a critical mass in a "gun-barrel" atomic bomb.

NEUTRON ACTIVATION ANALYSIS

Napoleon Bonaparte was born in August of 1769. Famous in his own time, he lived a life filled with adventure. He died in 1821 when he was 51 years old. Someone poisoned him with arsenic. But that was not known until 140 years later! The incredible story of this piece of detective work involves neutrons.

During the 1940s and 1950s workers in several countries were rapidly expanding knowledge of the nucleus. As special techniques were perfected, each nuclear isotope was systematically studied. The masses of the isotopes were measured—ever more precisely. And the decay scheme of each radioisotope was investigated.

As this information was collected and sorted, one fact became increasingly clear. Each radioisotope had a unique set of decay properties. For different nuclei there are different half-lives. Particles are emitted from radioactive nuclei with different energies. A list of the nuclear properties of each radioisotope provides decay schemes for all the radioisotopes. This knowledge of decay schemes can be used to detect the presence of various elements in an unknown sample.

When a sample of material is bombarded with neutrons, some of the isotopes present become radioisotopes. The neutron bombardment produces some radioactive substances. Then the decay schemes can be compared with the known list to see which elements are present in the unknown sample. This technique is called neutron activation analysis.

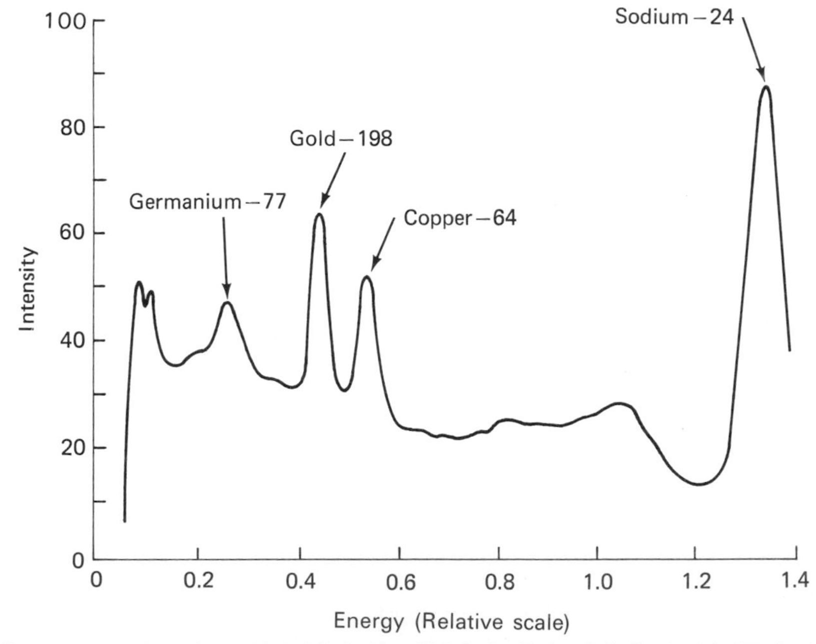

(Gamma ray spectrum of normal hair. Adapted from W. R. Corliss, Neutron Activation Analysis, U.S. Atomic Energy Commission, 1964(rev.) p. 9.)

In 1961 this method was applied to some hair from Napoleon Bonaparte's head. The arsenic content of his hair was found to be much higher than is normal (see graph on opposite page). By examining different parts of each hair (see photograph below) it was even discovered that he had been poisoned little by little over a period of many weeks.

(Photograph by Sten Forshufvud, courtesy of DOE.)

Neutron activation analysis has been applied to many other areas. It has been used to detect mineral ores in the ground and to determine the authenticity of paintings. It has been used medically to determine trace amounts of certain elements which are vital to proper health. Also, the use of neutron activation analysis to obtain legal evidence is gaining popularity. One technique is to examine the hand of a suspect in an armed robbery. Neutron activation analysis is sensitive enough to detect small traces of gunpowder. You might see it mentioned in the newspaper in connection with a trial.

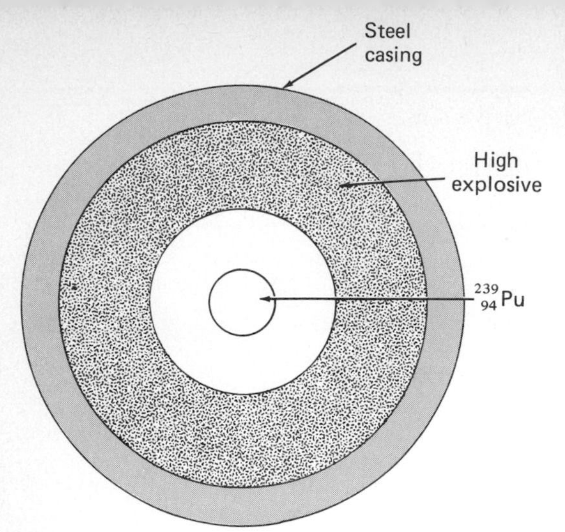

Fig. 10-15
High explosives are used to compress a subcritical mass of plutonium-239 into a critical mass in an "implosion" atomic bomb.

is compressed into a supercritical condition. An implosion device using plutonium-239 was dropped on the city of Nagasaki, Japan, during World War II. The explosion of those two nuclear bombs used up all the fissionable material available at that time.

The difficulties encountered by any nation attempting to produce its first atomic bomb are formidable. The fissionable material, whether uranium-235 or plutonium-239, is extremely difficult to obtain. If a nation has uranium ore, there is the problem of separating the $^{235}_{92}U$ radioisotope from the much more common $^{238}_{92}U$ radioisotope. The $^{239}_{94}Pu$ radioisotope is artificial and is produced by bombarding $^{238}_{92}U$ with neutrons, assuming that the uranium is available. (An adequate amount of fissionable material is typically a few kilograms for one nuclear fission bomb.) Next is the problem of holding a supercritical mass together long enough for a chain reaction to occur. If, however, a nation has the technology to produce an atomic bomb, it is a relatively easy step to escalate to a nuclear fusion bomb. Figure 10-16 shows the size of a typical atomic bomb.

A *hydrogen bomb* is based on the fusion of hydrogen nuclei. The thermonuclear fusion reaction for hydrogen nuclei occurs only at very high temperatures. These temperatures are typically in the range of 10 to 100 million degrees. Such high temperatures can be achieved by the explosion of an atomic bomb. In other words, an atomic bomb is needed to trigger the fusion reaction of a hydrogen bomb. A hydrogen bomb is constructed by surrounding an atomic bomb with a supply of the hydrogen isotope $^{2}_{1}H$. (The much more plentiful $^{1}_{1}H$ isotope of hydrogen does not participate in a fusion reaction.)

The destructiveness of nuclear weapons is awesome. In an attempt to rate the destructiveness of atomic or hydrogen bombs, it has become customary to describe an equivalent in tons of TNT. TNT is a chemical explosive mostly used in road building and mining. One *pound* of TNT can easily pulverize a

Fig. 10-16
An atomic bomb casing, similar to the one used in World War II, is on display for visitors to a government laboratory. (*Photograph courtesy of Los Alamos Scientific Laboratory.*)

boulder. Typical atomic bombs, such as the two used in the Second World War, are as destructive as 20,000 *tons* of TNT. (In the military the 20-kiloton figure is usually abbreviated. You might see this referred to as 20 kT.) A typical hydrogen bomb used for military and civil defense planning purposes is as destructive as 20 million tons of TNT, abbreviated 20 MT (for megaton). Tests have been made with hydrogen bombs rated at more than 50 megatons of TNT.

For a hydrogen bomb, about 95 percent of the total energy output is emitted at the time of the explosion. This percentage is distributed in the following way: 50 percent blast waves, 40 percent thermal radiation (heat and light), and 5 percent initial nuclear radiation, as indicated in Fig. 10-17. Such a distribution of energy is fairly typical of an explosion of TNT, except that there is no nuclear radiation with a conventional chemical explosive. The remaining 5 percent of the total energy released in a thermonuclear explosion

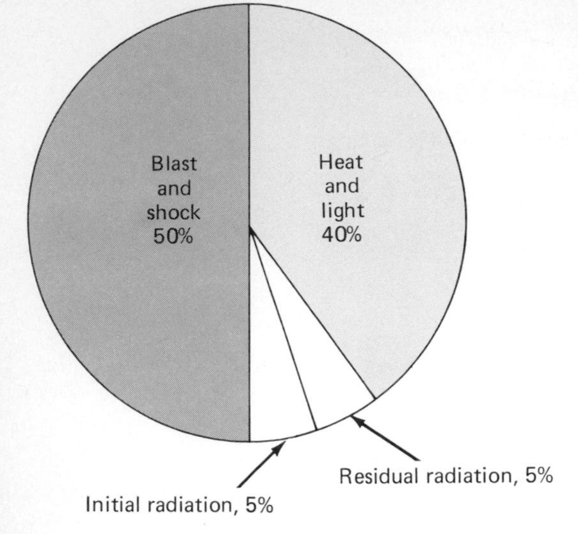

Fig. 10-17
The distribution of energy released in a hydrogen bomb.

is in the form of a lingering nuclear radiation which occurs long after the explosion.

The *initial* nuclear radiations consist mainly of gamma rays and neutrons. These radiations, especially gamma rays, can travel great distances through air and can penetrate considerable thicknesses of various material. Although they cannot be seen or felt, gamma rays and neutrons can produce very harmful effects. The *residual* nuclear radiations arise from the radioactive fission products. These products are called *fallout.* Fallout typically emits alpha particles, beta particles, and gamma rays in the course of the decay of the radioisotopes. Beta particles are much less penetrating than gamma rays. Beta particles are dangerous if emitted from fallout that has landed on your skin or that is taken in with food and drink. Alpha particles are the least penetrating form of nuclear radiation and are dangerous only if taken in with food or drink.

10.7 REACTORS

Nuclear fission, when occurring in an uncontrolled chain reaction, is the basis of nuclear weapons. But nuclear fission can be maintained as a *controlled* chain reaction in a *nuclear reactor.* In a nuclear reactor the energy and neutrons are released gradually to produce useful heat and artificial radioisotopes.

One type of nuclear reactor is constructed of carbon blocks containing long cylindrical holes. Containers of fissionable material, either uranium-235 or plutonium-239, are inserted in some of the holes in the carbon blocks (see Fig. 10-18). The carbon slows down neutrons, which enhances the chain reaction. The energy released in the chain reaction can be used to generate steam, which in turn produces electricity by driving an electric power plant.

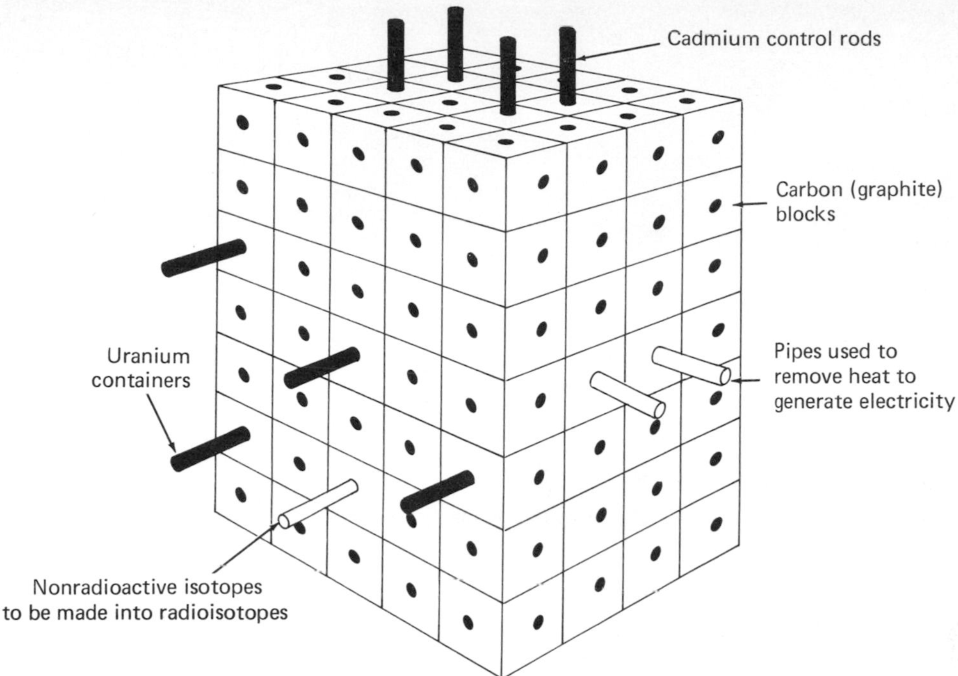

Fig. 10-18
A schematic diagram of a nuclear reactor. Artificial radioisotopes can be produced, as well as heat for the production of electric power.

Such nuclear power reactors are in use by many electric companies that have sought fuel supplies other than coal, oil, or natural gas.

Of course, the nuclear reactor must be kept under control. To prevent the temperature from rising too high, control rods are used to absorb neutrons. As neutrons are absorbed by the control rods, the fission rate is slowed down.

Some people are concerned that a malfunction or sabotage of the control rods might result in a nuclear explosion. This is a needless fear. Even if all the control rods were deliberately removed, a nuclear reactor still would *not* explode like an atomic bomb. Recall that a nuclear weapon functions by holding together all the fissionable material by means of high explosives. Nuclear reactors absolutely do not have the capability of producing an uncontrolled chain reaction of the fissionable material.

The uranium in the reactor could rise in temperature if there were a loss of coolant because a pipe burst. In the event of such an accident, other pipes would flood the reactor with water and chemicals from several different safety reservoirs. The resulting steam and released radioactivity would escape from the reactor itself, but not the specially constructed dome-topped building (see Fig. 10-20). The building housing the reactor is not only strong enough to hold the steam released in an accident, but is also strong enough to withstand a direct hit by a crashing airplane.

The supply of uranium, either as fissionable uranium-235 or as uranium-238 to produce fissionable plutonium-239, is known to be limited.

Fig. 10-19
The nuclear fuel is loaded into the holes on the face of the reactor. (*Photographs courtesy of Los Alamos Scientific Laboratory.*)

Current estimates suggest that uranium will probably be used up early in the next century. To provide the vast amounts of energy that industrial nations use, research is now underway to control the nuclear fusion process. The basic material for a fusion reaction is the ^2_1H isotope of hydrogen, and this isotope is quite plentiful in nature. However, a fusion reaction requires that the hydrogen be raised to a temperature of millions of degrees.

Unfortunately, there are exceedingly complicated problems in control-

Fig. 10-20
The nuclear power reactor that produces the steam to drive electrical generators is housed in the domelike structure. (*Photograph by Paolo Koch. From Photo Researchers, Inc.*)

ling such a thermonuclear reaction. Any material used to house the thermonuclear reaction would be vaporized by these high temperatures. So the reaction must be somehow suspended. One method being investigated suspends the fusion reaction with magnetic fields. But even when the problem of holding the fusion reaction is solved, there will still be others. Useful energy must be extracted from the process. And this must be done without stopping the reaction. There is probably little hope of controlling the fusion reaction and producing usable energy until the next century.

In addition to its use for the generation of electricity, a nuclear reactor can also be used to produce artificial radioisotopes. Most of the elements with atomic numbers less than 82 do not have radioisotopes that occur naturally in nature. With a nuclear reactor, radioisotopes of many elements can be made artificially. Of course, the half-life and type of particle or ray emitted from a selected radioisotope cannot be changed. These physical properties can only be measured and cataloged for possible future uses.

The fact that normally nonradioactive elements can be made radioactive opens the door to entirely new areas of scientific research and application. For example, chemical and biological processes can be examined using radioisotopes. In agricultural research it is important to know how nutrients are taken into plants. Any radioisotope behaves chemically exactly like a nonradioactive isotope of the same element. For example, consider phosphorus. Radioactive phosphorus in a fertilizer will behave chemically just like nonradioactive phosphorus until it emits a beta particle. A Geiger counter can be used to follow the progress of the radioactive phosphorus as it is taken in by a plant from the fertilizer. In other words, radioisotopes can be used as *tracers* in biochemical processes.

10.8 NUCLEAR MEDICINE

Radioisotopes have two primary functions in medicine: *diagnosis* and *therapy.* Diagnosis refers to the *identification* of a disease or improper body function. A radioisotope serves as an identifying and easily detected tracer. Therapy refers to the *treatment* of a disease or improper body function. Radioisotopes in nuclear therapy selectively destroy abnormal cells or tissues. For both diagnosis and therapy, radioisotopes are directed to a specific part of the body. The radioactive compounds used in *diagnosis* are of such small quantities that they are relatively harmless to nearby cells and organs. The radioactive doses used in *therapy* are large enough to destroy cells and tumors, and so care must be taken to avoid damage to healthy parts of the body.

An example of a tracer used in diagnosis is the radioisotope iodine-131. Iodine is used in studies of the thyroid gland. Since normal thyroid operation requires iodine, this is a natural procedure, except that iodine-131 is radioactive.

Thyroxin, an iodine compound, is manufactured in the thyroid gland. After manufacture it is transferred by the blood to the body tissues. Thyroxin helps to govern oxygen utilization in the body. In addition, thyroxin is essential for the proper utilization of nutrients. Thyroxin is monitored by substituting radioactive iodine-131 for the natural nonradioactive iodine-127.

Iodine-131 goes to the same spots and enters into the same chemical reactions as iodine-127, but a Geiger counter is used to detect the radiation from iodine-131 and monitor the progress of the thyroxin (see Fig. 10-21).

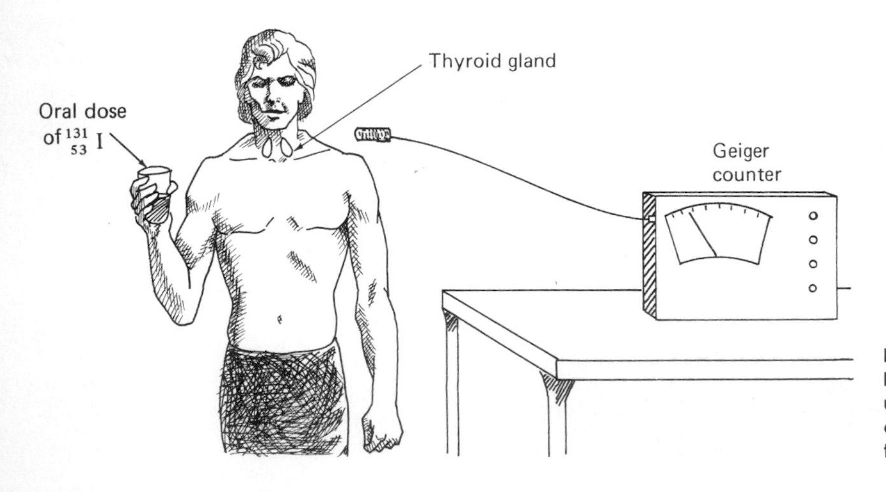

Fig. 10-21
Radioactive tracers can be used to detect many diseases and malfunctions of the body.

One of the common diagnostic procedures for checking the thyroid is to measure the percentage of a dose of iodine-131 that is taken up by the gland. To begin this process, the patient drinks a very small dose of radioactive iodine. The iodine is part of a harmless chemical compound. About 2 hours later the amount of iodine in the thyroid is determined by measuring the radiation coming from the neck area. This amount can then be compared with amounts known to be normal.

Another diagnostic test using radioactive iodine-131 is a check for the spread of thyroid cancer. If cancer is present, pieces of cancerous thyroid tissue may move to other parts of the body. If further growth takes place at new locations, these new cancers are a signal of an advanced state of the disease. Even complete surgical removal of the original cancer may not save the patient. When a thyroid cancer is discovered, a physician may check for the spread of the disease before deciding to operate.

The same radioisotope iodine-131 can also be used to *treat* thyroid malfunction. For therapy, the concentrations are much higher than those used in diagnostic tests. The higher levels of radiation will damage thyroid cells and thus reduce the activity of an overactive thyroid.

In therapy the energy is released within the diseased gland and much of the energy is absorbed there. Iodine-131 has a half-life of 8 days. In contrast, iodine-132 has a half-life of about 2 hours. As a result, the same mass of iodine-132 will give up its radiation quicker. This is often preferred, since the

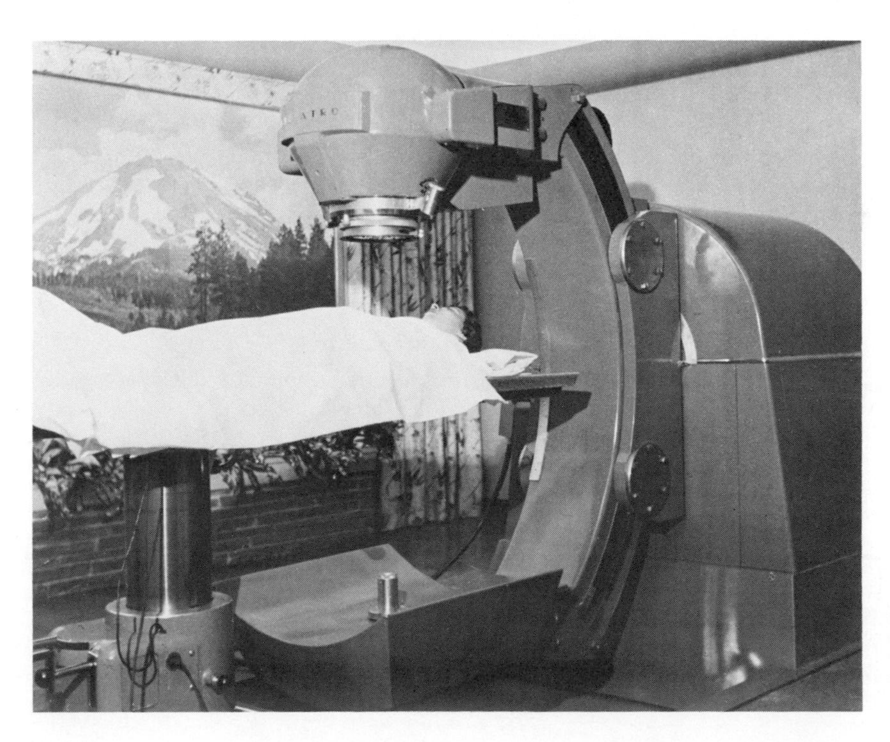

Fig. 10-22
Gamma rays, which are released from the container above the patient, destroy cancerous cells. (*Photograph courtesy of American Cancer Society.*)

radiation can be delivered to the desired organ before the radioisotope has passed through the body. Therefore, iodine-132 is often preferred for treatment of this sort.

Another form of nuclear therapy has the source of radiation outside the patient. The use of radioactive cobalt-60 is an example of placing the radioisotope outside the patient. Cobalt-60 emits very energetic gamma rays, which can be directed toward a cancerous region of the body. The cobalt-60 is stored in a thick lead box having only one small opening through which the gamma rays may pass. The patient is then put in position so that the gamma rays can strike a cancerous part of the body (see Fig. 10-22). Gamma rays are directed to the patient in repeated daily doses until the cancerous tissue is killed. Nearby healthy tissue and organs are protected from the gamma rays with sheets of lead-impregnated cloth.

Radioactive tracers and radiation sources are necessary in modern medicine. The frightening use of nuclear energy for nuclear weapons is balanced by the lifesaving techniques developed in nuclear medicine.

REVIEW You should be able to define or describe the following terms: mass number, isotopes, half-life, nuclear fission, nuclear fusion, critical mass, and containment time.

You should be able to explain and give examples of the release of nuclear energy.

PROBLEMS

1. How many protons are in the nucleus of $^{15}_{8}O$? Repeat for $^{16}_{8}O$.
2. For each of the nuclei $^{35}_{17}Cl$, $^{13}_{6}C$, and $^{14}_{7}N$, identify the atomic number.
3. How many neutrons are in $^{238}_{92}U$?
4. Calculate the number of neutrons in $^{137}_{53}I$.
5. Some nuclear reactions only take place at very high temperatures. One such reaction suggested to occur in dense stars is

$$^{4}_{2}He + ^{8}_{4}Be \longrightarrow ?$$

If the two nuclei are assumed to undergo fusion, what larger nucleus is created?

6. A value for the mass of a neutron has been determined by considering the nuclear reaction

$$^{11}_{5}B + ^{4}_{2}He \longrightarrow ? + ^{1}_{0}n$$

Identify the missing nucleus.

7. What percent of a sample of radioisotope undergoes decay during three half-lives?
8. The half-life of radon is 3.82 days. What fraction of a sample decays in two half-lives?

DISCUSSION QUESTIONS

1. How can the ionizing radiation coming from a mineral be detected?
2. What part of an atom releases energy in the process of radioactivity?

3. What is a proton? A neutron? What electrical charge does each particle have?

4. Can a stable nucleus contain only protons?

5. What is the atomic number of an element? What is the mass number of an element?

6. What is an alpha particle?

7. Why are most alpha particles that are emitted from radioisotopes considered harmless to humans?

8. When an alpha particle is emitted from a uranium-235 nucleus, does the nucleus change?

9. What is a beta particle?

10. Do beta particles exist in the nucleus?

11. Given the same kinetic energy, which are more penetrating, alpha particles or beta particles?

12. On what feature of the nucleus is the periodic chart of the elements based?

13. Is it possible for two isotopes of the same element to have the same number of protons? Neutrons?

14. In what sense is the Geiger counter a device that cannot be made more sensitive?

15. Define half-life for a sample of a single radioactive isotope.

16. Can a radioactive isotope's half-life be altered by ordinary pressure and temperature changes?

17. What is carbon-14 dating?

18. Why are several different elements used for radioactive dating?

19. When the isotopes uranium-235 and plutonium-239 undergo fission, is energy released?

20. What is meant by a nuclear chain reaction?

21. Are all isotopes radioactive?

22. Is it possible for a nuclear fission reactor to explode like a fission bomb?

23. What is a thermonuclear reaction?

24. What obstacles must be overcome to develop a useful fusion reactor?

25. What is meant by kT and MT in describing nuclear weapons?

ANNOTATED BIBLIOGRAPHY

A reference book mentioned in Chap. 9, *The Atom and Its Nucleus,* by G. Gamow, is also appropriate here. Chapter 6, pages 76 to 89, on natural radioactivity is particularly well done. Also see the diverse material in M. Bennett, *The Intelligent Woman's Guide to Atomic Radiation,* Penguin Books, Inc., Baltimore, 1964.

For a longer discussion of quantum physics, G. Gamow is an excellent author. His brief book *Thirty Years That Shook Physics* is subtitled *The Story of Quantum Theory.* It was published by Anchor Books, Doubleday & Company, Inc., Garden City, N.Y., 1966.

An authoritative source for information on nuclear bombs is S. Glasstone (ed.), *The Effect of Nuclear Weapons,* U.S. Atomic Energy Commission, 1962. Photographs showing various levels of damage to houses and machinery are scattered throughout the book.

Although many books have been written on the people involved with weapons development, one book stands out. *Brighter Than a Thousand Suns,* subtitled *The Story of the Men Who Made the Bomb,* was written by a professional journalist, R. Jungk. It won international acclaim when released by Grove Press, Inc., New York, 1958.

The role of nuclear power reactors in society is under constant discussion. A good collection of articles is H. Foreman (ed.), *Nuclear Power and the Public,* Anchor Books, Doubleday & Company, Inc., Garden City, N.Y., 1972. Another interesting article is H. Bethe's "The Necessity of Fission Power," *Scientific American,* January 1976, p. 21.

Another source of information on nuclear physics is a large series of pamphlets published by the Division of Technical Information, U.S. Energy Research and Development Administration. Write to ERDA at Oak Ridge, Tenn. 37830. Ask for a listing of titles. If you wish to read one, we suggest *Radioisotopes in Medicine* by E. Phelan.

Theories of the nucleus and fundamental particles are constantly changing in this very active area of research. Two articles presenting experimental results as well as speculations on future developments are D. Cline's "The Search for New Families of Elementary Particles," *Scientific American,* January 1976, p. 44, and R. Schwitter's "Fundamental Particles with Charm," *Scientific American,* October 1977, p. 56.

ANSWERS TO PROBLEMS

CHAPTER 1

1. The problem states that the distance d is 2,100 miles and the elapsed time t is 4 hours. Average speed v is given by the formula

$$v = \frac{d}{t}$$

and so
$$v = \frac{2{,}100 \text{ miles}}{4 \text{ hours}}$$

$$v = 525 \text{ mph}$$

3. The problem states that the average speed v is 1.5 feet per second and the elapsed time t is 30 seconds. From the formula for average speed we solve for distance d, which is

$$d = vt$$

and so $d = (1.5 \text{ feet per second})(30 \text{ seconds})$

$$d = 45 \text{ ft}$$

5. The distance d from the sun to the earth is 149,900,000 kilometers (about 93,000,000 miles), and the speed v is about 300,000 kilometers per second. From the formula for speed we solve for time t, which is

$$t = \frac{d}{v}$$

and so
$$t = \frac{149{,}900{,}000 \text{ kilometers}}{300{,}000 \text{ kilometers per second}}$$

$$t = 500 \text{ seconds}$$

$$t = 8 \text{ min } 20 \text{ sec}$$

7. It is known that a force F_1 of 500 pounds causes an acceleration a_1 of 5 ft/sec^2 and an unknown force F_2 causes an acceleration a_2 of 10 ft/sec^2. The mass of the car and driver is unknown. Newton's second law can be written twice:

$$F_1 = ma_1$$

and

$$F_2 = ma_2$$

Dividing the second equation by the first eliminates the unknown mass.

$$\frac{F_2}{F_1} = \frac{ma_2}{ma_1} = \frac{a_2}{a_1}$$

Thus,

$$F_2 = F_1\left(\frac{a_2}{a_1}\right)$$

$$F_2 = 500 \text{ pounds}\left(\frac{10 \text{ ft/sec}^2}{5 \text{ ft/sec}^2}\right)$$

$$F_2 = 1{,}000 \text{ lb}$$

9. From Newton's universal law of gravitation, the force of gravity can be calculated as

$$F_1 = \frac{Gm_E\, m_m}{r^2}$$

where F_1 = present force of gravity
m_E = mass of the earth
m_m = mass of the moon
r = distance from the earth to the moon

If the distance were doubled, then the distance could be expressed as $2r$, and the force of gravity would be

$$F_2 = \frac{Fm_E\, m_m}{(2r)^2}$$

$$= \frac{1}{4}\frac{Gm_E\, m_m}{r^2}$$

But

$$\frac{Gm_E\, m_m}{r^2} = F_1$$

and so

$$F_2 = \frac{1}{4}F_1$$

Notice that doubling the distance r has decreased the force of gravity to one-quarter of its original value.

CHAPTER 2

1. The weight of the elevator is 5,000 pounds, and so 5,000 pounds is the force F needed to lift the elevator. (The force of friction is neglected in this problem.) The elevator moves a distance d of 30 feet from the first to fourth floors. The work $\mathcal{W}$ can be calculated:

$$\mathcal{W} = Fd$$

$$= (5{,}000 \text{ pounds})(30 \text{ feet})$$

$$= 150{,}000 \text{ ft}\cdot\text{lb}$$

3. The problem states that the work $\mathcal{W}$ is 1,080,000 joules. In order to calculate power P, the time t must be expressed in seconds, namely, 1 hour equals 3,600 seconds. Substitute the numerical values into the formula for power and obtain

$$P = \frac{\mathcal{W}}{t}$$

$$= \frac{1{,}080{,}000 \text{ joules}}{3{,}600 \text{ seconds}}$$

$$= 300 \text{ W}$$

5. The weight w of the elevator is 5,000 pounds, and the elevator is raised a height h equal to 30 feet. The increase in the potential energy is calculated from the formula

$$PE = wh$$

$$= (5{,}000 \text{ pounds})(30 \text{ feet})$$

$$= 150{,}000 \text{ ft}\cdot\text{lb}$$

Notice the similarity in calculations between this problem and problem 1. The work $\mathcal{W}$ done in problem 1 is exactly the increase in potential energy PE of this problem, because the work caused potential energy to be increased.

7. The mass m of the truck is 5,000 kilograms, and the speed v is 20 meters per second. The formula for kinetic energy is:

$$KE = \tfrac{1}{2}mv^2$$

$$= \tfrac{1}{2}(5{,}000 \text{ kilograms})(20 \text{ meters per second})^2$$

$$= (2{,}500)(400) \text{ joules}$$

$$= 1{,}000{,}000 \text{ J}$$

9. No calculations are needed for this problem, only reasoning. The law of conservation of energy states that energy cannot be created or destroyed, only changed in form. The 150,000 foot-pounds of potential energy is changed to kinetic energy of exactly the same numerical value (assuming no loss of energy to friction), namely, 150,000 foot-pounds.

CHAPTER 3

1. The total area of contact A for the four tires is 128 square inches. The weight of the car is 3,600 pounds, which is the force F. The pressure p is calculated from the formula:

$$p = \frac{F}{A}$$

$$= \frac{3{,}600 \text{ pounds}}{128 \text{ square inches}}$$

$$= 28 \text{ lb/in}^2$$

This is the pressure exerted by each tire on the ground beneath.

3. The total surface area A is 240 square inches for the rectangular aquarium bottom. The pressure p is given as 0.4 pound per square inch. The force F can be calculated from the formula for pressure, which is rewritten in the form:

$$F = (p)(A)$$

$$= (0.4 \text{ pound per square inch})(240 \text{ square inches})$$

$$= 96 \text{ lb}$$

5. The volume V of the cube must be calculated using the fact that 6 inches equals $\frac{1}{2}$ foot. The volume V of a cube having side of length ℓ is

$$V = \ell^3$$

$$= (\tfrac{1}{2} \text{ foot})(\tfrac{1}{2} \text{ foot})(\tfrac{1}{2} \text{ foot})$$

$$= \tfrac{1}{8} \text{ ft}^3$$

The weight of the cube is given as 18.75 pounds. The weight density is determined from the formula

$$d_w = \frac{w}{V}$$

$$= \frac{18.75 \text{ pounds}}{\frac{1}{8} \text{ cubic foot}}$$

$$= 150 \text{ lb/ft}^3$$

7. The mass m is given as 7.9 kilograms, and the volume V is given as 0.001 cubic meter. The mass density d_m is calculated to be

$$d_m = \frac{m}{V}$$

$$= \frac{7.9 \text{ kilograms}}{0.001 \text{ cubic meter}}$$

$$= 7{,}900 \text{ kg/m}^3$$

CHAPTER 4

1. The length L is 100 meters, and the change of temperature ΔT is -50 C deg. The coefficient of thermal expansion α is 11×10^{-6} per degree, and so the change of length ΔL is

$$\Delta L = \alpha \, L \, \Delta T$$

$$= \left(\frac{11 \times 10^{-6}}{\text{degree}}\right)(100 \text{ meters})(-50 \text{ degrees})$$

$$= -55 \times 10^{-3} \text{ meter}$$

$$= -55 \text{ millimeters}$$

The minus sign means the bridge contracted in length. A positive sign would have meant that an expansion occurs. Notice that the contraction is about $2\frac{1}{4}$ inches, which shows why movable joints are important (see Fig. 4-4).

3. The work $\mathcal{W}$ is 1350 joules, and the heat Q is 650 joules. The change in internal energy ΔU is calculated from the first law of thermodynamics:

$$\Delta U = \mathcal{W} + Q$$
$$= 1350 \text{ joules} + 650 \text{ joules}$$
$$= 2000 \text{ J}$$

5. Refer to Table 4-4 and you will find that the wind-chill factor is 5°C. This means that the motorcyclist is cooled *as if* at rest in still air with a temperature of 5°C instead of the actual 15°C.

7. The relative humidity is calculated from the data in Table 4-5.

$$\text{Relative humidity} = \left(\frac{\text{actual humidity density}}{\text{saturation density}}\right)(100\%)$$
$$= \left(\frac{0.0020 \text{ kilogram per cubic meter}}{0.0022 \text{ kilogram per cubic meter}}\right)(100\%)$$
$$= 91\%$$

The outside relative humidity is 91 percent. Repeating the calculation using a saturation density 0.0173 kilogram per cubic meter yields a relative humidity of 12 percent in the living room. A relative humidity of 12 percent is too low for comfortable living, and so a humidifier should be used to increase the relative humidity to a comfortable level.

CHAPTER 5

1. The frequency f is 85 hertz, and the wavelength λ is 4 meters. The speed of the wave (that is, speed of sound in air) can be calculated from the wave formula

$$v = f\lambda$$
$$= (85 \text{ Hz})(4 \text{ m})$$
$$= 340 \text{ m/sec}$$

Notice that the hertz (Hz) is a unit equivalent to one cycle per second, so that the product of the hertz and the meter is the meter per second.

3. The speed of sound v is 340 meters per second, and the frequency f is 262 hertz. To calculate the wavelength λ, rewrite the wave formula as

$$\lambda = \frac{v}{f}$$
$$= \frac{340 \text{ m/sec}}{262 \text{ Hz}}$$
$$= 1.3 \text{ m}$$

The distance from one compression to the next compression in a longitudinal sound wave is 1.3 meters, or about $4\frac{1}{4}$ feet.

5. To solve this problem you need to recall from Chap. 1 that an object (or wave) with a speed v moves a distance d in a time t given by the formula $d = vt$. In this problem the time t is 12 seconds and the speed of sound v is 340 meters per second. The lightning is a distance d away, given by

$$d = vt$$
$$= (340 \text{ m/sec})(12 \text{ sec})$$
$$= 4{,}080 \text{ m}$$

The distance is 4.08 kilometers, or about $2\frac{1}{2}$ miles.

7. The frequency of the twenty-first harmonic is 21 times the fundamental frequency, or 21 times 196 Hz. The product of these two numbers is 4,116 Hz.

CHAPTER 6

1. Since each cell (battery) has a voltage of 1.5 volts, the flashlight runs on a voltage V of 4.5 volts. The resistance R is 20 ohms. Use Ohm's law to calculate the current,

$$I = \frac{V}{R}$$

$$= \frac{4.5 \text{ volts}}{20 \text{ ohms}}$$

$$= 0.225 \text{ ampere}$$

3. The voltage V delivered to the starter can be taken as 12 volts, and the current I is 48 amperes. Rewrite Ohm's law in the following form to compute the resistance:

$$R = \frac{V}{I}$$

$$= \frac{12 \text{ volts}}{48 \text{ amperes}}$$

$$= 0.25 \text{ ohm}$$

5. It is stated that the voltage V is 25,000 volts and the current I is 0.001 ampere. Substitute these values in the formula for power:

$$P = IV$$

$$= (0.001 \text{ ampere})(25{,}000 \text{ volts})$$

$$= 25 \text{ W}$$

7. The power of the light bulb P is 100 watts, and the voltage V is 120 volts. Use the formula for power in a rearranged form to obtain

$$I = \frac{P}{V}$$

$$= \frac{100 \text{ watts}}{120 \text{ volts}}$$

$$= 0.83 \text{ ampere}$$

CHAPTER 7

1. Broadcasts on AM radio occur on assigned carrier frequencies between 540 kilohertz (540,000 cycles per second) and 1,600 kilohertz (1,600,000 cycles per second). The number 11 is an abbreviation for 1,100 kilohertz (1,100,000 cycles per second), which is the frequency f of the carrier wave. The wavelength λ can be calculated from the wave formula for electromagnetic waves

$$c = f\lambda$$

where c is the speed of light (186,000 miles per second). Solving for wavelength you obtain

$$\lambda = \frac{c}{f}$$

$$= \frac{186{,}000 \text{ miles per second}}{1{,}100{,}000 \text{ cycles per second}}$$

$$= 0.169 \text{ mile}$$

$$= 893 \text{ feet}$$

3. It is stated that the wavelength λ of the red light is 650×10^{-9} meters. From the wave formula for electromagnetic waves you can solve for the frequency f and obtain

$$f = \frac{c}{\lambda}$$

where c is the speed of light (300,000 kilometers per second). Substitution of numerical values yields

$$f = \frac{300{,}000 \text{ kilometers per second}}{650 \times 10^{-9} \text{ meters}}$$

$$= \frac{3 \times 10^{5} \text{ kilometers per second}}{6.5 \times 10^{-7} \text{ meters}}$$

$$= \frac{3 \times 10^{8} \text{ meters per second}}{6.5 \times 10^{-7} \text{ meters}}$$

$$= 0.46 \times 10^{15} \text{ hertz}$$

$$= 4.6 \times 10^{14} \text{ Hz}$$

1. The energy lost by the electron ΔE is given as 3.96×10^{-19} joules. This energy loss can be related to the frequency f of the emitted radiation through the formula

$$\Delta E = hf$$

$$3.96 \times 10^{-19} \text{ joules} = (6.6 \times 10^{-34} \text{ joule-seconds})\, f$$

Notice that the value of Planck's constant h has been used. This constant has the value 6.6×10^{-34} joule-seconds. The same value of h is used in every problem of this kind. By dividing the numbers above, we obtain

$$f = \frac{3.96 \times 10^{-19} \text{ joules}}{6.6 \times 10^{-34} \text{ joule-seconds}}$$

$$f = 0.6 \times 10^{15} \text{ hertz}$$

3. The energy change ΔE is given as 7.8×10^{-19} joules, and so the frequency f can be found from the formula $\Delta E = hf$. However, we are asked to determine the wavelength λ. Thus, we will first calculate the frequency f and then find the wavelength λ from the formula $c = f\lambda$.

$$\Delta E = hf$$

$$7.8 \times 10^{-19} \text{ joules} = (6.6 \times 10^{-34} \text{ joule-seconds})\, f$$

$$f = 1.2 \times 10^{15} \text{ hertz}$$

Now that the frequency has been calculated, the wavelength can also be calculated.

$$c = f\lambda$$

$$3 \times 10^{8} \text{ meters per second} = (1.2 \times 10^{15} \text{ hertz})\lambda$$

$$\lambda = \frac{3 \times 10^{8} \text{ meters per second}}{1.2 \times 10^{15} \text{ hertz}}$$

$$\lambda = 2.5 \times 10^{-7} \text{ meters}$$

5. We are given two time intervals $t = 6.5 \times 10^{-6}$ seconds and $t_0 = 0.7 \times 10^{-6}$ seconds to be used in the relativistic time formula $t_0 = \sqrt{1 - v^2/c^2}\, t$. For any speed v (less than the speed of light) the quantity $\sqrt{1 - v^2/c^2}$ will be less than 1. Therefore the time interval t_0 will be less than t. The speed v of the muons can be calculated.

$$t_0 = \sqrt{1 - \frac{v^2}{c^2}}\, t$$

$$0.7 \times 10^{-6} \text{ sec} = \sqrt{1 - \frac{v^2}{(3 \times 10^{8} \text{ m/sec})^2}}\,(6.5 \times 10^{-6} \text{ sec})$$

$$\frac{0.7 \times 10^{-6} \text{ sec}}{6.5 \times 10^{-6} \text{ sec}} = \sqrt{1 - \frac{v^2}{(3 \times 10^{8} \text{ m/sec})^2}}$$

$$0.108 = \sqrt{1 - \frac{v^2}{(3 \times 10^{8} \text{ m/sec})^2}}$$

$$0.012 = 1 - \frac{v^2}{(3 \times 10^{8} \text{ m/sec})^2}$$

$$\frac{v^2}{(3 \times 10^{8} \text{ m/sec})^2} = 0.988$$

$$v^2 = (0.988)(3 \times 10^{8} \text{ m/sec})^2$$

$$v = 0.994(3 \times 10^{8} \text{ m/sec})$$

$$v = 2.98 \times 10^{8} \text{ m/sec}$$

7. We are given an amount of mass m equal to 6×10^{24} kilograms. This mass is equivalent to an amount of energy E which can be calculated from the formula $E = mc^2$.

$$E = mc^2$$

$$E = (6 \times 10^{24} \text{ kg})(3 \times 10^{8} \text{ m/sec})^2$$

$$E = (6 \times 10^{24} \text{ kg})(9 \times 10^{16} \text{ m}^2/\text{sec}^2)$$

$$E = 54 \times 10^{40} \text{ J}$$

9. If the radiant energy E is given as 2.4×10^{28} joules, the equivalent mass m can be found from $E = mc^2$.

$$E = mc^2$$

$$2.4 \times 10^{28} \text{ J} = m(3 \times 10^{8} \text{ m/sec})^2$$

$$\frac{2.4 \times 10^{28} \text{ joules}}{9 \times 10^{16} \text{ m}^2/\text{sec}^2} = m$$

$$m = 0.26 \times 10^{12} \text{ kg}$$

1. The number of protons in any nucleus is called the atomic number Z. In symbols the atomic number appears as a subscript to the left of the chemical symbol. Thus, ${}^{15}_{8}O$ has eight protons. The nucleus ${}^{16}_{8}O$ also has eight protons as given by the subscript. *Every* nucleus of oxygen, regardless of the isotope, has eight protons.

3. In the nucleus ${}^{238}_{92}U$, the number 238 is the mass number A and the number 92 is the atomic number Z. Therefore, the formula for the number of neutrons, N, may be used.

$$N = A - Z$$

$$N = 238 - 92$$

$$N = 146$$

There are 146 neutrons in the nucleus of ${}^{238}_{92}U$.

5. If the two nuclei ${}^{4}_{2}He$ and ${}^{8}_{4}Be$ fuse into a single nucleus, the new nucleus will contain all the protons and all the neutrons from the helium and the beryllium. The ${}^{4}_{2}He$ contains two protons and two neutrons. The ${}^{8}_{4}Be$ contains four protons and four neutrons. Thus, the new nucleus contains altogether six protons and six neutrons. The unknown nucleus may therefore be represented by ${}^{12}_{6}?$. By looking in the periodic chart we see that the element with atomic number 6 is carbon (symbol C). The fusion process yields ${}^{12}_{6}C$.

7. The half-life is the time for one-half of a sample to undergo radioactive decay. Thus, after one half-life, half of the sample has decayed.

$$\begin{matrix}\text{Fraction decayed}\\ \text{after one half-life}\end{matrix} = \frac{1}{2}$$

If we wait for a second interval of time to pass, equal to the half-life, one-half of the *remainder* decays. That is, *one-fourth* of the original sample decays during this second time interval. After two half-lives a *total* of three-quarters of the sample has decayed.

$$\begin{matrix}\text{Fraction decayed}\\ \text{after two half-lives}\end{matrix} = \frac{1}{2} + \frac{1}{4}$$

For each succeeding half-life, one-half of the remainder decays.

$$\begin{matrix}\text{Fraction decayed}\\ \text{after three half-lives}\end{matrix} = \frac{1}{2} + \frac{1}{4} + \frac{1}{8}$$

Adding the fractions, we conclude that the total fraction decayed after three half-lives is $\frac{7}{8}$. ($\frac{1}{2} + \frac{1}{4} + \frac{1}{8} = \frac{7}{8}$). Expressed as percent, $\frac{7}{8}$ is equal to 87.5 percent.

ANSWERS TO DISCUSSION QUESTIONS

CHAPTER 1

1. The speed of the growing tip is the distance the tip moves divided by the elapsed time. The estimated speed would be approximately 2 feet per month. Various answers would be obtained, depending on whether the plant was young or mature, in fertile soil or not, properly irrigated or not, healthy or not, and so on.

2. Measure the distance from the pitcher's hand when the ball is released to the catcher's mitt. Divide this distance by the elapsed time, which could be measured by a stopwatch. Some inaccuracy would occur because the ball follows a curved path rather than the straight-line distance, and stopwatch reading would be affected by your reflex time.

3. If you pull the needle with some vigor, the speed will be typically from 1 to 2 feet per second. In the case of a gentle pull, the speed might be a few inches per second. The choice of units for the speed is a matter of convenience depending on the distance moved in 1 second.

4. You push the toaster handle in lowering the bread, the drawers in closing them, an egg in cracking it, a swinging door in leaving the room. You pull on drawers in opening them, a pickle in removing it from a jar, spaghetti in removing it from a package, string when tying or untying a knot.

5. There are three "action" forces caused by the fingers on the tweezers, the tweezers on the splinter, and the splinter on the skin. According to Newton's third law, there will be three "reaction" forces, namely, forces caused by the tweezers on the fingers, the splinter on the tweezers, and the skin on the splinter.

6. In the absence of mortar the bricks would come in contact due to the weight of the upper brick. When mortar is present, the upper brick exerts a force that compresses the mortar. The mortar, however, pushes to hold the bricks apart.

7. Yes, your weight changes. The change of weight

can be easily measured on any bathroom scale capable of showing a quarter of a pound change of weight.

8. Place two bathroom scales side by side and stand with one foot on each scale. Each scale will give a reading of the weight on that foot. The sum of the two readings will yield the same weight that would occur if both feet were on one scale.

9. No, weight depends only on the mass of the snake at the same location on the earth's surface, not the configuration of the snake.

10. No, weight depends only on the mass of the automobile at the same location on the earth's surface, not the speed or acceleration of the automobile.

11. You can increase the scale reading by holding a heavy book, by eating a meal, or by pushing against a low ceiling or set of bars above your head.

12. No, shoelaces need the friction between the fibers in order for the laces to remain tied in a knot. Similarly, a nail in a board needs friction to remain in place.

13. The flour covering the raisins increases the friction between the raisins and the dough. Therefore, the raisins do not sink to the bottom while the dough is being cooked. If the dough were left uncooked for several weeks, the raisins would slowly settle, but the dough would also spoil!

14. Air resistance can be neglected in studying falling rocks because this friction is a small force in comparison to the weight of the rocks. But air resistance is not a small force in comparison to the weight of a parachute or a falling feather and must be included in a study of their motion.

15. No, 1 kilogram of any object has the same mass as 1 kilogram of another object. Of course, for lead and feathers, there is a difference in the size and shape, and also the air resistance experienced as these masses fall.

16. The mass of the pencil decreases as lead and wood are removed; the mass of the automobile decreases as its fuel is used; the mass of the book is unchanged; the mass of the blade of grass increases as it takes in water and nutrients.

17.——.

18. The cause of this difference in behavior is the inertia of the entire roll of toilet paper. In the first case the state of motion of the entire roll is gently changed from rest to motion as momentum is slowly transferred to the roll. In the second case the force is momentarily very large, causing a single sheet to tear before the entire roll has had time to begin unrolling.

19. The horse will win the 1-mile race, but a good sprinter will probably win a 10-foot race. The important factor in the 10-foot race is that a person has much less inertia (or mass) than a horse and can accelerate faster. In 10 feet the horse is still building up speed (which will ultimately be faster than a person's speed) while the person is already moving almost at his or her maximum speed.

20. The car, including the seats, is pushed forward. The body of the person is also pushed forward by the seat. A head unsupported by a cushion tends to remain at rest, according to Newton's first law, and the head is put in motion only because the neck is pulling forward. The neck cannot withstand the forces as it is stretched and consequently is injured. To prevent neck injuries, a headrest is mounted on the back of the seat.

21. The center of gravity of the earth pulls on the center of gravity of the head and vice versa. The head pushes against the pillow and vice versa. The pillow pushes against the bed and vice versa. And the bed pushes against the house and vice versa.

22. The skater exerts a backward force on the ice, perhaps using the serrated edge on the front of the blade or pushing with the side of the blade. An equal and opposite force is transmitted to the skater's body through the legs, and the skater moves forward. If the skate does not exert a force against the ice (in other words, the skate simply moves over the slippery ice), the ice skater cannot be propelled forward.

23. Yes, but the motion would be very slight. Newton's second law $F\,\Delta t = \Delta(mv)$ shows that your small force expended over a long time will increase the momentum of the building. Since the mass of the skyscraper is so large, the change (increase) in velocity will be very small, but will occur.

24. A change in direction, even at constant speed, is a change in velocity during some time interval, hence, an acceleration. This acceleration is not in the forward direction, but from the side in order to change direction. Accompanying the sideward acceleration is a sideways force, according to Newton's second law. Hence a force is felt from the side of the seat belt. For a left turn the force is felt on the right side of your body, pushing toward the left.

25. No. If the bean is motionless, then Newton's first

law applies. If the bean is in motion, its momentum must change because of forces, which is in accord with Newton's second law. When a force is exerted, there is also an equal and opposite reaction force, which is in agreement with Newton's third law. Although the internal action of a jumping bean is usually not visible, the physics of a jumping bean is no different than the physics of a person jumping in a potato sack at a picnic.

26. Yes, the force can change the direction by causing the object to move at constant speed in a circle. A merry-go-round is a good example.

27. Your weight is opposed by a force of support from the earth, and so no *net* force is exerted on your body. Newton's second law requires a net force to cause a velocity change (for an unchanging mass).

28. A batted ball will usually be given a larger velocity than a thrown ball. The two important factors, according to Newton's second law, are the size (also referred to as magnitude) of the force and the length of time that the force is exerted. For a thrown ball the force is moderate while the time is short, whereas for a batted ball the time is even shorter, but the force is very large.

29. In either case the same amount of momentum must be absorbed by the floor. But for a carpeted floor the time is longer (fibers are being compressed). Consequently the magnitude of the force is less, which may not exceed the strength of the eggshell.

CHAPTER 2

1. Your head and body move to shift your center of gravity directly over the foot supporting your weight. No, you could not prevent this sideward motion if you did not touch a nearby object. You could prevent the motion if you raised one arm and leaned against the wall, but leaning against the wall would introduce an external force as you push against the wall.

2. Yes. Now each half will have a new balance point at the center of that piece.

3. When your feet are under the chair, your center of gravity is located over the seat of the chair. By leaning very far forward, you could get off a chair even though your feet were not under the seat of the chair.

4. You fall, unless you hold onto something, because your center of gravity is not located over your feet. Your center of gravity must be over your feet (or between them, if separated) in order to stand unassisted.

5. You are more stable standing on the bottom step because the center of gravity of you plus the ladder is nearer to the points of contact with the floor.

6. A cat is more stable lying on its side, because its center of gravity is closer to the surface of support, namely, the ground. An elephant is more stable on its side for the same reason. A turtle is more stable on its feet, not only because its center of gravity is lower than when it is balanced on its side, but also because its four feet provide four surfaces of contact rather than one.

7. A tall, thin vase would be more stable if it were laid on its side (which is not the usual way to display flowers). The vase could be made more stable in an upright position by gluing the base to a wooden pedestal, or by putting gravel or coins in the base, or by breaking off the top of the vase (which is admittedly not very practical). The idea is to lower the center of gravity.

8. The everyday definition of work includes time (e.g., an 8-hour day or a 40-hour week), whereas the scientific definition does not involve time. Furthermore, the scientific definition involves motion (that is, distance moved), whereas the everyday definition usually does not involve motion or distance. Both definitions involve force, but in the everyday definition force is usually considered to be "muscular effort."

9. You do work on a suitcase in lifting it because you exert a force on the suitcase, which moves the distance from the floor to the tabletop. When the suitcase is returned to the floor, the earth does the work by means of the force of gravity. No work is done on a suitcase which is neither lifted nor lowered, because the suitcase was not moved in the direction of the force. Note, however, that you might expend considerable muscular effort, become tired, and claim "to have done a lot of work." You become tired in continuing to exert a force to hold up the suitcase, in turn using more oxygen and affecting your metabolism. Even though these bodily changes occur, you do no work on the suitcase if it moves no distance in the direction of the force, according to the technical definition of work.

10. Leather is not as porous as cotton flannel, and so it

takes more force to push a needle through leather. Consequently, for the same-length needles it takes more work to sew leather than cotton flannel.

11. Other examples of objects capable of storing elastic potential energy are springs (a Slinky), rubber bands, and panty hose.

12. Internal springs, which were compressed by your weight to give a weight reading, release their stored compressional potential energy and give the top of the scale some small amount of gravitational potential energy, and the rest of the formerly stored compressional energy is given to you.

13. A car will only begin rolling if it is not on level ground. The kinetic energy comes from the gravitational potential energy that the car had at the higher elevation.

14. The kinetic energy is 4 times larger when the speed is doubled and 9 times larger when the speed is tripled. If a motorcycle must be suddenly stopped, the kinetic energy is transformed into heat, which raises the temperature of the brakes and the tires.

15. No, a squirt gun cannot store potential energy in the water. The kinetic energy of the ejected water is acquired directly from the hand as the trigger is squeezed.

16. The kinetic energy comes from the potential energy of the elevated weights.

17. Drawing a bowstring stores elastic potential energy in the bow and string. When the string is released, potential energy is converted into kinetic energy of the arrow, which is propelled forward.

18. The water at rest in the reservoir has gravitational potential energy. When water is released, the potential energy is converted to kinetic energy as the water falls and acquires a speed. The water is then used to drive a turbine, thereby converting the kinetic energy of the water to electrical energy.

19. No. A rising rock would be acquiring gravitational potential energy and would need a source for this energy. "Spontaneous" eliminates the possibility of explosives or another object (volcano). Moreover, a rock has no internal source of energy, such as a living being.

20. A perpetual motion machine would violate the law of conservation of energy. Such a violation is not impossible, just so unlikely that the Patent Office demands a working model, not just plans and diagrams.

21. The jellyfish could not be made into a perpetual motion machine. Extracting useful energy would stop the impulse.

22. Linear momentum and kinetic energy are similar insofar as both involve motion, that is, mass and speed. Linear momentum depends upon velocity and is a *vector* quantity (has direction), whereas kinetic energy depends upon speed and is a *scalar* quantity (does not have a direction).

23. When the velocity is doubled, the linear momentum is doubled. When the velocity is tripled, the linear momentum is tripled. If a motorcycle must be suddenly stopped, the linear momentum is transferred to the earth (with no perceptible change in the earth's motion).

24. Your linear momentum could be increased either by walking faster, which would increase your velocity, or by eating while walking, which would increase your mass.

25. Both yes and no, depending on what is occurring. While the ferris wheel is rotating with constant speed, the angular momentum is constant and unchanging. However, while the ferris wheel is slowing down or speeding up in order to change riders, the angular momentum is not constant and unchanging. (*Note:* If someone fell out while the ferris wheel was rotating with constant speed, this accident would result in a change in angular momentum—mass loss.)

26. If the earth were an isolated body for all time (no collisions with other objects that would exert forces at a point other than through the earth's center of gravity), the law of conservation of angular momentum would apply. The rate of rotation could change as the earth solidified and material moved inward or outward in the earth's core, but the direction of the axis of rotation could not be changed. If the earth had collided with large, massive objects, the law of conservation of angular momentum would not apply, and both the rate of rotation and the direction of the axis of rotation may have changed.

27. Your arm feels a twisting movement at the moment the electric mixer is turned on. As the mixer acquires angular momentum in one direction, your arm acquires angular momentum in the opposite direction in order to conserve the initial zero angular momentum.

28. If both propellers were to start rotating in the same direction, they would acquire angular momentum in one direction and the boat would acquire angular momentum in the opposite direction in order to conserve the

zero angular momentum of the boat. Consequently, the boat would tilt to one side. (Also, the boat would tilt to the other side when the propellers are stopped.) Therefore, boats with two propellers are engineered to have one propeller rotate in one direction and the other propeller to rotate in the opposite direction. As one propeller acquires angular momentum in one direction, the other propeller acquires equal angular momentum in the other direction. The two angular momenta cancel one another, and the boat is unaffected by any tilting.

29. The small propeller on the tail of the helicopter keeps the body from rotating by pushing the body in a direction opposite to the direction that the body would otherwise rotate.

CHAPTER 3

1. Syrup is a fluid, even when quite cold. If frozen, it would behave as a solid. The borderline between fluid and solid is often controlled by temperature and time factors.

2. Yes. It would be reasonable to describe salt as both a solid and a fluid if we recognize that the salt has different conditions (block or grains). Salt can be considered to be a fluid in the sense that it flows from a shaker.

3. A jar filled with popcorn has visible voids between the popcorn. The voids contain molecules of air which are not visible. For an inflated balloon there is still empty space between the moving molecules of air.

4. The force on the floor is the same whether standing flat-footed or on tiptoe. Both are equal to your weight. But the pressure on the floor is higher when standing on tiptoe, because the area of contact is smaller. The force can be made the same cutting with the dull knife or a sharp knife. But the sharp knife will exert a larger pressure, because the area of contact is smaller.

5. Because the area of contact of the needle (stylus) is *very* small, even a small force exerts a large pressure.

6. The paper edge acts like the cutting edge of a knife. The very small area at the edge of the paper produces a large pressure even with a small force.

7. The same force is exerted on your body by either a regular mattress or a water bed. The water bed exerts less pressure because it has a larger area of contact. Consequently, a water bed is more comfortable (if you do not mind occasional waves and the sound of "sloshing" water).

8. Yes, if there were a large number of nails and the nails were not too sharp (a little dullness goes a long way for comfort and a short way into your skin).

9. The pressures at the three depths are 21.5 pounds per square inch, 2.1 pounds per square inch, and 0.21 pound per square inch.

10. A barometer measures pressure. Units must be force divided by area and could be pounds per square inch. However, it is often expressed as millimeters of mercury, which then refers to the pressure at the bottom of a mercury column of this height.

11. A tornado is associated with lowered pressure. Thus opening windows may help to equalize pressure inside and outside.

12. No, blood pressure varies with each beat of the heart.

13. The Ping-Pong balls float, and so they would provide you with more buoyant force and you would float more easily. Since marbles do not float, they would pull you downward and make floating more difficult.

14. No. Although helium provides enough buoyant force to lift the small weight of a rubber balloon, helium does not exert enough buoyant force to lift the weight of the teakettle.

15. There would be the same scale reading of 13 ounces in each case. The marble pushes on the bottom of the cup and increases the weight of the water-filled cup by 1 ounce. The wood is buoyed up by the water with a force of 1 ounce, which makes the wood float. But the wood pushes down on the water with an equal and opposite force of 1 ounce. Thus the weight of the water-filled cup is increased again by 1 ounce.

16. The important factor for two objects of the same weight is the air resistance. Freely floating feathers would fall very slowly. But if the feathers were compressed into a smooth ball, the feathers would fall just as fast as a smooth ball of lead.

17. Hot air has a smaller weight density than room

temperature air, and so hot air rises, according to Archimedes' principle. On a hot summer day, the balloon would rise only if its air had a higher temperature than the surrounding air. On a cold winter day the balloon would rise much faster than on a normal day, but again the balloon's air must be kept warmer than the surrounding air to keep afloat.

18. You would float extremely high in the mercury, almost completely out of the liquid, because mercury has such a high weight density (13.6 times higher than water). A similar experience of swimming in a dense liquid is possible in the Great Salt Lake in Utah or in the Dead Sea.

19. You would find it very difficult, if not impossible, to float or swim because gasoline has a very low weight density for a liquid (only .66–.69 that of water).

20. No, Pascal's principle applies only to *liquids.*

21. Pressure applied at one point is transmitted throughout the tube.

22. No, a hydraulic press requires a liquid. Gases are too easily compressed.

23. Compressible fluids (most gases) are air and steam. Incompressible fluids (most liquids) are water, cigarette lighter fluid, and beverages.

24. Units must be volume divided by time and thus could be gallons per hour, quarts per second, liters per minute, cubic meters per second, and so forth.

25. Volume flow rate is inversely related to resistance and directly related to pressure drop.

26. Volume flow rate is halved. Volume flow rate is multiplied by one-third.

27. Friction from all causes determines the resistance. This could include, for example, a dependence on pipe length, pipe diameter, and fluid viscosity.

28. It would experience a downward force in addition to its weight if the tail were at the same height as the nose. Consequently, a plane flying upside down has its tail lower than its nose in order to have air pushed downward and the plane lifted upward.

29. Streamlines in air can be marked with smoke, even ordinary tobacco smoke. Streamlines in air can also be marked with swirling dust, leaves, feathers, or bits of paper. Streamlines in liquids can be observed with bubbles or suspended pieces of metal, such as aluminum paint. In cake batter the streamlines made by a spoon can be observed clearly and for several minutes using markers of food dye.

30. Glider pilots look for rising updrafts of wind to keep aloft. If the air is very clear, the updrafts are not visible and only found by chance. If, however, the updrafts carry leaves or dust, the streamlines are easily seen.

31. Add some floating markers, such as pepper, parsley, or cracker crumbs, to make the streamlines more easily detected.

CHAPTER 4

1. The common household thermometer, such as those used to measure the temperature of a sick person or found mounted on a wall, is filled with mercury (note the shiny appearance of the mercury). The lowest temperature any mercury thermometer can measure is −39°F, the temperature at which mercury freezes. Any lower temperature would produce negligible change in the temperature reading. Alcohol, however, does not freeze until −202°F, and so an alcohol thermometer could be used to measure much colder temperatures in cold climates such as Alaska and in the Arctic.

2. Pyrex expands or contracts much less than normal glass. Consequently, Pyrex can be put into an oven and then suddenly cooled without fracturing because of the temperature variations.

3. No, the giraffe is a warm-blooded animal whose body temperature remains the same in summer and winter.

4. Quartz, like Pyrex, has a very small temperature coefficient of expansion. It expands or contracts very little when the temperature is greatly increased or decreased. Consequently, there are only small internal forces, and these are incapable of fracturing quartz.

5. The small particles that make up smoke rise from the end of a cigarette because the smoke is carried aloft by the rising hot air. But if smoke is blown gently from your mouth (at a temperature of about 98°F) through a straw, the smoke will settle on a tabletop in a warm room with no wind or air movement.

6. No, such a self-winding clock obtains the potential energy that winds the clock from the internal energy of

the surrounding air. The law of conservation of energy is not violated.

7. Glass is a poor heat conductor. Thin walls permit heat to be more easily transferred into the test tube.

8. The bracelet would not become snug, because both the bracelet and statue would contract in the freezer, thereby retaining the relative sizes. If the bracelet alone could be cooled, it would fit snugly.

9. The markings are made for mercury, which has a certain rate of expansion as the temperature rises. If alcohol were put in this thermometer, the alcohol would have a different rate of expansion and give false readings for the temperature. New markings would be necessary.

10. Cold air contracts and has a greater weight density than air at room temperature. Consequently, cold air sinks and warm air rises.

11. Theoretically, you would expect the water to have a higher temperature at the base of the falls because the water's potential energy is transformed into internal energy. But as a practical matter, the falls would have to be hundreds of feet high to observe significant temperature changes. (Water would have to fall 773 feet for a temperature change of 1 F degree.)

12. No, the cooling of the refrigerator interior and the cooling of any air circulated through the interior are done at the expense of raising the temperature of the air in the back of the refrigerator. (Window air conditioners expel the hotter air to the outside of the house.) A refrigerator also raises the air temperature even more from the electrical heating of the compressor motor.

13. Although the frost on the refrigerator is cold, the frost is a fairly good insulator to the even colder cooling coils. When the thermostat turns on the cooling mechanism, the refrigerator has to cool all the frost (an expensive waste) before there is any effect on the air in the rest of the refrigerator.

14. Objects come to thermal equilibrium by establishing a common temperature. A hot-water bottle would experience a temperature decrease, and the temperature of nearby objects would increase. In terms of internal energy, the internal energy of the hot-water bottle decreases and the internal energy of the nearby objects would increase. This process of changing internal energies is called *transfer of heat.*

15. Glass has a very low heat conductivity, and so the high internal energy at one end is poorly transferred to the other end. But iron has a very high heat conductivity. The high internal energy at one end is easily transferred to all parts of the iron rod. The end opposite to the red-hot end may not be glowing hot, but it will still be too hot to handle with bare hands.

16. Tile has a larger heat conductivity than a rug. Therefore, tile conducts more heat away from your feet than does a rug, which gives the sensory impression of the tile having a lower temperature than the rug.

17. A double-paned window or storm window traps a layer of air between the window panes. Since air has a small heat conductivity, heat is poorly transferred through these special windows. In the summer your air-conditioning unit need only cool the air in the house without much heat transfer from the outside. In the winter your furnace need only warm the air in the house without much heat transfer to the outside.

18. Snow has a small heat conductivity. The temperature could be below 0°F in the air above the snow, but under the snow the temperature might be 25°F.

19. The coffee in a porcelain cup cools by conduction of heat through the cup and by convection and evaporation from the exposed surface of the coffee. Styrofoam is a very poor conductor of heat, and so the coffee in a Styrofoam cup can only cool by convection and evaporation. Consequently, coffee stays hot longer in a Styrofoam cup than in a porcelain cup.

20. Aluminum has a large heat conductivity. Consequently, the temperature of all parts of an aluminum saucepan would be high when the pan is on the stove. An aluminum handle would burn your bare hands. Thus, better-designed aluminum pans are constructed with plastic or ceramic handles that have small heat conductivity and are not "too hot to handle."

21. Place your hand very near the metal without touching it. Your hand can sense the infrared radiation passing through the space between very hot metal and your hand, even though you cannot see the infrared radiation.

22. Cupcakes have a small heat conductivity, whereas the metal pan has a large heat conductivity. The cupcakes will be cool on the crust but remain quite hot inside the cupcakes. The metal pan will be at nearly the same high temperature both on the surface and inside.

23. A single-walled boiler containing water can be very hot on the lower surface, even much above the boiling point of water. When this happens, there is violent boiling as steam bubbles are formed at the bot-

tom of the water. Some soups and milk (all of which contain water) get burnt under these conditions. A double boiler has a second pan, which is heated by the steam from the violently boiling water. But the steam maintains a constant temperature of 212°F, which is less than the surface of the outer boiler. The advantage of the double boiler is a lower cooking temperature, which does not burn the food. The disadvantage of the double boiler is that it cooks slower than a single-walled boiler.

24. A pressure cooker holds much of the evaporated water vapor in the cooker, thereby increasing the pressure on the water surface and increasing the temperature needed to cause boiling. Therefore, many foods with higher heat conductivities can be cooked at temperatures greater than 212°F. The advantage of the pressure cooker is that food cooks faster. The disadvantage is that not all foods (eggs or dough, for example) have a large enough heat conductivity to be cooked in a pressure cooker.

25. Heat is transferred into ice in order to change the structure of the H_2O from a solid to a liquid. Melting ice is important in cooling the surroundings, such as in a beverage that is being cooled.

26. Heat is transferred away from water in order to change the structure of H_2O from a liquid to a solid. Freezing water is important in heating the surroundings, such as in snow or hail formation which warms the upper atmosphere.

27. Cold water freezes faster because it has less internal energy to be removed than hot water. Ice skating rinks may be flooded with hot water, but this is done only to melt down ridges and produce a smooth surface, not to speed up the freezing process (on the contrary, the freezing process is slower).

28. All the bodily fluids contain water and expand when frozen. This expansion ruptures all the delicate structures of your brain, kidneys, and so on, thereby causing irreversible damage to these organs. Try freezing a fresh tomato and observe its condition after it has been thawed out. Therefore, the idea has little merit.

29. No, water vapor does not have much internal energy to be transferred to her body. However, your friend could not sit in a tub of water at 160°F, because there would be enough internal energy to overcome her body's cooling mechanisms.

30. When perspiration evaporates, internal energy is removed from your body. This loss of internal energy through the process of heat transfer is felt as a cooling effect. In a very humid room (or outside) the air is already nearly saturated with water vapor and very little evaporation occurs. Consequently, a fan is not very effective when the humidity is high. In the winter, when the humidity in a house is always lower than that in the summer, any perspiration evaporates easily and you feel cool, even though the temperature may be considered comfortable.

31. In the winter the humidity is always lower in a house in comparison to the summer. The water on your body easily evaporates, thereby cooling you.

32. The windchill factor applies only to objects which are being heated. All the wind does to a car in 10°F weather is cool it more quickly down to 10°F than would occur on a windless day. Putting a blanket on the radiator will lengthen the time required to cool to 10°F, but it surely will not keep the radiator warm. Even for human beings the windchill factor does not mean that the temperature is going to drop to −40°F. It will only mean that you feel more uncomfortable, as if it were −40°F and no wind were blowing.

33. The internal energy can be increased by two different processes: (*a*) transfer heat into the air by the use of a flame or any high-temperature object or (*b*) do work on the air by compressing it, as in a tire pump.

CHAPTER 5

1. Visible waves might be a wave on a clothesline (amplitude of about 6 inches), coiled cord of telephone (amplitude of about 4 inches), Slinky (amplitude of about 3 inches), bathtub water (amplitude of about 2 inches), and dishwater (amplitude of about 1 inch).

2. Your heart probably beats 70 to 90 times each minute, or *about* once each second, and a Ping-Pong ball probably bounces 1 to 5 times per second. The frequency of your heartbeat is affected by exercise and rest. The frequency of the bouncing Ping-Pong ball is af-

fected by the amplitude of the bounce, that is, the height between bounces.

3. Estimate how many meters the crest of a wave moves in comparison to a fixed object, such as a pier, in 10 seconds. Then divide this distance by 10 seconds for the speed in units of meters per second.

4. The wave shape is best examined from the side of the wave at the same height as the wave, perhaps by swimming and looking at the cross section.

5. From the macroscopic point of view a sound wave is alternate regions of pressure slightly above atmospheric pressure and slightly below atmospheric pressure. These regions move together in a direction, known as the *direction of propagation.* From the microscopic point of view there is a vibratory motion of the air molecules superimposed on their existing random motion.

6. No, sound does not travel through a vacuum. A vacuum, including outer space, is essentially a region free of any atoms or molecules. Therefore, there is no material medium to transmit the pressure variations of sound. Sound can also travel through a solid or liquid, not just a gas such as air.

7. The sound of your friend eating comes to your ears through the air separating you. When you eat, the sound is mostly transmitted from your teeth to your jawbone, to your skull, and then to your ear. In other words, the sound travels through various bones, which are solid.

8. Sound, which is vibrating air, in one room causes an adjoining wall to vibrate. This vibrating wall causes air in the next room to vibrate—that is, a sound is produced in the next room. In order to soundproof a wall, nonvibrating materials, such as glass wool or loose plastic fibers, are stuffed into the spaces between the surfaces of the wall.

9. An echo is the reflection of sound off a large, fairly smooth surface. There can be multiple echoes if there are enough properly located sound-reflecting surfaces.

10. An echo could never contain more total acoustical energy than the sound possessed at its source. If the echo could contain more acoustical energy, the law of conservation of energy would be violated. Echoes can meet at some distant point so that the sound would be louder than if the sound came directly from the source to that spot. But either way, the sound would not be as loud as the original sound.

11. No, even though the walls reflect much of the sound, some acoustical energy is absorbed by the walls. This absorbed acoustical energy results in an increase in the internal energy (and temperature) of the walls, according to the law of conservation of energy.

12. Usually, you are not in the correct position, which is above the water, to hear echoes from a lake. If you do hear echoes at a lake, it is usually because the smooth nearby mountains reflect the sound, not the water of the lake.

13. An average young adult can probably hear sounds from a frequency of 20 cycles per second to 20,000 cycles per second. The upper limit of the frequency range of human hearing decreases as one grows older. Deafness or hearing problems further decrease the frequency range.

14. Ultrasonic refers to phenomena with frequencies just above the range of human hearing, hence, above about 20,000 cycles per second. Supersonic deals with speeds greater than the speed of sound, hence, above about 775 miles per hour at room temperature.

15. Infrasonic refers to phenomena with frequencies just below the range of human hearing, hence, below about 20 cycles per second. Subsonic deals with speeds less than the speed of sound, hence, below about 775 miles per hour at room temperature.

16. No, the sounds of working ants are well within the range of human hearing. The sounds are not very loud, however. If you want to hear them, simply put your ear close to them.

17. Pitch is the subjective idea of a note or tone and depends upon the observer. Frequency, however, is a technical, objective definition of the number of cycles per second of a periodic motion and can be measured with an instrument.

18. Frequency would be unchanged, because it is a technical definition and can be measured. Pitch, however, would depend upon the ability of the other intelligent being to understand and appreciate music.

19. He drops the Ping-Pong ball on the table and listens to the pitch of the sound which results. The pitch will be different than that of an unbroken ball.

20. Baseball players drop their bats on a hard surface and listen to the pitch of the sound given off. A broken bat will produce a different sound than a good bat. Yes, broken keys could be checked in the same way, but experienced players can hear the different pitch of the marimba's or xylophone's keys without dismantling the instrument.

21. Resonance is a phenomenon exhibited by a system acted upon by an external, periodic driving force at the same frequency as the natural frequency of the system. Consequently, the amplitude of oscillation of the system becomes large.

CHAPTER 6

1. Two positive charges ("like" charges) or two negative charges (also "like" charges) repel each other. A positive charge and a negative charge ("unlike" charges) attract each other. The magnitude of the electric force, whether attractive or repulsive, can be made larger either by using more charge or by bringing the charges closer together.

2. The dust is charged with one polarity and passes a plate charged with the opposite polarity. The attractive force between the unlike-charged objects thus traps and collects the dust on the plate.

3. An electric field exists in a region of space if a charge placed in that region experiences an electric force.

4. Yes, but some materials conduct electricity *much* better than other materials. Some materials conduct electricity so slightly that they are called *insulators.*

5. A flashlight battery usually is rated at $1\frac{1}{2}$ volts. A wall outlet in the home usually is rated at 110 volts. An automobile battery usually is rated at 12 volts.

6. Yes, such a battery, if it existed, could power a perpetual motion machine. However, such a battery would violate the law of conservation of energy by providing an unending supply of electrical energy. On a more practical level, such a battery does not exist. All batteries run down.

7. Most outlets in the home are capable of delivering 15 amperes if no other current is being used in that circuit. The kitchen, however, usually has outlets capable of delivering 20 amperes (again, if no other current is being drawn) because of the many electrical devices usually found in the kitchen. Special outlets serving only air conditioners or electrical clothes dryers may deliver 30 amperes.

8. Too much current in a circuit may cause a fire. Therefore, when the current exceeds a predetermined safe value (such as 15 amperes), the current melts a special wire in the fuse. The burnt-out fuse now opens the circuit, thereby stopping all current in that circuit.

9. Various appliances have different resistances and therefore use different currents. According to Ohm's law, the smaller the resistance, the larger the current for the same voltage.

10. The power can be calculated from the current and voltage ($P = IV$), or from the current and the resistance ($P = I^2R$), or from the voltage and resistance ($P = V^2/R$).

11. The kilowatt is a unit of power (1 kilowatt equals 1,000 watts equals 1000 joules per second). The kilowatthour is a unit of energy (1 kilowatthour equals 3,600,000 joules).

12. No, you pay for energy (kilowatthours), not power. A high-power appliance that uses electricity only for a short time is often less expensive to operate than a lower-power appliance that uses electricity continuously.

13. Direct current (dc) electricity has current directed always in one direction, because the voltage does not change at the terminals. Alternating current (ac) electricity has current directed alternately in one direction and then in the other, because the voltage changes sign typically 60 times per second at the terminals.

14. A 1-inch spark is generated when about 75,000 volts exists between two blunt voltage terminals. If the terminals are sharply pointed, then only about 28,000 volts between the voltage terminals will cause a 1-inch spark.

15. All magnetic fields are generated by electric currents, and all electric currents generate magnetic fields. Some electric currents, such as those in a flashlight, generate magnetic fields that are weak and difficult to detect. Other magnetic fields, such as found in a permanent horseshoe magnet, do not appear to be the result of an electric current. But each spinning electron can be viewed as a microscopic current, which contributes to the macroscopic magnetic field.

16. There is some warming of a wire whenever it carries an electric current. But the rubber or plastic insulation will make any warming difficult to detect. If an extension cord is warm, that is a danger signal that

the extension cord is carrying too much current. Such "overloads" can be prevented by a fuse, but if they are not, the extension cord may become hot enough to set the floor or rug on fire.

17. The current in a refrigerator runs the compressor motor, and indeed the motor and wires do become warm. But these warm parts are not inside the interior of the cooling compartment of the refrigerator. Instead, the compressor is connected to tubing carrying a fluid to the inside of the refrigerator. The warm air from the motor and the compressor flows into the room and somewhat warms the room.

18. The critical factor in an electric shock is the current that passes through a body. The current is affected by skin resistance, the voltage experienced, and whether or not a complete electrical circuit is formed.

19. An uninterrupted current of 0.1 to 0.3 amperes for several seconds across the chest disrupts the ability of the heart to pump blood (so-called ventricular fibrillation). Larger currents across the chest disrupt the heart by paralyzing it. A paralyzed heart can be more easily made to start pumping blood than can a fibrillating heart.

20. No, lightning usually strikes either from one part of a cloud to another part of the same cloud or from a cloud to the ground, but rarely does lightning strike from one cloud to another.

21. Consider a lightning bolt that strikes between a cloud and the ground. The leader stroke is the invisible movement of charge from the cloud to the ground. Once the leader stroke makes contact with the ground, the violent discharge of the channel of charge produces the visible return stroke, which appears in high-speed photography to be a stroke going from the ground to the cloud.

CHAPTER 7

1. The spectrum of colors is by agreement subdivided into six colors: red, orange, yellow, green, blue, and violet. Black is not included because black is the total absence of color. (A "flat" black object reflects no light, and such an object is identified by its shape, surroundings, and size.) White is not included because white is the color identified by the eye when equal amounts of all colors of the spectrum are received simultaneously. Many other colors are not included in the spectrum of colors. These colors are composed of varying amounts of the six spectral colors, for example, brown, magenta, turquoise, mauve, puce, and so on.

2. Yes, all spectral colors come from the sun, and green light is mixed in with the others to form "white" light. If there were no green light in sunlight, there would be no green color striking plants, which reflect green light.

3. Infrared radiation consists of electromagnetic waves with wavelengths longer than those of red light. Infrared radiation, while not visible to the human eye, can be detected by some animals, for example, snakes.

4. A forged document may easily appear to have the same shade of ink, although there were two different inks used (one for the original lettering, a second for the forgery). The different inks undoubtedly have different reflecting characteristics for ultraviolet radiation. Thus a picture taken using ultraviolet-sensitive film will easily detect the forgery.

5. No, ultraviolet radiation cannot be seen. This radiation, which is electromagnetic waves having wavelengths shorter than those of violet light, is not sensed by the eye.

6. No, only the ultraviolet radiation causes a chemical change in your skin that leads to a darkening of the skin's pigment.

7. Large doses of ultraviolet radiation kill bacteria and thus sterilize medical equipment. Ultraviolet light causes certain minerals to emit visible light (such a use of ultraviolet leads it to be called *blacklight*). Sunlamps, which can produce an indoor tan, emit ultraviolet light (as well as visible light to indicate that radiation is being emitted).

8. A typical human eye can see electromagnetic waves with wavelengths from 0.4 millionths of a meter to 0.7 millionths of a meter. Window glass (borosilicate crown) has above 98% transmission for visible light, but effectively drops to zero transmission for wavelengths of 0.3 millionths of a meter or less. Almost all glasses are not transparent to ultraviolet radiation with wavelengths less than 0.1 millionths of a meter.

9. An electromagnetic wave is a propagating disturbance which consists of vibrating electric and magnetic fields and is moving at the speed of light. The electric and magnetic fields are at right angles to one another and to the direction of motion.

10. Microwaves or any electromagnetic waves travel at the speed of light, namely, 186,000 miles per second or 300,000 kilometers per second in air or vacuum.

11. Microwaves penetrate throughout raw food and cause the molecules of the food to vibrate, which is equivalent to warming the food throughout without having the food placed in a high-temperature oven. Microwave ovens are sometimes called *radar* ovens, since there is no difference between microwaves and radar as far as electromagnetic waves are concerned. However, the word microwave usually indicates a part of the electromagnetic spectrum, while the word radar usually suggests a technique to detect objects.

12. Amplitude modulation (AM) is a type of radio communication where information (such as sounds and music) is coded as variations in the *amplitude* of the electromagnetic wave (the frequency of the wave is unchanged). AM radio uses low-frequency electromagnetic waves (540,000 to 1,600,000 cycles per second) and can be interrupted by static caused by nearby motors (for example, electric shavers).

13. Frequency modulation (FM) is a type of radio communication where information (sounds, music) is coded as variations in the *frequency* of the electromagnetic wave (the amplitude of the wave is unchanged). FM radio uses higher-frequency electromagnetic waves (88 million to 108 million cycles per second), but cannot be interrupted by static.

14. The visual portion of a TV signal is essentially AM, which controls bursts of electrons in the picture tube that map out the "dots" of the picture. The audio portion of a TV signal is simply an FM radio communication.

15. Stars emit part of their energy in the form of invisible electromagnetic waves, namely, radiowaves. These waves are detected by radio telescopes that are tuned to the electromagnetic waves being investigated. Detectors other than radio telescopes have found energy is also being released in the form of ultraviolet rays, x-rays, and gamma rays.

16. A radio telescope is essentially a giant antenna. Unlike an optical telescope, which intercepts visible light, a radio telescope is a receptor of radio waves. You could not "see" through a radio telescope, because you do not "see" radio waves.

17. Both UV rays and x-rays are electromagnetic waves with the same speed in air or vacuum. Ultraviolet rays have less energy than x-rays. So UV rays are stopped by the skin, whereas x-rays are only slightly stopped by flesh. Medical and dental x-rays are stopped by bone and teeth.

18. Both x-rays and gamma rays are electromagnetic waves capable of penetrating matter (gamma rays are even more penetrating than x-rays). In their useful applications, most x-rays are man-made, whereas most gamma rays are emitted from radioactive nuclei.

CHAPTER 8

1. Light can be seen coming from an object when the object actually generates the light (for example, a candle, a light bulb, the sun) or when the object merely reflects the light (for example, water, a mirror, a painted surface).

2. The mirror reflects light onto the glass slide holding the substance being investigated.

3. The silver coating on the back surface of the glass is not thick enough to completely reflect all light, if the light is bright enough. A person in a darkened room can see light from a brightly lit room through a one-way mirror. But in a brightly lit room a person would see only his or her own image, since no appreciable light has passed through the one-way mirror from the dimly lit room.

4. Light refraction is caused by a change in the speed of light for light that passes obliquely (that is, at angles other than right angles) to the surface between two transparent materials.

5. Dispersion of light is the separation of light into its various color components. The dispersion of white light would produce the rainbow, or spectrum of colors.

6.——.

7. When the sun *appears* on the horizon, the sun is actually below the horizon. The reason that the sun is still visible is that refraction of the sunlight bends the sunbeams down from above our heads to our eyes. So we see the sun longer than if no refraction occurred. The sun appears to set a few minutes later than it really did. And it appears to rise a few minutes before it actually does.

8. Heated air has different refractive characteristics for light than air at room temperature. The mixture of hot and normal air above a toaster jumbles the light beams passing through this mixture, thereby producing a "shimmer" to objects viewed through this mixture. If the toaster were very cold, the air would be cooled quite a bit and also have different refractive characteristics for light than air at room temperature. This mixture would also make objects appear to shimmer. But such objects would have to be viewed *below* the toaster, because cold air falls, whereas hot air rises.

9. A mirage is caused by the refraction of light through layers of air having different temperatures.

10. A magnifying lens alters the direction of the light beams emitted from the object.

11. The surface of the water is detected by reflection from the surface. In addition, refraction through the water will alter the background scene, indicating the presence of water.

12. A common household mirror called a "shaving mirror" is concave. When held near the face it produces an enlarged image useful for shaving or applying cosmetics.

13. When a ray of pure spectral green light shines through a prism, refraction takes place but all the light stays together. When a ray of white light enters a prism, it is spread out into the full spectrum of colors.

14. A lens could certainly be made of ice—or, indeed, any clear material. If small bubbles and dirt particles were avoided, the lens could focus sunlight to start a fire.

15. The bright spots and lines of light which are focused can often be seen in shallow water on the bottom of a sandy lake. Also, they can be seen on the bottom of a bathtub. Look the next time you take a bath!

16. Refraction takes place as light passes through the lens as well as when the light first enters the eye from the air.

17. The eye has muscles which affect the thickness of the lens.

18. Bifocal glasses have two separate regions, one above the other. They can be ground with different curvatures to allow a defective eye to see objects clearly at different distances.

19. Particles in the atmosphere, whether molecules of air or very small particles of dust or pollution, preferentially interact with light with shorter wavelengths (colors nearer the blue or violent end of the spectrum). Blue and violet light are absorbed from the light beams coming from the sun and are reradiated in every direction. Consequently, the light beam traveling on to your eyes is depleted of blue and violet. The remaining red and orange become the main components of the light, which now takes on a reddish-orange color at sunset. The sun at noon appears white because there is relatively little atmosphere to travel through and relatively little depletion of blue and violet from the white light.

20. Any white light passing through the smoke will have its blue light scattered in every direction. Consequently, the smoke will appear blue, even though the light illuminating a room is pure white.

21. A rainbow is formed when white sunlight interacts with a droplet of water. As the light both enters and exits a droplet, the light beam is refracted at the air-water surface. While the light beam is inside the droplet, it is reflected off the inside back surface of the droplet.

22. The first number in the rating of a pair of binoculars is the magnification. The second number is the diameter (measured in millimeters) of the objective lens, and so the bigger the diameter, the more light that is admitted. The 7X50 binoculars admit more light than the 10X35 binoculars. However, the 10X35 binoculars have a larger magnification than the 7X50 binoculars.

CHAPTER 9

1. A quantum is a discrete amount of a physical quantity. When something is *quantized,* it occurs in discrete amounts or increments.

2. The total energy of the hydrogen atom is quantized. Only certain discrete values of energy are allowed.

3. Every atom has a series of allowed energy levels for its electrons. The larger of these levels have the larger energies.

4. The frequency of an emitted photon is determined by the energy change of the atom. The quantitative relationship is given by the Bohr energy law.

5. Various applications of laser light are (*a*) cutting metal forms, (*b*) drilling holes in very hard materials (quartz), (*c*) attaching detached retinas in human eyes, (*d*) establishing reference lines for surveying or construction projects, (*e*) constructing holograms, and (*f*) determining distances and ranges of distant objects.

6. A hologram is a photographic negative that is a picture taken with laser light under special conditions. When the hologram is illuminated with laser light, a true three-dimensional image is produced.

7. An energy level of a laser temporarily stores electrons. These electrons are stimulated to return to the ground-state energy level, whereupon photons are emitted which contribute to the laser light.

8. Monochromatic blue light is pure blue light. No other colors are present (the ultimate in blue hue). Blue light from a lake is not a pure color, since small amounts of green, yellow, violet, and so on are present.

9. Coherent light is composed of visible electromagnetic waves of the same wavelength. For such light there is a definite relationship between the same points on different waves. *The waves are in phase.* A laser produces coherent light.

10. In the TV tube, electrons strike the phosphor-coated viewing screen. The molecules in the phosphor absorb the kinetic energy of the electrons and subsequently emit light of a certain color (energy). In a color TV the screen is made up of a matrix of three different phosphors, each of which emits a different color. The three different phosphors emit red, green, and blue light.

11. Photons are created, for example, in the emission of light from a hot wire, from a burning match, or from a television screen. Photons are destroyed, for example, in the absorption of light as when green light shines on a red surface.

12. No, a mirror does not create photons in the sense of a burning match, which obviously does not store photons. In the process of reflection photons are absorbed at the surface of the mirror's aluminum backing. The smooth aluminum surface reradiates photons in a preferential direction, namely, such that the path of the photon produces an equal angle for the incident and reflected photons.

13. Ionizing electromagnetic radiation is electromagnetic waves with sufficient energy to remove an electron from an atom or molecule. Some radiation, which is not ionizing, lacks enough energy to remove electrons and only raises the temperature of the absorbing substance. Ionizing electromagnetic energy involves the complete liberation of the electron from an atom or molecule.

14. The complete circuit involves not only the wires connecting various motors and electrical components, but also the electric eye that generates a *photocurrent.* If the light is interrupted, then the electric eye ceases to generate a photocurrent. This is comparable to opening a switch and stopping the flow of current. The interruption of the current automatically triggers a mechanism to open the door and to prevent the elevator from moving.

15. Electrons in lower energy levels are more tightly bound to the nucleus. Consequently, more energy is liberated in transitions between inner levels than in transitions between outer levels. When an electron in an outer level makes a transition to an inner level and the atom liberates energy, the energy usually takes the form of a photon of ultraviolet radiation, visible light, or infrared radiation. However, when an electron in a lower level makes a transition to a still lower level and liberates more energy than in the previous case, the energy takes the form of a photon of x-radiation.

16. X-rays are (1) capable of passing through solid objects if the x-rays have sufficient energy and the solid is not too dense or thick, (2) capable of causing fluorescence, that is, certain materials emit visible light when bombarded with x-rays, (3) capable of ionizing atoms and molecules, and (4) capable of causing a chemical reaction in photographic film, that is, x-rays can "expose" film.

17. A "hard" x-ray has a higher frequency and, correspondingly, more energy than a "soft" x-ray. A "hard" x-ray can penetrate more deeply than a "soft" x-ray, and so, for example, a "hard" x-ray easily escapes the glass container of a vacuum tube, whereas a "soft" x-ray is stopped by the glass.

18. Your internal organs, which all have approximately the same density as flesh, are difficult to identify in an x-ray photograph (unlike bones, which are much denser). In order to identify a specific organ, such as the colon or stomach, the individual must swallow a liquid containing chemicals that are opaque to x-rays, such as

barium sulfate mixed with water. The barium sulfate absorbs the x-rays, thereby indicating the size and shape of the colon and stomach. Since the barium sulfate does not immediately enter the kidneys, liver, or bladder, these organs do not absorb the x-rays and are not confused with the colon or stomach.

19. The word *relative* refers to relative motion. We might speak of the speed of a car relative to the earth or the speed of one rocket ship relative to another rocket ship.

20. An *absolute* quantity in relativity is one which does *not* depend on relative motion, such as the speed of light.

21. Time is an absolute quantity in Newtonian mechanics but a relative quantity in relativity theory.

CHAPTER 10

1. The ionizing radiation coming from a radioactive mineral can be detected by a Geiger counter or photographic film (which becomes exposed).

2. Energy released in the process of radioactivity comes from the nucleus of an atom.

3. *Nucleon* is the general name for a proton or a neutron. Such particles are the constituents of atomic nuclei. The proton is positively charged, and the neutron lacks electrical charge.

4. Yes, the most common isotope of the hydrogen atom contains only a single proton. However, no nucleus containing more than one particle could contain only protons because the positively charged protons would exert repulsive forces on one another, thereby making the nucleus unstable. All other nuclei, including the other isotopes of hydrogen, contain neutrons in addition to protons.

5. The atomic number is the number of protons in the nucleus. The mass number is the sum of the numbers of protons and neutrons in the nucleus of an atom.

6. An alpha particle is a positively charged particle consisting of two protons and two neutrons and is identical with the nucleus of a helium atom.

7. Most alpha particles have very low penetrating ability. They are incapable of passing through the skin when coming from outside your body. Alpha particles *are* harmful if they get into your lungs or stomach, where there is no protective skin.

8. Yes, the nucleus changes to that of thorium-231, which is a different element.

9. A beta particle is an electron emitted from certain radioactive nuclei.

10. No, beta particles do not exist in the nucleus, but beta particles are emitted *from* the nucleus. (This situation is analogous to having sound coming from a speaking person, but the sound does not exist in the person. Sound is *produced* in the act of speaking.)

11. Beta particles are more penetrating. If both an alpha particle and a beta particle have 100 keV of energy, the alpha particle will penetrate $\frac{1}{1,000}$ of a millimeter of human tissue, but a beta particle will penetrate 180 times this far. (One eV = 1.60×10^{-19} joule.)

12. The periodic chart of the elements is based on the atomic numbers of the nuclei of the different elements.

13. Isotopes of the same element always have the same number of protons in their nuclei, since an element is uniquely defined by the number of protons. But the number of neutrons is not the same for different isotopes of the same element. In fact, it is the variation in the number of neutrons that determines different isotopes of the same element.

14. A Geiger counter is capable of detecting the passage of a single charged particle through the detection tube. Since there is no such thing as *part* of a single charged particle, the Geiger counter cannot be made more sensitive.

15. Half-life is the time interval required for one-half of any quantity of a single radioactive isotope to undergo radioactive decay.

16. No, any pressure or temperature changes made in a laboratory are incapable of affecting the rate of radioactive decay.

17. Carbon-14 dating is the determination of the age of a sample containing naturally occurring carbon, which also contains the naturally occurring carbon-14 radioactive isotope.

18. Radioactive carbon-14 is useful in determining the age of a specimen if it is between 1,000 and 20,000 years old. With careful measurements this range can be ex-

tended to perhaps 35,000 years. But for ages of rocks from 100,000 years upward, radioactive potassium-40 is used. Ages of 5 million years upward use rubidium-57. Ages of 200 million years and upward use uranium-238. The various isotopes are used because of the varying half-lives.

19. Yes, the fissioning of a high atomic number nucleus releases energy. The energy appears mostly as kinetic energy of the fission fragments plus the radiant energy of the gamma ray photons.

20. A nuclear chain reaction is a succession of nuclear fissions. The neutrons that are set free in the splitting of one nucleus cause additional nuclei to also split (for example, using uranium-233, uranium-235, and plutonium-239).

21. No, not all isotopes are radioactive. Isotope simply means that various nuclei of the same element have different numbers of neutrons, but always the same number of protons. Within a given family of isotopes, some will be stable and some will be radioactive.

22. No, nuclear reactors are designed *without* the capability of sustaining a runaway chain reaction. Instead, a nuclear reactor's walls will crumble in a runaway chain reaction, thereby separating parts of the fissionable fuel rods.

23. A thermonuclear reaction is a fusion reaction (*not* a fission reaction) which occurs between various nuclei of low atomic number at high temperatures. This sort of reaction is continually occurring in the sun and the other stars.

24. A useful fusion reactor must be able to control and contain a thermonuclear reaction, as well as extract energy from the fusion reaction without destroying the reaction or liberating too much energy. None of these obstacles has been overcome.

25. The abbreviation kT stands for kilotons of TNT, which means that a 20-kT nuclear weapon is equivalent to 20,000 tons of TNT in destructive capability. The abbreviation MT stands for megatons of TNT, which means a 20-MT nuclear weapon is equivalent to 20 million tons of TNT in destructive capability.

APPENDIX I

Conversion factors for Metric and English units (Concluded)

Length:

1 cm = 0.394 in	1 in = 2.54 cm
1 m = 3.28 ft	1 ft = 0.305 m
1 km = 0.62 mi	1 mi = 1.61 km

Area:

1 cm^2 = 0.155 in^2	1 in^2 = 6.45 cm^2
1 m^2 = 10.76 ft^2	1 ft^2 = 0.093 m^2

Volume:

1 cm^3 = 0.061 in^3	1 in^3 = 16.4 cm^3
1 m^3 = 35.3 ft^3	1 ft^3 = 0.028 m^3
1 m^3 = 264 gal	1 gal = 3.79 liter
1 liter = 1.06 qt	1 qt = 0.95 liter

Time:

1 hr = 3600 s
1 day = 86,400 s
1 year = 3.16×10^7 s

Speed:

1 m/s = 3.28 ft/s = 2.24 mi/hr
1 km/hr = 0.911 ft/s = 0.62 mi/hr
1 ft/s = 0.68 mi/hr = 0.305 m/s
1 mi/hr = 1.47 ft/s = 1.61 km/hr

Mass:

1 kg = 0.069 slug
1 slug = 14.6 kg
1 amu = 1.66×10^{-27} kg
1 kg = 6.02×10^{26} amu

Force:

1 N = 0.225 lb
1 lb = 4.45 N

Conversion factors for Metric and English units (Concluded)

Energy (Work, Heat):

1 J = 0.738 ft-lb
1 kcal = 3.97 Btu
1 cal = 4.186 J
1 eV = 1.6×10^{-19} J
1 kW-hr = 3413 Btu

1 ft-lb = 1.36 J
1 Btu = 0.252 kcal
1 Btu = 778 ft-lb
1 Btu = 1055 J
1 Btu = 2.93×10^{-4} kW-hr

Power:

1 W = 0.738 ft-lb/s
1 W = 1.34×10^{-3} hp
1 W = 3.41 Btu/hr

1 ft-lb/s = 1.36 W
1 hp = 550 ft-lb/s
1 hp = 746 W

Pressure:

1 N/m^2 = 1.45×10^{-4} lb/in^2
1 atm = 1.01×10^5 N/m^2
1 atm = 76 cm Hg

1 lb/in^2 = 6.9×10^3 N/m^2
1 atm = 14.7 lb/in^2
1 atm = 407 in H_2O

Mass Density:

1 kg/m^3 = 1.94×10^{-3} $slug/ft^3$

1 $slug/ft^3$ = 515 kg/m^3

Weight Density:

1 N/m^3 = 6.36×10^{-3} lb/ft^3

1 lb/ft^3 = 157 N/m^3

Relativistic Mass Energy (from $E = mc^2$):

1 amu = 931 Mev
1 kg = 8.99×10^{16} J

1 MeV = 1.07×10^{-3} amu
1 J = 1.11×10^{-17} kg

INDEX